Mathematical Logic

A Course with Exercises

Mathematical Logic

A Course with Exercises
Part II: Recursion theory, Godel's Theorems, Set theory, Model theory

---

Rene Cori and Daniel Lascar

Equipe de Logique Mathematique
Universite Paris VII

Translated by

Donald H. Pelletier

York University, Toronto

OXFORD
UNIVERSITY PRESS

OXFORD
UNIVERSITY PRESS

Great Clarendon Street, Oxford OX2 6DP

Oxford University Press is a department of the University of Oxford.
It furthers the University's objective of excellence in research, scholarship,
and education by publishing worldwide in

Oxford New York

Athens Auckland Bangkok Bogota Buenos Aires Calcutta
Cape Town Chennai Dar es Salaam Delhi Florence Hong Kong Istanbul
Karachi Kuala Lumpur Madrid Melbourne Mexico City Mumbai
Nairobi Paris São Paulo Singapore Taipei Tokyo Toronto Warsaw

with associated companies in Berlin Ibadan

Oxford is a trade mark of Oxford University Press
in the UK and in certain other countries

Published in the United States
by Oxford University Press, Inc., New York

First published in English 2001

Reprinted 2004

A catalogue record for this book is available from the British Library

Library of Congress Cataloging in Publication Data
(Data available)

ISBN 0 19 850051 3 (Hbk)
ISBN 0 19 850050 5 (Pbk)

Typeset by Newgen Imaging Systems (P) Ltd., Chennai
using the translator's LaTeX files
Printed in Great Britain
on acid free paper by
Biddles Ltd.,
Guildford & King's Lynn

Foreword to the Original French edition

Jean-Louis Krivine

In France, the discipline of logic has traditionally been ignored in university-level scientic studies. This follows, undoubtedly, from the recent history of mathematics in our country which was dominated, for a long while, by the Bourbaki school for whom logic was not, as we know, a strong point. Indeed, logic originates from reecting upon mathematical activity and the common gut-reaction of the mathematician is to ask: `What is all that good for? We are not philosophers and it is surely not by cracking our skulls over modus ponens or the excluded middle that we will resolve the great conjectures, or even the tiny ones ...' Not so fast!

A new ingredient, of some substance, has come to settle this somewhat byzantine debate over the importance of logic: the explosion of computing into all areas of economic and scientic life, whose shock wave nally reached the mathematicians themselves.

And, little by little, one fact dawns on us: the theoretical basis for this nascent science is nothing other than the subject of all this debate, mathematical logic.

It is true that certain areas of logic were put to use more quickly than others. Boolean algebra, of course, for the notions and study of circuits; recursiveness, which is the study of functions that are computable by machine; Herbrand's theorem, resolution and unication, which form the basis of `logic programming' (the language PROLOG); proof theory, and the diverse incarnations of the Completeness theorem, which have proven themselves to be powerful analytical tools for mature programming languages.

But, at the rate at which things are going, we can imagine that even those areas that have remained completely `pure', such as set theory, for example, will soon see their turn arrive.

As it ought to be, the interaction is not one-way, far from it; a ow of ideas and new, deep intuitions, arising from computer science, has come to animate all these sectors of logic. This discipline is now one of the liveliest there is in mathematics and it is evolving very rapidly.

So there is no doubt about the utility and timeliness of a work devoted to a general introduction to logic; this book meets its destiny. Derived from lectures for the Diplôme d'Etudes Approfondies (DEA) of Logic and the Foundations of Computing at the University of Paris VII, it covers a vast panorama: Boolean algebras, recursiveness, model theory, set theory, models of arithmetic and Godel's theorems.

The concept of model is at the core of this book, and for a very good reason since it occupies a central place in logic: despite (or thanks to) its simple, and even elementary, character, it illuminates all areas, even those that seem farthest from it. How, for example, can one understand a consistency proof in set theory without rst mastering the concept of being a model of this theory? How can one truly grasp Godel's theorems without having some notion of non-standard models of Peano arithmetic? The acquisition of these semantic notions is, I believe, the mark of a proper training for a logician, at whatever level. R. Cori and D. Lascar know this well and their text proceeds from beginning to end in this direction. Moreover, they have overcome the risky challenge of blending all the necessary rigour with clarity, pedagogical concern and refreshing readability.

We have here at our disposal a remarkable tool for teaching mathematical logic and, in view of the growth in demand for this subject area, it should meet with a marked success. This is, naturally, everything I wish for it.

Foreword to the English edition

Wilfrid Hodges
School of Mathematical Sciences
Queen Mary and Westeld College
University of London

In the 1930s two young logicians, Kurt Godel and Alan Turing, proved theorems that eventually gave both of them cult status among twentieth century thinkers. Both theorems say, from different viewpoints, that there is no nite set of instructions that mathematicians can write down, which will lead to the solution of all problems in arithmetic (let alone all problems in mathematics).

The volume that you have in your hand, the second part of the text of Rene Cori and Daniel Lascar, contains full and lucid proofs of both of these theorems. In fact the two theorems are not as similar as my informal statement of them suggests. Godel's theorem is about proofs and Turing's is about calculations. Cori and Lascar make the difference very clear by putting the two theorems in their appropriate settingsóGodel's theorem in formal arithmetic (Chapter 6) and Turing's in recursion theory (the undecidability of the halting problem, in Chapter 5).

In spite of their breadth and depth, the theorems of Godel and Turing had little direct inuence on mathematical practice. One reason for this was that, by and large, twentieth century mathematicians came to accept ZermeloñFraenkel set theory as a summary of the starting axioms that they were prepared to use, even though many mathematical problems are known not to be settled by these axioms. Chapter 7 introduces you to ZermeloñFraenkel set theory, with some examples of how it is used in mathematics.

The book nishes with a chapter on model theory. It's hard to say in a few words what model theory does for us; but basically it is the framework within which mathematical logicians study truth, denition and classication. It's a lovely branch of logic, with many applications.

For English students this book is probably best suited to Masters or fourth-year undergraduate studies, or for students working largely on their own. As in the rst volume, the authors have included full solutions to the exercises. The rst volume contains all the background that you need, and more. But in fact you can read this second volume with prot and pleasure if you have studied enough logic to know what rst-order logic is, say as far as a proof calculus and a statement of the compactness theorem.

It remains only for me to repeat my closing comment on the rst part. This book comes from the famous Equipe de Logique Mathematique at the University of Paris, a research team that has had an enormous inuence on the development of mathematical logic and its links with other branches of mathematics. Read it with condence.

Preface

This book is based upon several years' experience teaching logic at the UFR of Mathematics of the University of Paris 7, at the beginning graduate level as well as within the DEA of Logic and the Foundations of Computer Science.

As soon as we began to prepare our rst lectures, we realized that it was going to be very difcult to introduce our students to general works about logic written in (or even translated into) French. We therefore decided to take advantage of this opportunity to correct the situation. Thus the rst versions of the eight chapters that you are about to read were drafted at the same time that their content was being taught. We insist on warmly thanking all the students who contributed thereby to a tangible improvement of the initial presentation.

Our thanks also go to all our colleagues and logician friends, from Paris 7 and elsewhere, who brought us much appreciated help in the form of many comments and moral support of a rare quality. Nearly all of them are co-authors of this work since, to assemble the lists of exercises that accompany each chapter, we have borrowed unashamedly from the invaluable resource that comprises the hundreds and hundreds of pages of written material that were handed out to students over the course of more than twenty-ve years during which the University of Paris 7, a pioneer in this matter, has organized courses in logic open to a wide public.

At this point, the reader generally expects a phrase of the following type: `they are so numerous that we are obviously unable to name them all'. It is true, there are very many to whom we extend our gratitude, but why shouldn't we attempt to name them all?

Thank you therefore to Josette Adda, Marouan Ajlani, Daniel Andler, Gilles Amiot, Fred Appenzeller, Jean-Claude Archer, Jean-Pierre Azra, Jean-Pierre Benejam, Chantal Berline, Claude-Laurent Bernard, Georges Blanc, Elisabeth Bouscaren, Albert Burroni, Jean-Pierre Calais, Zoe Chatzidakis, Peter Clote, Francois Conduche, Jean Coret, Maryvonne Daguenet, Vincent Danos, Max Dickmann, Patrick Dehornoy, Francoise Delon, Florence Duchêne, Jean-Louis Duret, Marie-Christine Ferbus, Jean-Yves Girard, Daniele Gondard, Catherine Gourion, Serge Grigorieff, Ursula Gropp, Philippe Ithier, Bernard Jaulin, Ying Jiang, Anatole Khelif, Georg Kreisel, Jean-Louis Krivine, Ramez Labib-Sami, Daniel Lacombe, Thierry Lacoste, Richard Lassaigne, Yves Legrandgerard, Alain Louveau, Francois Lucas, Kenneth MacAloon, Gilles Macario-Rat, Sophie Malecki, Jean Malifaud, Pascal Manoury, Francois Metayer, Marie-Helene Mourgues, Catherine Muhlrad-Greif, Francis Oger, Michel Parigot, Donald

Pelletier, Marie-Jeanne Perrin, Bruno Poizat, Jean Porte, Claude Precetti, Christophe Raffalli, Laurent Regnier, Jean-Pierre Ressayre, Iegor Reznikoff, Philippe Royer, Paul Roziere, Gabriel Sabbagh, Claire Santoni, Marianne Simonot, Gerald Stahl, Jacques Stern, Anne Strauss, Claude Sureson, Jacques Van de Wiele, Francoise Ville.

We also wish to pay homage to the administrative and technical work accomplished by Mesdames Sylviane Barrier, Gisele Goeminne, and Claude Orieux.

May those whom we have forgotten forgive us. They are so numerous that we are unable to name them all.

September 1993

The typographical errors in the rst printing were so numerous that even Alain Kapur was unable to locate them all. May he be assured of all our encouragement for the onerous task that still awaits him.

We also thank Edouard Dorard and Thierry Joly for their very careful reading.

Contents

Contents of Part I

Notes from the translator

In everyday mathematical language, the English word `contains' is often used indifferently, sometimes referring to membership of an element in a set, $\in$, and sometimes to the inclusion relation between sets, $\subseteq$. For a reader who is even slightly familiar with the subject, this is not a serious issue since the meaning is nearly always clear from the context. But because this distinction is precisely one of the stumbling blocks encountered by beginning students of logic and set theory, I have chosen to consistently use the word `contains' when the meaning is $\in$ and the word `includes' when the meaning is $\subseteq$.

It is perhaps more common in mathematical English to use the phrases `one-to-one' and `onto' in place of the more formal-sounding `injective' and `surjective'. I have none the less retained `injective' and `surjective' as more in keeping with the style of the original; even those who object must admit that `bijective' has the advantage over `one-to-one and onto'.

Where the original refers the reader to various standard texts in French for some basic facts of algebra or topology, I have replaced these references with suitable English-language equivalents.

It is useful to distinguish between bold zero and one (**0** and **1**) and plain zero and one (0 and 1). The plain characters are part of the metalanguage and have their usual denotations as integers. The bold characters are used, by convention, to denote the truth values of two-valued logic; they are also used to denote the respective identity elements for the operations of addition and multiplication in a Boolean algebra.

April, 2000 Donald H. Pelletier

Notes to the reader

The book is divided into two parts. The rst consists of Chapters 1 through 4; Chapters 5 through 8 comprise the second. Concepts presented in a given chapter presume knowledge from the preceding chapters (but Chapters 2 and 5 are exceptions to this rule).

Each of the eight chapters is divided into sections, which, in turn, are composed of several subsections that are numbered in an obvious way (see the Contents).

Each chapter concludes with a section devoted to exercises. The solutions to these are grouped together at the end of the corresponding volume.

The solutions, especially for the rst few chapters, are rather detailed.

Our reader is assumed to have acquired a certain practice of mathematics and a level of knowledge corresponding, roughly, to classical mathematics as taught in high school and in the rst years of university. We will refer freely to what we have called this `common foundation', especially in the examples and the exercises.

None the less, the course overall assumes no prior knowledge in particular.

Concerning the familiar set-theoretical (meta-)language, we will use the terminology and notations that are most commonly encountered: operations on sets, relations, maps, etc., as well as $\mathbb{N}$, $\mathbb{Z}$, $\mathbb{Z}/n\mathbb{Z}$, $\mathbb{Q}$, $\mathbb{R}$ for the sets we meet every day. We will use $\mathbb{N}^*$ to denote $\mathbb{N} - \{0\}$.

If E and F are sets and if f is a map dened on a subset of E with values in F, the **domain** of f is denoted by $\mathsf{dom}(f)$ (it is the set of elements in E for which f is dened), and its **image** is denoted by $\mathsf{Im}(f)$ (it is the set of elements y in F for which $y = f(x)$ is true for at least one element x in E). If A is a subset of the domain of f, the **restriction** of f to A is the map from A into F, denoted by $f \restriction A$, which, with each element x in A, associates $f(x)$. The image of the map $f \restriction A$ is also called the **direct image of** A **under** f and is denoted by $f[A]$. If B is a subset of F, the **inverse image of** B **under** f, denoted by $f^{-1}[B]$, consists of those elements x in E such that $f(x) \in B$. In fact, with any given map f from a set E into a set F, we can associate, in a canonical way, a map from $\wp(E)$ (the set of subsets of E) into $\wp(F)$: this is the `direct image' map, denoted by $\overline{f}$ which, with any subset $A$ of $E$, associates $f[A]$, which we could then just as well denote by $\overline{f}(A)$. In the same way, with this given map $f$, we could associate a map from $\wp(F)$ into $\wp(E)$, called the `inverse image' map and denoted by $\overline{f}^{\,-1}$, which, with any subset B of F, associates $f^{-1}[B]$, which we could then just as well denote by $\overline{f}^{\,-1}(B)$. (See also Exercise 19 from Chapter 2.)

Perhaps it is also useful to present some details concerning the notion of word on an alphabet; this concept will be required at the outset.

Let E be a set, nite or innite, which we will call the **alphabet.** A **word**, w, on the alphabet E is a nite sequence of elements of E (i.e. a map from the set $\{0, 1, \dots, n-1\}$ (where n is an integer) into E); $w = (a_0, a_1, \dots, a_{n-1})$, or even $a_0 a_1 \dots a_{n-1}$, represents the word whose domain is $\{0, 1, \dots, n-1\}$ and which associates a_i with i (for $0 \leq i \leq n-1$). The integer n is called the **length** of the word w and is denoted by $\mathsf{lg}[w]$. The set of words on E is denoted by $\mathcal{W}(E)$.

If $n = 0$, we obtain the **empty word**. We will adopt the abuse of language that consists in simply writing a for the word (a) of length 1. The set $\mathcal{W}(E)$ can also support a binary operation called **concatenation**: let $w_1 = (a_0, a_1, \dots, a_{n-1})$ and $w_2 = (b_0, b_1, \dots, b_{m-1})$ be two words; we can form the new word $w = (a_0, a_1, \dots, a_{n-1}, b_0, b_1, \dots, b_{m-1})$, i.e. the map w dened on $\{0, 1, \dots, n+m-1\}$ as follows:

$$w(i) = \begin{cases} a_i & \text{if } 0 \leq i \leq n-1; \\ b_{i-n} & \text{if } n \leq i \leq n+m-1. \end{cases}$$

This word is called the **concatenation** of w_1 with w_2 and is denoted by $w_1 w_2$. This parenthesis-free notation is justied by the fact that the operation of concatenation is associative.

Given two words w and w_1, we say that w_1 is an **initial segment** of w if there exists a word w_2 such that $w = w_1 w_2$. To put it differently, if $w = (a_0, a_1, \dots, a_{n-1})$, the initial segments of w are the words of the form $(a_0, a_1, \dots, a_{p-1})$ where p is an integer less than or equal to n. We say that w_1 is a **nal segment** of w if there exists a word w_2 such that $w = w_2 w_1$; so the nal segments of $(a_0, a_1, \dots, a_{n-1})$ are the words of the form $(a_p, a_{p+1}, \dots, a_{n-1})$ where p is an integer less than or equal to n. In particular, the empty word and w itself are both initial segments and nal segments of w. A segment (initial or nal) of w is **proper** if it is different from w and the empty word.

When an element b of the alphabet `appears' in a word $w = a_0 a_1 \dots a_{n-1}$, we say that it has an **occurrence in** $w$ and the various `positions' where it appears are called the **occurrences of** b **in** w. We could, of course, be more precise and more formal: we will say that b **has an occurrence in** w if b is equal to one of the a_i for i between 0 and $n-1$ (i.e. if b belongs to the image of w). An **occurrence of** b **in** w is an integer k, less than $\mathsf{lg}[w]$, such that $b = a_k$. For example, the third occurrence of b in w is the third element of the set $\{k : 0 \leq k \leq n-1 \text{ and } a_k = b\}$ in increasing order. This formalism will not be used explicitly in the text; the idea sketched at the beginning of this paragraph will be more than adequate for what we have to do.

The following facts are more or less obvious and will be in constant use:

- for all words w_1 and w_2, $\mathsf{lg}[w_1w_2] = \mathsf{lg}[w_1] + \mathsf{lg}[w_2]$;
- for all words w_1, w_2, and w_3, the equality $w_1w_2 = w_1w_3$ implies the equality $w_2 = w_3$ (we call this **left cancellation**);
- for all words w_1, w_2, and w_3, the equality $w_1w_2 = w_3w_2$ implies the equality $w_1 = w_3$ (we call this **right cancellation**);
- for all words w_1, w_2, w_3, and w_4, if $w_1w_2 = w_3w_4$, then either w_1 is an initial segment of w_3 or else w_3 is an initial segment of w_1. Analogously, under the same assumptions, either w_2 is a nal segment of w_4 or else w_4 is a nal segment of w_2;
- if w_1 is an initial segment of w_2 and w_2 is an initial segment of w_1, then $w_1 = w_2$.

We will also use the fact that $\mathcal{W}(E)$ is countable if E is either nite or countable (this is a theorem from Chapter 7).

Introduction

There are many who consider that logic, as a branch of mathematics, has a somewhat special status that distinguishes it from all the others. Curiously, both its keenest adversaries and some of its most enthusiastic disciples concur with this conception which places logic near the margin of mathematics, at its border, or even outside it. For the rst, logic does not belong to `real' mathematics; the others, on the contrary, see it as the reigning discipline within mathematics, the one that transcends all the others, that supports the grand structure.

To the reader who has come to meet us in this volume seeking an introduction to mathematical logic, the rst advice we would give is to adopt a point of view that is radically different from the one above. The frame of mind to be adopted should be the same as when consulting a treatise in algebra or differential calculus. It is a mathematical text that we are presenting here; in it, we will be doing mathematics, not something else. It seems to us that this is an essential precondition for a proper understanding of the concepts that will be presented.

This does not mean that the question of the place of logic in mathematics is without interest. On the contrary, it is enthralling, but it concerns problems external to mathematics. Any mathematician can (and we will even say must) at certain times reect on his work, transform himself into an epistemologist, a philosopher or historian of science; but the point must clearly be made that, in doing this, he temporarily ceases his mathematical activity. Besides, there is generally no ambiguity: when reading a text in analysis, what the student of mathematics expects to nd there are denitions, theorems, and proofs for these theorems. If the author has thought it appropriate to add some comments of a philosophical or historical nature, the reader never has the slightest difculty separating the concerns contained in these comments from the subject matter itself.

We would like the reader to approach the course that follows in this way and to view logic as a perfectly ordinary branch of mathematics. True, it is not always easy to do this.

The major objection surfaces upon realizing that it is necessary to accept simultaneously the following two ideas:

(1) logic is a branch of mathematics; and

(2) the goal of logic is to study mathematics itself.

Faced with this apparent paradox, there are three possible attitudes. First, one may regard it as so serious that to undertake the study of logic is condemned in advance; second, one may deem that the supposed incompatibility between (1) and (2) simply compels the denial of (1), or at least its modication, which leads to the belief that one is not really doing mathematics when one studies logic; the third attitude, nally, consists in dismantling the paradox, becoming convinced that it is not one, and situating mathematical logic in its proper place, within the core of mathematics.

We invite you to follow us in this third path.

Those for whom even the word paradox is too weak will say: `Wait a minute! Aren't you putting us on when you nally get around, in your Chapter 7, to providing denitions of concepts (intersection, pair, map, ordered set, . . .) that you have been continually using in the six previous chapters? This is certainly paradoxical. You are surely leading us in a vicious circle'.

Well, in fact, no. There is neither a paradox nor a vicious circle.

This text is addressed to readers who have already `done' some mathematics, who have some prior experience with it, beginning with primary school. We do not ask you to forget all that in order to rebuild everything from scratch. It is the opposite that we ask of you. We wish to exploit the common background that is ours: familiarity with mathematical reasoning (induction, proof by contradiction, . . .), with everyday mathematical objects (sets (yes, even these!), relations, functions, integers, real numbers, polynomials, continuous functions, . . .), and with some concepts that may be less well known (ring, vector space, topological space, . . .). That is what is done in any course in mathematics: we make use of our prior knowledge in the acquisition of new knowledge. We will proceed in exactly this way and we will learn about new objects, possibly about new techniques of proof (but caution: the mathematical reasoning that we habitually employ will never be called into question; on the contrary, this is the only kind contemplated here).

If we simplify a bit, the approach of the mathematician is almost always the same whether the subject matter under study is measure theory, vector spaces, ordered sets, or any other area of so-called classical mathematics. It consists in examining structures, i.e. sets on which relations and functions have been dened, and correspondences among these structures. But, for each of these classical areas, there was a particular motivation that gave birth to it and nurtured its development. The purpose was to provide a mathematical model of some more or less `concrete' situation, to respond to an expressed desire arising from the world outside mathematics, to furnish a useful mathematical tool (as a banal illustration of this, consider that vector spaces arose, originally, to represent the physical space in which we live).

Logic, too, follows this same approach; its particularity is that the reality it attempts to describe is not one from outside the world of mathematics, but rather the reality that is mathematics itself.

This should not be awkward, provided we remain aware of precisely what is involved. No student of mathematics confuses his physical environment with an oriented three-dimensional Euclidean vector space, but the knowledge of this environment assists one's intuition when it comes to proving some property of this mathematical structure. The same applies to logic: in a certain way, we are going to manufacture a copy, a prototype, we dare say a reduced model of the universe of mathematics, with which we are already relatively familiar. More precisely, we will build a whole collection of models, more or less successful (not every vector space resembles our physical space). In addition to a specimen that is truly similar to the original, we will inevitably have created others (at the close of Chapter 6, we should be in a position to understand why), often rather different from what we imagined at the outset. The study of this collection teaches us many lessons; notably, it permits those who undertake this study to ask themselves interesting questions about their perceptions and their intuitions of the mathematical world. Be that as it may, we must understand that it is essential not to confuse the original that inspired us with the copy or copies. But the original is indispensable for the production of the copy: our familiarity with the world of mathematics guides us in fabricating the representation of it that we will provide. But at the same time, our undertaking is a mathematical one, within this universe that we are attempting to better comprehend.

So there is no vicious circle. Rather than a circle, imagine a helix (nothing vicious there!), a kind of spiral staircase: we are on the landing of the nth oor, where our mathematical universe is located; call this the `intuitive level'. Our work takes us down a level, to the $(n-1)$st oor, where we nd the prototype, the reduced model; we will then be at the `formal' level and our passage from one level to the other will be called `formalization'. What is the value of n? This makes absolutely no difference; there is no rst nor last level. Indeed, if our model is well constructed, if in reproducing the mathematical universe it has not omitted any detail, then it will also contain the counterpart of our very own work on formalization; this requires us to consider level $n-2$, and so on. At the beginning of this book, we nd ourselves at the intuitive level. The souls that inhabit it will also be called intuitive objects; we will distinguish these from their formal replica by attaching the prex `meta' to their names (meta-integers, meta-relations, even meta-universe since the word `universe' will be given a precise technical meaning in Chapter 7). We will go so far as to say that for any value of n, the nth level in our staircase is intuitive relative to level $n - 1$ and is formal relative to level $n + 1$. As we descend, i.e. as we progress in our formalization, we could stop for a rest at any moment, and take the opportunity to verify that the formal model, or at least what we can see of it, agrees with the intuitive original. This rest period concerns the meta-intuitive, i.e. level $n + 1$.

So we must face the facts: it is no more feasible to build all of mathematics `*ex nihilo*' than to write an EnglishñEnglish dictionary that would be of use to a Martian who knows nothing of our lovely language. We are faced here with a question that had considerable importance in the development of logic at the beginning of the century and about which it is worth saying a few words.

Set theory (it matters little which theory: ZF, Z, or some other), by giving legitimacy to innite objects and by allowing these to be manipulated just like `real' objects (the integers, for example), with the same logical rules, spawned a fair amount of resistance among certain mathematicians; all the more because the initial attempts turned out to be contradictory. The mathematical world was then split into two clans. On the one hand, there were those who could not resist the freedom that set theory provided, this `Cantorian paradise' as Hilbert called it. On the other hand were those for whom only nite objects (the integers, or anything that could be obtained from the integers in a nite number of operations) had any meaning and who, as a consequence, denied the validity of proofs that made use of set theory.

To reconcile these points of view, Hilbert had imagined the following strategy (the well-known `Hilbert programme'): rst, proofs would be regarded as nite sequences of symbols, hence, as nite objects (that is what is done in this book in Chapters 4 and 6); second, an algorithm would be found that would transform a proof that used set theory into a nitary proof, i.e. a proof that would be above all suspicion. If this programme could be realized, we would be able to see, for example, that set theory is consistent: for if not, set theory would permit a proof of $0 = 1$ which could then, with the help of the algorithm suggested above, be transformed into a nitary proof, which is absurd.

This hope was dashed by the second incompleteness theorem of Godel: surely, any set theory worthy of this name allows the construction of the set of natural numbers and, consequently, its consistency would imply the consistency of Peano's axioms. Godel's theorem asserts that this cannot be done in a nitary way.

The conclusion is that even nitary mathematics does not provide a foundation for our mathematical edice, as presently constructed.

The process of formalization involves two essential stages. First, we x the context (the structures) in which the objects evolve while providing a syntax to express their properties (the languages and the formulas). Here, the important concept is the notion of satisfaction which lies at the heart of the area known as semantics. It would be possible to stop at this point but we can also go further and formalize the reasoning itself; this is the second stage in the formalization. Here, we treat deductions or formal proofs as mathematical objects in their own right. We are then not far from proof theory, which is the branch of logic that specializes in these questions.

This book deliberately assigns priority to the rst stage. Despite this, we will not ignore the second, which is where the most famous results from mathematical logic (Godel's theorems) are situated. Chapter 4 is devoted to the positive results in this area: the equivalence between the syntactic and semantic points of view in the context that we have selected. This equivalence is called `completeness'. There are several versions of this simply because there are many possible choices for a formal system of deduction. One of these systems is in fashion these days because of its use in computer science: this is the method of resolution. We have chosen to introduce it after rst presenting the traditional completeness theorem.

The negative results, the incompleteness theorems, will be treated in Chapter 6, following the study of Peano's arithmetic. This involves, as we explained above, abandoning our possible illusions.

The formalization of reasoning will not occur outside the two chapters that we have just mentioned.

Chapter 1 treats the basic operations on truth values, `true' and `false'. The syntax required is very simple (propositional formulas) and the semantics (well-known truth tables) is not very complicated. We are interested in the truth value of propositions, while carefully avoiding any discussion of the nature of the properties expressed by means of these propositions. Our concern with what they express, and with the ways they do it, is the purpose of Chapter 3. We see immediately that the operators considered in the rst chapter (the connectives `and', `or', `implies', and so on) do not sufce to express familiar mathematical properties. We have to introduce the quantiers and we must also provide a way of naming mathematical objects. This leads to formulas that are sequences of symbols obeying rather complicated rules. Following the description of a syntax that is considerably more complex than that for propositional calculus, we dene the essential concept: satisfaction of a formula in a structure. We will make extensive use of all this, which is called predicate calculus, in Chapters 4 and 6, to which we referred earlier, as well as in Chapters 7 and 8. You will have concluded that it is only Chapter 5 that does not require prior knowledge of predicate calculus. Indeed, it is devoted to the study of recursive functions, a notion that is absolutely fundamental for anyone with even the slightest interest in computer science. We could perfectly well begin with this chapter provided we refer to Chapter 1 for the process of inductive definition, which is described there in detail and which is used as well for recursive functions.

In Chapter 7 we present axiomatic set theory. It is certainly there that the sense of paradox to which we referred will be most strongly felt since we purport to construct mathematical universes as if we were dening a eld or a commutative group. But, once a possible moment of doubt has passed, one will nd all that a mathematician should know about the important notions of cardinals and ordinals, the axiom of choice, whose status is generally poorly understood, and, naturally, a list of the axioms of set theory.

Chapter 8 carries us a bit further into an area of which we have so far only caught a glimpse: model theory. Its ambition is to give you a taste for this subject and to stimulate your curiosity to learn more. In any case, it should lead you to suspect that mathematical logic is a rich and varied terrain, where one can create beautiful things, though this can also mean difcult things.

Have we forgotten Chapter 2? Not at all! It is just that it constitutes a singularity in this book. To begin with, it is the only one in which we employ notions from classical mathematics that a student does not normally encounter prior to the upper-level university curriculum (topological spaces, rings, and ideals). Moreover, the reader could just as well skip it: the concepts developed there are used only in some of the exercises and in one section of the last chapter. But we have included it for at least three reasons: the rst is that Boolean algebras are the `correct' algebraic structures for logic; the second is that it affords us an opportunity to display how perfectly classical mathematics, of a not entirely elementary nature, could be linked in a natural way with the study of logic; the third, nally, is that an exposure to Boolean algebras is generally absent from the mathematical literature offered to students, and is even more rarely proposed to students outside the technical schools. So you should consider Chapter 2, if you will, as a little supplement that you may consult or not, as you wish.

We will probably be criticized for not being fair, either in our choice of the subjects we treat or in the relative importance we accord to each of them. The domain of logic is now so vast that it would have been absolutely impossible to introduce every one of its constituents. So we have made choices: as we have already noted, proof theory is barely scratched; lambda calculus and algorithmic complexity are absent despite the fact that they occupy an increasingly important place in research in logic (because of their applications to the theory of computing which have been decisive). The following are also absent: non-classical logics (intuitionist . . .), second-order logic (in which quantications range over relations on a structure as well as over its elements), or so-called `innitary' logics (which allow formulas of innite length). These choices are dictated, rst of all, by our desire to present a basic course. We do not believe that the apprentice logician should commence anywhere else than with a detailed study of the rst-order predicate calculus; this is the context that we have set for ourselves (Chapter 3). Starting from this, we wished to present the three areas (set theory, model theory, recursive function theory and decidability) that seem to us to be the most important. Historically speaking, they certainly are. They also are because the `grand' theorems of logic are all found there. Finally, it is our opinion that familiarity with these three areas is an indispensable prerequisite for anyone interested in any other area of mathematical logic. Having chosen this outline, we still had the freedom to modify the relative importance given to these three axes. In this matter, we cannot deny that we allowed our personal preferences to guide us; it is clear that Chapter 8 could just as well have been devoted to something other than model theory.

These lines were drafted only after the book that you are about to read was written. We think that they should be read only after it has been studied. As already pointed out, we can only truly speak about an activity, describe it (formalize it!), once we have acquired a certain familiarity with it.

Until then.

5 Recursion theory

Recursive functions are certain functions from $\mathbb{N}^p$ *(a Cartesian power of the set of natural numbers) into* $\mathbb{N}$. *They are functions that can, intuitively speaking, be effectively computed or, if one prefers, those whose values can be calculated by some algorithm or some machine. One should note that we are considering here only the theoretical possibility of a mechanical calculation (the actual calculation may well require far too much time to be reasonably undertaken).*

In Section 5.1, we will dene a class of functions, the primitive recursive functions, which manifestly satisfy the criterion in the previous paragraph. We will attempt to convince the reader that this class is already extremely broad by showing that all the functions that immediately come to mind are primitive recursive. Unfortunately, the class of primitive recursive functions does not exhaust the class that we wish to describe: in Section 5.2, we will construct a function, Ackerman's function, which is not primitive recursive although it is effectively computable. So we will dene a wider class, the class of recursive functions. But as it happens, for reasons that will also appear in Section 5.4, we will have to dene a class that is more complicated and, a priori, less natural: the class of partial recursive functions. A partial function f *of p variables is a map from a subset E of* $\mathbb{N}^p$ *into* $\mathbb{N}$ *and such a function is recursive if there exists an algorithm that computes it in the following sense: when the algorithm is applied to compute* $f(n_1, n_2, \ldots, n_p)$, *if* $(n_1, n_2, \ldots, n_p)$ *does belong to E, then it will compute the function value; and if* $(n_1, n_2, \ldots, n_p)$ *does not belong to E, the algorithm will never halt. There is strong evidence that the notion of an effectively computable function has been correctly circumscribed: no one has ever found an effectively computable function that could not be proven to be either recursive or partial recursive.*

Section 5.3 presents the concept of a Turing machine; this is a mathematical idealization of a calculating machine or computer. We will show that the functions calculated by Turing machines are precisely the partial recursive functions. Many other mathematical machines have been dened but we have preferred Turing machines because they are interesting for several reasons: in the rst instance, historical, for these were the earliest mathematical models of machines to be introduced; then, pedagogical, for we can see how they function in a practically mechanical way; and nally, theoretical, for they allow us to prove the important enumeration and xed point theorems. This will be done in Section 5.4.

5.1 Primitive recursive functions and sets

5.1.1 Some initial denitions

We will dene the set of primitive recursive functions by induction using a process analogous to the one we used to dene the sets of formulas of propositional calculus and of predicate calculus: it will be the smallest class that contains certain specic functions and that is closed under certain operations. Before giving the denition, we need to supply some notation and explanations.

- Let p be an integer. We will denote the set of mappings from $\mathbb{N}^p$ into $\mathbb{N}$ by $\mathcal{F}_p$. We will agree, by convention, that if $p = 0$ then the only element of $\mathbb{N}^p$ is the empty sequence and that, consequently, the elements of $\mathcal{F}_0$ can be identied with the elements of $\mathbb{N}$. The set $\bigcup_{p \in \mathbb{N}} \mathcal{F}_p$ will be denoted by $\mathcal{F}$.
- If i is an integer from 1 to p inclusive, the ith **projection** P^i_p is the function in $\mathcal{F}_p$ dened by

$$P^i_p(x_1, x_2, \ldots, x_p) = x_i.$$

- In this chapter, we will make use of the following notation whose origin is in lambda-calculus: with this notation, the function P^i_p is written

$$P^i_p = \lambda x_1 x_2 \ldots x.x_i.$$

  More generally, if t is an expression that involves the variables $x_1, x_2, \ldots, x_p$, then $\lambda x_1 x_2 \ldots x_p.t$ will denote the function that assigns the value $t(n_1, n_2, \ldots, n_p)$ to the arguments $n_1, n_2, \ldots, n_p$. This notation can also be used for functions from $\mathbb{N}^p$ to $\mathbb{N}^q$: for example, $\lambda xy.(x + y, 3x + 2y)$ is the function from $\mathbb{N}^2$ into itself which, to the pair (m, n), assigns the pair $(m + n, 3m + 2n)$.
- By denition, the **successor function** S is the function $\lambda x.x+1$, i.e. the function in $\mathcal{F}_1$ whose value at an integer n is $n + 1$.
- If $f_1, f_2, \ldots, f_n$ belong to $\mathcal{F}_p$ and g belongs to $\mathcal{F}_n$, then the **composite function** or **composition** $h = g(f_1, f_2, \ldots, f_n)$ is the element of $\mathcal{F}_p$ that is equal to

$$\lambda x_1 x_2 \ldots x_p.g(f_1(x_1, x_2, \ldots, x_p), f_2(x_1, x_2, \ldots, x_p), \ldots,$$
$$f_n(x_1, x_2, \ldots, x_p)).$$

Denition 5.1. *(Denitions by recursion)* *This is a procedure for dening functions which is justied by the following obvious fact: if $g \in \mathcal{F}_p$ and $h \in \mathcal{F}_{p+2}$, then there is one and only one function $f \in \mathcal{F}_{p+1}$ that satises the following conditions:*

for all $x_1, x_2, \ldots, x_p$ and y in $\mathbb{N}$,

(i) $f(x_1, x_2, \ldots, x_p, 0) = g(x_1, x_2, \ldots, x_p)$,

(ii) $f(x_1, x_2, \ldots x_p, y + 1) = h(x_1, x_2, \ldots, x_p, y, f(x_1, x_2, \ldots, x_p, y))$.

We say that f is **the function dened by recursion** from g (the **initial condition**) and h (the **recursion step**).

Remark We must convince ourselves that a function dened by recursion will be effectively computable. More precisely, suppose that g and h are two functions, as above, which, in addition, can be effectively computed by algorithms $\mathcal{A}_1$ and $\mathcal{A}_2$, respectively. Then it is not difcult to imagine another algorithm that will compute the function f that is dened by recursion from g and h: to compute $f(n_1, n_2, \ldots, n_p, m)$, one must rst compute $f(n_1, n_2, \ldots, n_p, 0)$ (which is equal to $g(n_1, n_2, \ldots, n_p)$ and is obtained using the algorithm $\mathcal{A}_1$), then $f(n_1, n_2, \ldots, n_p, 1)$ (using the denition of f and the algorithm $\mathcal{A}_2$), and so on until the desired value is obtained.

Denition 5.2 *The set of* ***primitive recursive functions*** *is the smallest subset E of $\mathcal{F}$ satisfying the following conditions:*

(i) *for every integer p, E contains all the constant functions from $\mathbb{N}^p$ into $\mathbb{N}$;*

(ii) *for all integers p and for all integers i such that $1 \leq i \leq p$, E contains all the projections P^i_p;*

(iii) *E contains the successor function S;*

(iv) *E is closed under composition; this means that if n and p are integers, if $f_1, f_2, \ldots, f_n$ are elements of $\mathcal{F}_p$ that belong to E and if g is an element of $\mathcal{F}_n$ that also belongs to E, then the* ***composite function*** $g(f_1, f_2, \ldots, f_n)$ *belongs to E;*

(v) *E is closed under recursion; this means that if p is an integer and if the functions g in $\mathcal{F}_p$ and h in $\mathcal{F}_{p+2}$ are both in E, then the function f dened by recursion from g and h is also in E.*

Remark As we did for the sets of formulas of propositional calculus or predicate calculus, we could have provided a denition `from below' of the set of primitive recursive functions. We set

$$R_0 = \{\gamma : p \in \mathbb{N} \text{ and } \gamma \text{ is a constant function from } \mathbb{N}^p \text{ into } \mathbb{N}\}$$
$$\cup \{P^i_p : 1 \leq i \leq p\} \cup \{S\},$$

and for every n,

$$R_{n+1} = R_n \cup \{\, h : h \text{ is obtained by recursion from two functions in } R_n\}$$
$$\cup \{h : h \text{ is obtained by composition from functions that are in } R_n\};$$

the set of primitive recursive functions is then equal to $\bigcup_{n \in \mathbb{N}} R_n$.

To prove that a function is primitive recursive, it sufces to show how to obtain it using clauses (iv) and (v) starting with functions described in (i)ñ(iii) or, more generally, starting with functions that are already known to be primitive recursive. We will see some examples very shortly.

On the other hand, to prove that all primitive recursive functions possess some property $\wp$, it sufces to show that the functions mentioned in (i)ñ(iii) have this property and that the class of functions having this property is closed under composition and recursion.

We can also see that, for each primitive recursive function f, there exists an algorithm that computes it: this is true for the functions in R_0 and, if it is true for the functions in R_n, then it is also true for those in R_{n+1}.

Denition 5.3 *A subset $A \subseteq \mathbb{N}^p$ is called* ***primitive recursive*** *if its characteristic function is primitive recursive.*

Recall that the **characteristic function** χ_A of the set A is dened by

$$\chi_A(n_1, n_2, \ldots, n_p) = \begin{cases} 1 & \text{if } (n_1, n_2, \ldots, n_p) \in A; \\ 0 & \text{otherwise.} \end{cases}$$

The characteristic function of the set A will be denoted by χ_A or by $\chi(A)$ depending on typographical requirements. If $\wp$ $(x_1, x_2, \ldots, x_p)$ is a property applicable to integers $n_1, n_2, \ldots, n_p$ (we will also use the phrase n-**ary predicate**), we will say that $\wp$ is primitive recursive if the set

$$\{(x_1, x_2, \ldots, x_p) : (x_1, x_2, \ldots, x_p) \text{ satises } \wp\}$$

is primitive recursive.

5.1.2 Examples and closure properties

- Addition $\lambda xy.x + y$ is primitive recursive: indeed, it can be dened in the following way by recursion:

$$x + 0 = x;$$
$$x + (y + 1) = (x + y) + 1.$$

  For this example (and for this one only), let us be rigorously precise. Let us denote the addition function by ad (i.e. $ad = \lambda xy.x + y$). Then

$$ad(x, 0) = P_1^1(x);$$
$$ad(x, y + 1) = S(P_3^3(x, y, ad(x, y))).$$

- Multiplication is also primitive recursive. It can be dened by recursion from addition:

$$x \cdot 0 = 0;$$
$$x \cdot (y + 1) = x \cdot y + x.$$

- The function $\lambda xy.x^y$ is also primitive recursive. It can be dened by

$$x^0 = 1;$$
$$x^{y+1} = x^y \cdot x.$$

- Let us agree that $x \dot{-} 1$ denotes the integer that is equal to $x - 1$ if $x > 0$ and is equal to 0 otherwise. The function $\lambda x.x \dot{-} 1$ is primitive recursive. It can be dened by recursion as follows:

$$0 \dot{-} 1 = 0;$$
$$(x + 1) \dot{-} 1 = x.$$

- More generally, let $x \dot{-} y$ denote the integer that is equal to $x - y$ if $x \geq y$ and is equal to 0 otherwise. The function $\lambda xy.x \dot{-} y$ is also primitive recursive:

$$x \dot{-} 0 = x;$$
$$x \dot{-} (y + 1) = (x \dot{-} y) \dot{-} 1.$$

- Let us dene the function sg by setting $sg(0) = 0$ and $sg(x) = 1$ if $x \neq 0$. The function sg is primitive recursive: indeed, $sg(x) = 1 \dot{-} (1 \dot{-} x)$.
- The predicate $x > y$ is primitive recursive (this means that the set

$$\{(x, y) : x > y\}$$

is primitive recursive). To see this, note that the characteristic function of this set is equal to $sg(x \dot{-} y)$. Similarly, the predicate $x \geq y$, whose characteristic function is $sg((x + 1) \dot{-} y)$, is primitive recursive.

We will now show that the collections of primitive recursive functions and primitive recursive predicates satisfy a certain number of closure properties.

- The set of primitive recursive functions is closed under substitution of variables: if $f \in \mathcal{F}_p$ is primitive recursive and if σ is any mapping of the set $\{1, 2, \ldots, p\}$ into itself, then the function

$$\lambda x_1 x_2 \ldots x_p . f(x_{\sigma(1)}, x_{\sigma(2)}, \ldots, x_{\sigma(p)})$$

is also primitive recursive. In fact, this function is equal to

$$f(P_p^{\sigma(1)}, P_p^{\sigma(2)}, \ldots, P_p^{\sigma(p)}).$$

- If $A \subseteq \mathbb{N}^n$ is primitive recursive and if $f_1, f_2, \ldots, f_n$ belong to $\mathcal{F}_p$ and are primitive recursive, then the set

$$\{(x_1, x_2, \ldots, x_p) : (f_1(x_1, x_2, \ldots, x_p), \ldots, f_n(x_1, x_2, \ldots, x_p)) \in A\}$$

is also primitive recursive [its characteristic function is $\chi_A(f_1, f_2, \ldots, f_n)$].

- From the preceding, we can easily deduce that if f and g are two primitive recursive functions in $\mathcal{F}_p$, then the sets

$$\{(x_1, x_2, \ldots, x_p) : f(x_1, x_2, \ldots, x_p) > g(x_1, x_2, \ldots, x_p)\},$$
$$\{(x_1, x_2, \ldots, x_p) : f(x_1, x_2, \ldots, x_p) = g(x_1, x_2, \ldots, x_p)\}, \quad \text{and}$$
$$\{(x_1, x_2, \ldots, x_p) : f(x_1, x_2, \ldots, x_p) < g(x_1, x_2, \ldots, x_p)\}$$

are primitive recursive. In particular, the set

$$\{(x_1, x_2, \ldots, x_p) : f(x_1, x_2, \ldots, x_p) > 0\}$$

is primitive recursive.

- For every integer p, the set of primitive recursive subsets of $\mathbb{N}^p$ is closed under Boolean operations: if A and B are primitive recursive subsets of $\mathbb{N}^p$, then so are $A \cap B$, $A \cup B$, and $\mathbb{N}^p - A$. Indeed, the characteristic functions of these new sets can be computed:

$$\chi(A \cap B) = \chi(A) \cdot \chi(B);$$
$$\chi(A \cup B) = sg(\chi(A) + \chi(B));$$
$$\chi(\mathbb{N}^p - A) = 1 \dot{-} \chi(A).$$

In particular, note that $A - B = A \cap (\mathbb{N}^p - B)$ is primitive recursive.

- **The schema of denition by cases**: let f and g be two primitive recursive functions in $\mathcal{F}_p$ and let A be a primitive recursive subset of $\mathbb{N}^p$; then the function h dened by

$$h(x_1, x_2, \ldots, x_p) = \begin{cases} f(x_1, x_2, \ldots, x_p) & \text{if } (x_1, x_2, \ldots, x_p) \in A, \\ g(x_1, x_2, \ldots, x_p) & \text{otherwise,} \end{cases}$$

is primitive recursive. It sufces to observe that

$$h = f \cdot \chi(A) + g \cdot \chi(\mathbb{N}^p - A).$$

We can generalize this ability to dene functions by cases: let $f_1, f_2, \ldots, f_{n+1} \in \mathcal{F}_p$ be primitive recursive functions and let $A_1, A_2, \ldots, A_n \subseteq \mathbb{N}^p$ be primitive recursive sets; then the function g dened by

$$g(x_1, x_2, \ldots, x_p) = f_1(x_1, x_2, \ldots, x_p)$$
$$\quad \text{if } (x_1, x_2, \ldots, x_p) \in A_1,$$
$$g(x_1, x_2, \ldots, x_p) = f_2(x_1, x_2, \ldots, x_p)$$
$$\quad \text{if } (x_1, x_2, \ldots, x_p) \notin A_1 \text{ and } (x_1, x_2, \ldots, x_p) \in A_2,$$

$$
\begin{aligned}
&g(x_1, x_2, \ldots, x_p) = f_3(x_1, x_2, \ldots, x_p)\\
&\quad \text{if } (x_1, x_2, \ldots, x_p) \notin A_1 \cup A_2 \text{ and } (x_1, x_2, \ldots, x_p) \in A_3,\\
&\vdots\\
&g(x_1, x_2, \ldots, x_p) = f_n(x_1, x_2, \ldots, x_p)\\
&\quad \text{if } (x_1, x_2, \ldots, x_p) \notin A_1 \cup A_2 \cup \cdots \cup A_{n-1} \text{ and } (x_1, x_2, \ldots, x_p) \in A_n,\\
&g(x_1, x_2, \ldots, x_p) = f_{n+1}(x_1, x_2, \ldots, x_p)\\
&\quad \text{if } (x_1, x_2, \ldots, x_p) \notin A_1 \cup A_2 \cup \cdots \cup A_n,
\end{aligned}
$$

is a primitive recursive function. To see this, we may observe that

$$
\begin{aligned}
g = f_1 \cdot \chi(A_1) &+ f_2 \cdot \chi(A_2 - A_1) + f_3 \cdot \chi(A_3 - (A_1 \cup A_2)) + \cdots\\
&+ f_n \cdot \chi(A_n - (A_1 \cup A_2 \cup \cdots \cup A_{n-1}))\\
&+ f_{n+1} \cdot \chi(\mathbb{N}^p - (A_1 \cup A_2 \cup \cdots \cup A_n)).
\end{aligned}
$$

- As a corollary, we see that the functions

$$
\begin{aligned}
&\lambda x_1 x_2 \ldots x_p . \sup(x_1, x_2, \ldots, x_p) \quad \text{and}\\
&\lambda x_1 x_2 \ldots x_p . \inf(x_1, x_2, \ldots, x_p)
\end{aligned}
$$

are primitive recursive. For example, $\sup(x_1, x_2, \ldots, x_p)$ can be dened as follows:

$$
\begin{aligned}
&\sup(x_1, x_2, \ldots, x_p)\\
&\quad = \begin{cases} x_1 & \text{if } x_1 \geq x_2 \text{ and } x_1 \geq x_3 \text{ and } \ldots \text{ and } x_1 \geq x_p;\\ x_2 & \text{if not and if } x_2 \geq x_3 \text{ and } \ldots \text{ and } x_3 \geq x_p, \text{ etc.} \end{cases}
\end{aligned}
$$

- **Bounded sums and products**: if $f \in \mathcal{F}_{p+1}$ is a primitive recursive function, then the functions

$$
\begin{aligned}
g &= \lambda x_1 x_2 \ldots x_p y . \sum_{t=0}^{t=y} f(x_1, x_2, \ldots, x_p, t) \quad \text{and}\\
h &= \lambda x_1 x_2 \ldots x_p y . \prod_{t=0}^{t=y} f(x_1, x_2, \ldots, x_p, t)
\end{aligned}
$$

are also primitive recursive. They are easily dened by recursion. The sum, for example, is given by

$$
\begin{aligned}
g(x_1, x_2, \ldots, x_p, 0) &= f(x_1, x_2, \ldots, x_p, 0);\\
g(x_1, x_2, \ldots, x_p, y+1) &= g(x_1, x_2, \ldots, x_p, y) + f(x_1, x_2, \ldots, x_p, y+1).
\end{aligned}
$$

In particular, the factorial function $\lambda x . x!$, which can be dened as a bounded product, is primitive recursive.

- **The bounded μ-operator**: let A be a primitive recursive subset of $\mathbb{N}^{p+1}$. Then the function f in $\mathcal{F}_{p+1}$, whose denition follows, is primitive recursive:

$$f(x_1, x_2, \ldots, x_p, z) = 0 \quad \text{if there does not exist an integer } t \leq z \text{ such that } (x_1, x_2, \ldots, x_p, t) \in A;$$
$$f(x_1, x_2, \ldots, x_p, z) = \text{the smallest integer } t \leq z \text{ such that } (x_1, x_2, \ldots, x_p, t) \in A \quad \text{otherwise.}$$

The denition of f uses recursion, the schema of denition by cases and bounded sums:

$$f(x_1, x_2, \ldots, x_p, 0) = 0;$$
$$f(x_1, x_2, \ldots, x_p, z+1)$$
$$= \begin{cases} f(x_1, x_2, \ldots, x_p, z) & \text{if } \sum_{y=0}^{y=z} \chi_A(x_1, x_2, \ldots, x_p, y) \geq 1; \\ z+1 & \text{if not and if } (x_1, x_2, \ldots, x_p, z+1) \in A; \\ 0 & \text{in all other cases.} \end{cases}$$

To denote this function, we will use the following notation:

$$f(x_1, x_2, \ldots, x_p, z) = \mu t \leq z\,[(x_1, x_2, \ldots, x_p, t) \in A].$$

Read this as

> `$f(x_1, x_2, \ldots, x_p, z)$ is the smallest integer t less than or equal to z such that $(x_1, x_2, \ldots, x_p, t) \in A$ if such a t exists; otherwise $f(x_1, x_2, \ldots, x_p, z) = 0$'.

When this scheme is used, the condition $(x_1, x_2, \ldots, x_p, t) \in A$ will often take the form `$g(x_1, x_2, \ldots, x_p, t) = 0$', where g is a primitive recursive function.

- The set of primitive recursive predicates is closed under **bounded quantification**. This means that if $A \subseteq \mathbb{N}^{p+1}$ is primitive recursive, then so are the following sets:

$$B = \{(x_1, x_2, \ldots, x_p, z) : \exists t \leq z\,(x_1, x_2, \ldots, x_p, t) \in A\} \quad \text{and}$$
$$C = \{(x_1, x_2, \ldots, x_p, z) : \forall t \leq z\,(x_1, x_2, \ldots, x_p, t) \in A\}.$$

To see this, observe that the characteristic function of B is given by the formula

$$\chi_B(x_1, x_2, \ldots, x_p, z) = sg\left(\sum_{t=0}^{t=z} \chi_A(x_1, x_2, \ldots, x_p, t)\right)$$

and that of C by

$$\chi_C(x_1, x_2, \ldots, x_p, z) = \prod_{t=0}^{t=z} \chi_A(x_1, x_2, \ldots, x_p, t).$$

Let us take advantage of our newly acquired knowledge to prove that certain specic functions and sets are primitive recursive.

- $\mathbb{N}$ is primitive recursive: its characteristic function is the constant function in $\mathcal{F}_1$ equal to 1;
- the set of even integers is also primitive recursive; its characteristic function is dened by recursion:

$$\chi(0) = 1 \quad \text{and} \quad \chi(n+1) = 1 \dot{-} \chi(n);$$

- the function $q(x, y)$ that is equal to the integer part of x/y if $y \neq 0$ and 0 otherwise is primitive recursive; it is dened by

$$q(x, y) = \mu t \leq x\, [(t+1) \cdot y > x];$$

- the set $\{(x, y) : y \text{ divides } x\}$ is primitive recursive; its characteristic function is equal to

$$1 \dot{-} sg(x \dot{-} y \cdot q(x, y));$$

- the set $\{x : x \text{ is prime}\}$ is primitive recursive: indeed, x is prime if and only if $x > 1$ and $\forall y \leq x (y \leq 1$ or $y = x$ or y does not divide $x)$;
- the function π whose value at n is the $(n+1)$st prime number is primitive recursive: it is dened by recursion using the bounded μ-scheme in the following way:

$$\pi(0) = 2;$$
$$\pi(n+1) = \mu z \leq (\pi(n)! + 1)\, [z > \pi(n) \text{ and } z \text{ is prime}].$$

  (Here, we are invoking the well-known fact that there always exists a prime number strictly between p and $p! + 2$.)
- The exercises contain many other examples of primitive recursive sets and functions.

5.1.3 Coding of sequences

The notion of computability is not applicable only to functions from integers to integers. The simplest and most useful generalization consists in considering functions which, with each nite sequence of integers, associate some integer or even some other nite sequence of integers. To be able to use the theory of recursive functions in this context, we will code the nite sequences of integers. To be precise, we are going to develop a mapping from the set of nite sequences of integers into the integers. It is clearly necessary that this mapping be computable; namely that we know how to calculate the integer that corresponds to a given nite sequence and that, conversely, we know how to recover a nite sequence from its code. There are many ways of doing this. Here, we will provide two such codings that will be used in the rest of the chapter.

Proposition 5.4 *For every non-zero integer p, there exist primitive recursive functions* $\alpha_p \in \mathcal{F}_p$ *and* $\beta_p^1, \beta_p^2, \ldots, \beta_p^p \in \mathcal{F}_1$ *that have the following property:* α_p *is a bijection from* $\mathbb{N}^p$ *onto* $\mathbb{N}$ *whose inverse mapping is* $\lambda x.(\beta_p^1(x), \beta_p^2(x), \ldots, \beta_p^p(x))$.

Proof We will begin by constructing α_2. To do this, we enumerate the set of pairs of integers according to the diagram below:

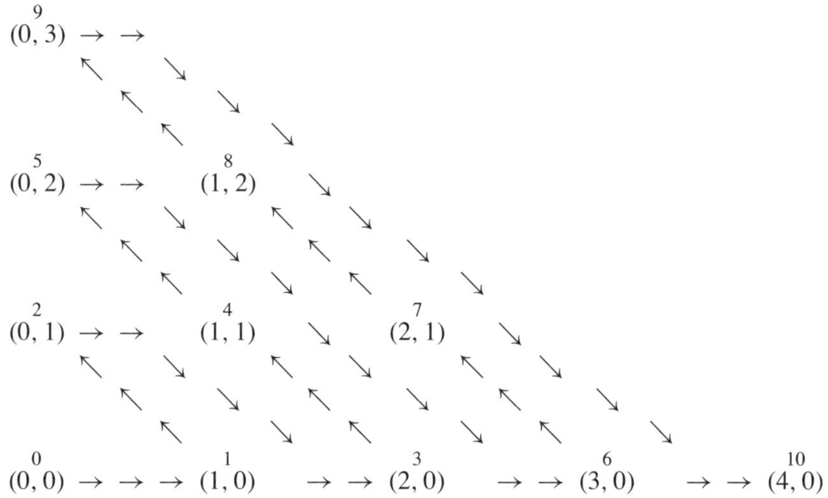

More precisely, we enumerate the pairs (x, y) by following the diagonals on which $x + y$ is constant. We start with the diagonal $x + y = 0$ [which contains only the single pair $(0, 0)$], then we proceed to the diagonal $x + y = 1$ starting from below, and so on. The value of $\alpha_2(x, y)$ is exactly the number of predecessors of (x, y) in this enumeration. Thus, preceding the pair $(p + n, 0)$, there are

$$1 + 2 + \cdots + (n + p) = \frac{1}{2}(n + p + 1)(n + p)$$

elements. The pair (p, n) is on the same diagonal as $(p + n, 0)$ and lies exactly n positions beyond it. Consequently,

$$\alpha_2(p, n) = \frac{1}{2}(n + p + 1)(n + p) + n.$$

Note that α_2 is clearly primitive recursive and is greater than or equal to both n and p. Because α_2 is bijective, we can recover n and p from $\alpha_2(p, n)$ by using one of the following functions:

$$\beta_2^1(x) = \mu z \leq x\ [\exists t \leq x\ \alpha_2(z, t) = x] \quad \text{and}$$
$$\beta_2^2(x) = \mu z \leq x\ [\exists t \leq x\ \alpha_2(t, z) = x].$$

Observe that β_2^1 and β_2^2 are also primitive recursive.

We can then dene α_3 by $\alpha_3(x, y, z) = \alpha_2(x, \alpha_2(y, z))$ and

$$\beta_3^1 = \beta_2^1, \qquad \beta_3^2 = \beta_2^1 \circ \beta_2^2, \qquad \beta_3^3 = \beta_2^2 \circ \beta_2^2$$

and, more generally,

$$\alpha_{p+1}(x_1, x_2, \ldots, x_p, x_{p+1}) = \alpha_p(x_1, x_2, \ldots, x_{p-1}, \alpha_2(x_p, x_{p+1}));$$
$$\beta_{p+1}^1 = \beta_p^1, \qquad \beta_{p+1}^2 = \beta_p^2, \quad \ldots, \quad \beta_{p+1}^{p-1} = \beta_p^{p-1},$$
$$\beta_{p+1}^p = \beta_2^1 \circ \beta_p^p, \qquad \beta_{p+1}^{p+1} = \beta_2^2 \circ \beta_p^p.$$

To conclude, we set $\alpha_1(x) = x$ and $\beta_1^1(x) = x$.

Notation We will use $\mathcal{S}$ to denote the set of nite sequences of integers $[\mathcal{S} = \mathcal{M}(\mathbb{N})]$.

In Exercise 3, we will show how to employ these functions to establish a single coding of the set of all nite sequences.

And here is a different coding, a true classic, that will be used subsequently.

Denition 5.5 *The function Ω is the map from $\mathcal{S}$ into $\mathbb{N}$ dened as follows:*

$$\Omega((x_0, x_1, \ldots, x_p)) = \pi(0)^{x_0} \cdot \pi(1)^{x_1} \cdots \pi(p)^{x_p}$$

[recall that π is the function whose value at n is the $(n+1)$st prime number].

We complete this denition by setting $\Omega(s) = 1$ if s is the empty sequence.

The function δ is the function in $\mathcal{F}_2$ dened by

$$\delta(i, x) = \mu z \leq x \ [x \text{ is not divisible by } \pi(i)^{z+1}]$$

[$\delta(i, x)$ is the exponent of $\pi(i)$ in the decomposition of x into a product of primes].

Observe that the function δ is primitive recursive. It is also not difcult to see that the range of Ω [i.e. $\{x : \text{there exists } s \in \mathcal{S} \text{ such that } x = \Omega(s)\}$] is the set $\mathbb{N} - \{0\}$. We do not have a perfect coding since the function Ω is not injective [clearly, if s and s' are in $\mathcal{S}$, then $\Omega(s) = \Omega(s')$ if and only if the longer of the two sequences s or s' is obtained from the other one by adjoining zeros at the end]. In fact, we could make this injective by adding 1 to each exponent, but then we would lose surjectivity. Also, the values assumed by Ω rapidly become enormous and thus it is useless except for computations `in theory'. But this is of no consequence for the purposes we have in mind for it.

Double recursion: We are given four functions, $g, g' \in \mathcal{F}_p$ and $h, h' \in \mathcal{F}_{p+3}$. With the help of these functions, we may dene simultaneously two new functions

f and f' in $\mathcal{F}_{p+1}$ by the conditions

$$\begin{aligned} f(x_1, x_2, \ldots, x_p, 0) &= g(x_1, x_2, \ldots, x_p); \\ f'(x_1, x_2, \ldots, x_p, 0) &= g'(x_1, x_2, \ldots, x_p); \\ f(x_1, x_2, \ldots, x_p, y+1) &= h(x_1, x_2, \ldots, x_p, y, f(x_1, x_2, \ldots, x_p, y), \\ &\qquad f'(x_1, x_2, \ldots, x_p, y)); \\ f'(x_1, x_2, \ldots, x_p, y+1) &= h'(x_1, x_2, \ldots, x_p, y, f(x_1, x_2, \ldots, x_p, y), \\ &\qquad f'(x_1, x_2, \ldots, x_p, y)). \end{aligned}$$

We will show that if all four of g, g', h, h' are primitive recursive, then so are f and f'. To do this, let us introduce the function $k = \alpha_2(f, f')$. This function is denable by recursion as follows:

$$\begin{aligned} &k(x_1, x_2, \ldots, x_p, 0) = \alpha_2(g(x_1, x_2, \ldots, x_p), g'(x_1, x_2, \ldots, x_p)); \\ &k(x_1, x_2, \ldots, x_p, y+1) \\ &\quad = \alpha_2(h(x_1, x_2, \ldots, x_p, y, \beta_2^1(k(x_1, x_2, \ldots, x_p, y)), \beta_2^2(k(x_1, x_2, \ldots, x_p, y))), \\ &\qquad h'(x_1, x_2, \ldots, x_p, y, \beta_2^1(k(x_1, x_2, \ldots, x_p, y)), \beta_2^2(k(x_1, x_2, \ldots, x_p, y)))). \end{aligned}$$

Thus the function k is primitive recursive; hence $f = \beta_2^1 \circ k$ and $f' = \beta_2^2 \circ k$ are as well.

5.2 Recursive functions

5.2.1 Ackerman's function

Our aim in this subsection is to give an example of a function that is effectively computable, in the intuitive sense of the word, but that is not primitive recursive. This will justify all the extra work that we will demand of the reader in the future. We dene a function (which we call **Ackerman's function** even though it is in fact a slight variant of the one Ackerman dened originally) of two variables that we will denote by ξ as follows:

(i) for every integer x, $\xi(0, x) = 2^x$;

(ii) for every integer y, $\xi(y, 0) = 1$;

(iii) for all integers x and y, $\xi(y+1, x+1) = \xi(y, \xi(y+1, x))$.

For each integer n, let ξ_n denote the function $\lambda x.\xi(n, x)$. Then $\xi_0(x) = 2^x$ and, by invoking clause (iii), it is easy to show that for all positive n, ξ_n is dened by recursion from ξ_{n-1} by

$$\xi_n(0) = 1 \quad \text{and} \quad \xi_n(x+1) = \xi_{n-1}(\xi_n(x)).$$

This shows, rst of all, that there is a unique function ξ satisfying the given conditions. Moreover, all the functions ξ_n are primitive recursive (this is proved by induction on n). On the contrary, nothing permits us to afrm that the function ξ

itself is primitive recursive; this is fortunate since we are about to show that it is not. However, we can effectively compute $\xi(x, y)$ for any values of x and y, as the reader should easily be convinced. We must next prove a few easy but annoying lemmas concerning the function ξ.

Lemma 5.6 *For every n and for every x, $\xi_n(x) > x$.*

Proof Our proof will involve two interleaved inductions. By induction on n, we will show that for all x, $\xi_n(x) > x$. This is clear for $n = 0$. Now x an $n > 0$ and assume that the assertion

$$\text{for every integer } x, \quad \xi_{n-1}(x) > x$$

is true. We must then prove the assertion

$$\text{for every integer } x, \quad \xi_n(x) > x.$$

To do this, we will now argue by induction on x. For $x = 0$, this is clear since $\xi_n(0) = 1$. Next, assuming that $\xi_n(x) > x$, we will prove that $\xi_n(x+1) > x+1$. We know that $\xi_n(x+1) = \xi_{n-1}(\xi_n(x))$ and so, by the rst induction hypothesis, we see that

$$\xi_n(x+1) > \xi_n(x), \text{ or, equivalently, } \xi_n(x+1) \geq \xi_n(x) + 1.$$

Now, according to the second induction hypothesis, $\xi_n(x) > x$; so the lemma is proved. ∎

Lemma 5.7 *For every integer n, the function ξ_n is strictly increasing.*

Proof This is clear for $n = 0$. For positive n, it follows immediately from the previous lemma and from the formula $\xi_n(x+1) = \xi_{n-1}(\xi_n(x))$. ∎

Lemma 5.8 *For all $n \geq 1$ and for all x, $\xi_n(x) \geq \xi_{n-1}(x)$.*

Proof This is clear for $x = 0$. For $x + 1$, since $\xi_n(x) \geq x + 1$ and since ξ_{n-1} is increasing, $\xi_{n-1}(\xi_n(x)) \geq \xi_{n-1}(x+1)$; it now sufces to apply the formula

$$\xi_n(x+1) = \xi_{n-1}(\xi_n(x)).$$ ∎

If k is an integer, let ξ_n^k denote the function ξ_n iterated k times (i.e. $\xi_n^0 = \lambda x.x$, $\xi_n^1 = \xi_n$, and $\xi_n^{k+1} = \xi_n \circ \xi_n^k$). The following lemma is now a collection of trivialities.

Lemma 5.9 *The functions ξ_n^k are all strictly increasing. Moreover, for all m, n, k, and x,*

$$\xi_n^k(x) < \xi_n^{k+1}(x), \qquad \xi_n^k(x) \geq x, \qquad \xi_n^k \circ \xi_n^h = \xi_n^{k+h}$$

and, if $m \leq n$, then $\xi_m^k(x) \leq \xi_n^k(x)$.

Next, let us give a denition.

Denition 5.10 *Suppose that $f \in \mathcal{F}_1$ and $g \in \mathcal{F}_p$. We say that f **dominates** g if there exists an integer A such that, for all $(x_1, x_2, \ldots, x_p)$,*

$$g(x_1, x_2, \ldots, x_p) \leq f(\sup(x_1, x_2, \ldots, x_p, A)).$$

In particular, when f is strictly increasing, f dominates g if and only if $g(x_1, x_2, \ldots, x_p) \leq f(\sup(x_1, x_2, \ldots, x_p))$ holds for all but nitely many p-tuples $(x_1, x_2, \ldots, x_p)$.

Let C_n denote the set of functions that are dominated by at least one iterate of ξ_n:

$$C_n = \{g : \text{there exists a } k \text{ such that } \xi_n^k \text{ dominates } g\}.$$

It is obvious that the following functions belong to C_0: the projection functions P_n^i, the constant functions, the successor function S, the function

$$\lambda x_1 x_2 \ldots x_p.\sup(x_1, x_2, \ldots, x_p),$$

the function $\lambda xy.x + y$, and the functions $\lambda x.kx$ where k is an arbitrary integer. Also, the function ξ_n belongs to C_n. Finally, if f and g both belong to $\mathcal{F}_p$, if $g \in C_n$, and if for all $x_1, x_2, \ldots, x_p$, $f(x_1, x_2, \ldots, x_p) \leq g(x_1, x_2, \ldots, x_p)$, then $f \in C_n$.

We will now establish

Lemma 5.11 *For every integer n, the set C_n is closed under composition.*

Proof Let $f_1, f_2, \ldots, f_m$ be functions of p variables and let g be a function of m variables and suppose all these functions are in C_n. We need to prove that $g(f_1, f_2, \ldots, f_m)$ is also in C_n. We know that there exist integers A, A_1, $A_2, \ldots, A_m$, k, $k_1, k_2, \ldots, k_m$ such that, for all $y_1, y_2, \ldots, y_m$,

$$g(y_1, y_2, \ldots, y_m) \leq \xi_n^k(\sup(y_1, y_2, \ldots, y_m, A)),$$

and for all $x_1, x_2, \ldots, x_p$ and for all i between 1 and m inclusive,

$$f_i(x_1, x_2, \ldots, x_p) \leq \xi_n^{k_i}(\sup(x_1, x_2, \ldots, x_p, A_i)).$$

Set $B = \sup(A, A_1, A_2, \ldots, A_m)$ and $h = \sup(k_1, k_2, \ldots, k_m)$. By invoking Lemma 5.9, we can now see that, for all $x_1, x_2, \ldots, x_p$,

$$\begin{aligned} &g(f_1(x_1, x_2, \ldots, x_p), f_2(x_1, x_2, \ldots, x_p), \ldots, f_m(x_1, x_2, \ldots, x_p)) \\ &\quad \leq \xi_n^k(\xi_n^h(\sup(x_1, x_2, \ldots, x_p, B))), \end{aligned}$$

and hence that

$$\begin{aligned} &g(f_1(x_1, x_2, \ldots, x_p), f_2(x_1, x_2, \ldots, x_p), \ldots, f_m(x_1, x_2, \ldots, x)) \\ &\quad \leq \xi_n^{k+h}(\sup(x_1, x_2, \ldots, x_p, B)). \end{aligned}$$ ■

Lemma 5.12 *For all integers n, k, and x,*

$$\xi_n^k(x) \leq \xi_{n+1}(x+k).$$

Proof The proof is by induction on k. For k equal to 0 or 1, it is obvious. Assume it is true for k; then it is also true for $k+1$ because

$$\begin{aligned} \xi_n^{k+1}(x) &= \xi_n(\xi_n^k(x)) \\ &\leq \xi_n(\xi_{n+1}(x+k)) \quad \text{(by the induction hypothesis)} \\ &= \xi_{n+1}(x+k+1) \quad \text{(by the denition of } \xi\text{)}. \end{aligned}$$ ■

Lemma 5.13 *Suppose that $g \in \mathcal{F}_p$, that $h \in \mathcal{F}_{p+2}$ and that g and h both belong to C_n $(n \geq 0)$. Then the function f dened by recursion from g and h belongs to C_{n+1}.*

Proof We begin by translating the hypotheses. First, the denition of f:

$$\begin{aligned} f(x_1, x_2, \ldots, x_p, 0) &= g(x_1, x_2, \ldots, x_p), \\ f(x_1, x_2, \ldots x_p, y+1) &= h(x_1, x_2, \ldots, x_p, y, f(x_1, x_2, \ldots, x_p, y)); \end{aligned}$$

next, the domination conditions:

there exist A_1, A_2, k_1, k_2 such that, for all $x_1, x_2, \ldots, x_p, y$,
$g(x_1, x_2, \ldots, x_p) \leq \xi_n^{k_1}(\sup(x_1, x_2, \ldots, x_p, A_1))$ and
$h(x_1, x_2, \ldots, x_p, y, z) \leq \xi_n^{k_2}(\sup(x_1, x_2, \ldots, x_p, y, z, A_2))$.

We will now prove by induction on y that, for all $x_1, x_2, \ldots, x_p, y$,

$$f(x_1, x_2, \ldots, x_p, y) \leq \xi_n^{k_1+yk_2}(\sup(x_1, x_2, \ldots, x_p, y, A_1, A_2)). \qquad (*)$$

For $y = 0$, this is clear. If it is true for y, then it is also true for $y+1$ because

$$f(x_1, x_2, \ldots, x_p, y+1) = h(x_1, x_2, \ldots, x_p, y, f(x_1, x_2, \ldots, x_p, y));$$
$$f(x_1, x_2, \ldots, x_p, y+1) \leq \xi_n^{k_2}(\sup(x_1, x_2, \ldots, x_p, y, f(x_1, x_2, \ldots, x_p, y), A_2)).$$

So, using the induction hypothesis $(*)$ and Lemma 5.9,

$$f(x_1, x_2, \ldots, x_p, y+1) \leq \xi_n^{k_2}(\xi_n^{k_1+yk_2}(\sup(x_1, x_2, \ldots, x_p, y, A_1, A_2))),$$

which proves the assertion. Now, we invoke Lemma 5.12 to get

$$f(x_1, x_2, \ldots, x_p, y) \leq \xi_{n+1}(\sup(x_1, x_2, \ldots, x_p, y, A_1, A_2) + k_1 + yk_2).$$

Note that the function

$$\lambda x_1 x_2 \ldots x_p y.\xi_{n+1}(\sup(x_1, x_2, \ldots, x_p, y, A_1, A_2) + k_1 + yk_2)$$

is obtained by composition from functions belonging to C_{n+1}; so it too belongs to C_{n+1} and so does f. ■

We are now in a position to assert:

Corollary 5.14 *The set $\bigcup_{n\in\mathbb{N}} C_n$ contains all primitive recursive functions.*

Proof Indeed, this set contains the constant functions, the projections, and the successor function; also, it is closed under composition, and under denitions by recursion. ■

This brings us to the main theorem of this subsection.

Theorem 5.15 *Ackerman's function is not primitive recursive.*

Proof Suppose, to the contrary, that Ackerman's function is primitive recursive; then so is the function $\lambda x.\xi(x, 2x)$. So, there exist integers n, k, and A such that for all $x > A$, $\xi(x, 2x) \leq \xi_n^k(x)$. Thus, for all $x > A$, we have

$$\xi(x, 2x) \leq \xi_n^k(x) \leq \xi_{n+1}(x + k)$$

(by Lemma 5.12), and, if $x > \sup(A, k, n + 1)$,

$$\xi_{n+1}(x + k) < \xi_{n+1}(2x) < \xi_x(2x) = \xi(x, 2x)$$

(by Lemma 5.9), which is absurd. ■

In fact, we can see that the function $\lambda x.\xi(x, x)$ dominates all the primitive recursive functions.

5.2.2 The μ-operator and the partial recursive functions

We must therefore dene a larger class which we will call the class of recursive functions. We will accomplish this by allowing a new denition scheme, the unbounded μ-operator. The idea is as follows: given a subset A of $\mathbb{N}^{p+1}$, this scheme permits us to dene the function $f \in \mathcal{F}_p$ which, with the p-tuple $(x_1, x_2, \ldots, x_p)$, associates the least integer z such that $(x_1, x_2, \ldots, x_p, z) \in A$. The problem with this is immediately apparent: what happens if there does not exist an integer z such that $(x_1, x_2, \ldots, x_p) \in A$? Observe that it is not possible in this situation to do what we did for the bounded μ-operator and simply set $f(x_1, x_2, \ldots, x_p) = 0$. Indeed, assuming, as we must, that we have an algorithm at our disposal which computes the characteristic function χ_A of A, the only way we can imagine for computing $f(x_1, x_2, \ldots, x_p)$ is to calculate $\chi_A(x_1, x_2, \ldots, x_p, 0)$. If the result is 1, we may stop; if not, then calculate $\chi_A(x_1, x_2, \ldots, x_p, 1)$, then $\chi_A(x_1, x_2, \ldots, x_p, 2)$, and so on, until the value 1 is obtained. But if, for every integer z, $(x_1, x_2, \ldots, x_p, z) \notin A$, this process will never halt and we will never know the value of $f(x_1, x_2, \ldots, x_p)$. In other words, we do not have an algorithm for computing f, so we may not allow this denition scheme. One possibility would be to restrict the scheme to the case in which, for every $(x_1, x_2, \ldots, x_p)$, there exists a z such that $(x_1, x_2, \ldots, x_p, z) \in A$ (we will call this the **total** μ**-operator**). We would obtain in this way, as we will see in later sections, all the recursive functions. It is preferable, however, to dene the class of partial recursive functions; the reason for this is that the enumeration and xed point theorems (see Theorems 5.32 and 5.52), which are essential in this subject, are true only for this latter class (see Exercise 22).

Our task now is to formalize this intuition. To begin with, here are some denitions that pertain to partial functions.

Denition 5.16 *A **partial function from** $\mathbb{N}^p$ **into** $\mathbb{N}$ is a pair (A, f) where $A \subseteq \mathbb{N}^p$ and f is a mapping from A into $\mathbb{N}$; A is called the **domain** of the function.*

Notation The set of partial functions from $\mathbb{N}^p$ into $\mathbb{N}$ will be denoted by $\mathcal{F}^*_p$ and we let $\mathcal{F}^* = \bigcup_{p \geq 0} \mathcal{F}^*_p$.

If $(a_1, a_2, \ldots, a_p) \notin A$, we will say that the function f is **undened at** $(a_1, a_2, \ldots, a_p)$ or that $f(a_1, a_2 \ldots, a_p)$ is **undened**. We will freely abuse notation by identifying (A, f) and f. We must insist on the fact that two partial functions f and g are equal if, rst, they have the same domain and, second, if they are identical on this domain. If the domain of a partial function $f \in \mathcal{F}^*_p$ is the whole of $\mathbb{N}^p$, we say that f is **total**. The word `function', by itself, will be reserved to refer to functions that are total.

Denition 5.17 *Suppose that $f_1, f_2, \ldots, f_n \in \mathcal{F}^*_p$ and $g \in \mathcal{F}^*_n$. The **composite function** or **composition** $h = g(f_1, f_2, \ldots, f_n)$ is the element of $\mathcal{F}^*_p$ dened as follows:*

- *$h(x_1, x_2, \ldots, x_p)$ is undened if any one of the $f_i(x_1, x_2, \ldots, x_p)$ is undened, or, should all these be dened, if*

$$g(f_1(x_1, x_2, \ldots, x_p), f_2(x_1, x_2, \ldots, x_p), \ldots, f_n(x_1, x_2, \ldots, x_p))$$

  is undened.
- *Otherwise, $h(x_1, x_2, \ldots, x_p)$ is dened and is equal to*

$$g(f_1(x_1, x_2, \ldots, x_p), f_2(x_1, x_2, \ldots, x_p), \ldots, f_n(x_1, x_2, \ldots, x_p)).$$

Remark When working with partial functions, one must be wary of reex actions. For instance, take two functions f and g in $\mathcal{F}^*_1$; it is not always true that $(f+g)-g$ and f are equal: if f is total and g is never dened, for example, then $(f+g)-g$ is never dened. In fact, an algorithm that attempted to calculate $(f+g)-g$ would begin by calculating $f+g$ and would never succeed.

Recursive denitions: Within the class of partial functions, we may dene functions by recursion, thanks to the following fact:

Proposition 5.18 *Suppose $g \in \mathcal{F}^*_p$ and $h \in \mathcal{F}^*_{p+2}$. Then there exists one and only one function $f \in \mathcal{F}^*_{p+1}$ satisfying the following conditions:*

- *For all $(x_1, x_2, \ldots, x_p) \in \mathbb{N}^p$, $f(x_1, x_2, \ldots, x_p, 0) = g(x_1, x_2, \ldots, x_p)$ [which, to be precise, means that $f(x_1, x_2, \ldots, x_p, 0)$ is dened if and only if $g(x_1, x_2, \ldots, x_p)$ is dened and, in this case, is equal to it].*
- *For all $(x_1, x_2, \ldots, x_p, y) \in \mathbb{N}^{p+1}$,*

$$f(x_1, x_2, \ldots, x_p, y + 1) = h(x_1, x_2, \ldots, x_p, y, f(x_1, x_2, \ldots, x_p, y))$$

[the same remark applies: $f(x_1, x_2, \ldots, x_p, y + 1)$ is dened if and only if $h(x_1, x_2, \ldots, x_p, y, f(x_1, x_2, \ldots, x_p, y))$ is dened].

We will say, in this context as well, that f is **dened by recursion from** g **and** h.

Denition 5.19 ***The** μ**-operator** (unbounded). Let $f \in \mathcal{F}^*_{p+1}$. Then we may dene the partial function*

$$g(x_1, x_2, \ldots, x_p) = \mu y\, [f(x_1, x_2, \ldots, x_p, y) = 0]$$

in the following way:

- *if there exists at least one integer z such that $f(x_1, x_2, \ldots, x_p, z)$ equals 0 and if, for every $z' < z$, $f(x_1, x_2, \ldots, x_p, z')$ is dened, then $g(x_1, x_2, \ldots, x_p)$ equals the least such z;*
- *in the opposite case, $g(x_1, x_2, \ldots, x_p)$ is undened.*
  If $A \subseteq \mathbb{N}^{p+1}$, then, by denition,

$$\mu y\, [(x_1, x_2, \ldots, x_p, y) \in A] = \mu y\, [1 \dot{-} \chi_A(x_1, x_2, \ldots, x_p, y) = 0].$$

One must take care that $z = \mu y\, [f(x_1, x_2, \ldots, x_p, y) = 0]$ implies that for all y less than z, $f(x_1, x_2, \ldots, x_p, y)$ is dened and not equal to zero. First of all, this is the denition that one must take to respect our intuition concerning effective computability; second, Exercise 24 will show that to neglect this precaution would lead to disaster.

We may now dene the set of partial recursive functions:

Denition 5.20 *The set of **partial recursive functions** is the smallest subset of $\mathcal{F}^*$ that*

- *contains all the (total) constant functions, the successor function S, the projections P^i_p (for $1 \leq i \leq p$),*
- *is closed under composition, denition by recursion and the (unbounded) μ-operator.*

A subset A of $\mathbb{N}^p$ is called **recursive** if its characteristic function is (total) recursive.

We see in particular that all primitive recursive functions are partial recursive (even total recursive). Also, it is not a problem to show that the closure properties stated in Section 5.1 for the primitive recursive functions are true for the partial recursive functions and the total recursive functions as well. We already know that the Ackerman function is not primitive recursive. We will have to await the end of this chapter (or do Exercise 11) to see that it is recursive; this will prove that the class of recursive functions is strictly larger than the class of primitive recursive functions. Also, it is easy to construct a partial recursive function that is not total: for example, the partial function $f(x) = \mu y\ (2y = x)$ is only dened for the even integers. In Exercise 20, we will provide examples of partial recursive functions whose `partial' character is innate: there are some that are impossible to extend to total recursive functions.

The reader should be convinced that the partial recursive functions are effectively computable in the sense mentioned in the introduction: for each of them, there exists an algorithm that, if the function is dened at the given point, will halt after a nite amount of time and yield the value of the function or, in the opposite case, will never halt. The next section will show how to mechanically compute a partial recursive function.

There remains the converse problem: is an intuitively computable function necessarily recursive? In other words, have we succeeded in our attempt to formalize the notion of an effectively computable function? The afrmative answer to this question is known as **Church's thesis**. It is clear that this afrmation is not subject to proof since we do not have a precise denition of what it means to be an effectively computable function. Indeed, the failure of our initial attempt via the primitive recursive functions should make us cautious. But, in fact, no counterexample to Church's thesis is known and, moreover, experience has shown that every time we have encountered a function that is intuitively effectively computable, this same intuition has enabled us to provide a proof that it is recursive. In this sense, the last theorems of this chapter (the xed point theorems) are strong arguments in favour of Church's thesis.

5.3 Turing machines

Turing machines are theoretical machines that are able to compute, in a sense that we will dene, certain functions in $\mathcal{F}_p^*$. The important fact in this section is that a partial function is computable by a Turing machine if and only if it is partial recursive.

5.3.1 Description of Turing machines

A Turing machine is composed of

- a **tape** consisting of a nite number of parallel **bands** placed horizontally; the tape is bounded to the left and innite to the right; each band is divided into a number of cells with the leftmost cell of each band numbered 1, followed at the right by cells numbered 2, and so on. Because the bands lie parallel on the tape, the cells bearing the same number are aligned vertically.

cell number 1	cell number 2		band number 1
cell number 1	cell number 2		band number 2
cell number 1	cell number 2		band number 3

↑

read head

- a head that we will call the **reading head** or **read head** though it is also able to write or erase symbols on the tape (limited to a single symbol per cell). The head is able to move horizontally; at each instant, it is pointed at a single vertical, i.e. to the sequence of cells bearing the same number n corresponding to the different bands and it can perform these operations (read, write, erase) on all these cells. There are three symbols that the head may be able to either read or write or erase: d, which denotes the debut of the band, the stroke, $|$, and the blank, b. We will use the notation $S = \{d, |, b\}$.

The above is common to all Turing machines. Individual machines are characterized by the following data:

- the number n of its bands;
- a nite set of **states**, E; at each instant, the machine will be in some given state. Every machine has two particular states: its **initial state**, e_i, and its **nal state**, e_f;
- a **table**, M; this is a map from $S^n \times E$ into $S^n \times E \times \{-1, 0, +1\}$. It is often called the machine's **transition table**.

The machine operates by changing its state, writing and erasing on the bands, and moving its head at each instant according to the following rules:

- at the instant $t = 0$, the head is situated on the leftmost cells (numbered 1), on each of which is written the symbol d; there is a symbol written on every cell of the tape; the machine is in the initial state e_i;
- at each instant, t, the machine reads the symbols $s_1, s_2, \ldots, s_n$ written in the cells on which the head is located; the table M indicates what it must do next: if it is in state e and $M(s_1, s_2, \ldots, s_n, e) = (s'_1, s'_2, \ldots, s'_n, e', \varepsilon)$, where $\varepsilon \in \{-1, 0, +1\}$, then the machine erases the symbols $s_1, s_2, \ldots, s_n$ and writes the symbols $s'_1, s'_2, \ldots, s'_n$ in their place; it moves its head one cell to the right if $\varepsilon = +1$, to the left if $\varepsilon = -1$, and does not move its head if $\varepsilon = 0$; nally, it changes to state e'. The instant t is then nished and, at instant $t + 1$, the machine repeats these same operations;
- when the machine reaches the nal state e_f, it stops functioning.

For the machine to operate properly, the table M must satisfy a certain number of constraints:

- the machine must halt as soon as it reaches the nal state; this is reected by the fact that, for all $s_1, s_2, \ldots, s_n \in S^n$, $M(s_1, s_2, \ldots, s_n, e_f) = (s_1, s_2, \ldots, s_n, e_f, 0)$;
- it must not be possible for the machine to write or erase the symbol d that marks the beginning of the tape; also, the head cannot move to the left if it reads the symbol d. Thus, for any state e, $M(d, d, \ldots, d, e) = (d, d, \ldots, d, e', \varepsilon)$, where e' is a state and ε is equal to 0 or $+1$; if $(s_1, s_2, \ldots, s_n) \neq (d, d, \ldots, d)$ and $M(s_1, s_2, \ldots, s_n, e) = (s'_1, s'_2, \ldots, s'_n, e', \varepsilon)$, then none of the s'_i is equal to d.

We will always assume implicitly that these conditions are satised. We will also assume that, at the instant $t = 0$, the symbol d is written at the beginning of each band and nowhere else and that only a nite number of cells contain a symbol other than the blank b. These hypotheses continue to remain valid at every instant. We should note also that the operation of the machine is completely determined: we can be certain of what will be written on the tape at any instant t if we know what is written on the tape at the initial instant $t = 0$.

We will now see how a Turing machine is able to compute a partial function and will show that the partial functions that can be computed in this way are precisely the partial recursive functions.

5.3.2 T-computable functions

For a machine to be able to compute the value of a function f at the point $(x_1, x_2, \ldots, x_n)$, it is obviously necessary to enter the values of the variables $(x_1, x_2, \ldots, x_n)$ in one way or another. This will be done in the initial conguration. To compute a function of p variables, we require a machine with at least $p + 1$ bands: the values of the inputs are entered on the rst p bands, the output is coded on the $(p + 1)$st band, and the remaining bands, if any, are used for intermediate calculations. Let us begin by seeing how to code an integer on a band.

Denition 5.21 *We will say that a band* ***represents*** *an integer x at a given instant if the symbols written on it at that instant are*

$$(d, \underbrace{|, |, \ldots, |}_{x\,strokes}, b, b, \ldots);$$

i.e. the symbol d is in the rst cell, the stroke is in cells numbered $2, 3, \ldots, x + 1$, and the blank character is in all subsequent cells. A band which represents the integer zero (so there is a d followed by blanks) will be called a ***clean band****.*

Denition 5.22 *Let f be a partial function of p variables and let M be a Turing machine with at least $p + 1$ bands. s that $\mathcal{M}$* ***computes*** *f if for every sequence* *of integers* $(x_1, x_2, \ldots, x_p)$, *if the machine is started in the initial conguration (in which bands* $1, 2, \ldots, p$ *represent* $x_1, x_2, \ldots, x_p$ *respectively, all other bands being clean), then*

- *if* $f(x_1, x_2, \ldots, x_p)$ *is undened, then the machine never halts (i.e. never reaches the nal state);*
- *if* $f(x_1, x_2, \ldots, x_p)$ *is dened, the machine will halt in a nite amount of time and, at that instant, the rst band will represent* x_1, *the second band will represent* x_2, *and so on through the pth band, and the* $(p+1)$*st band will represent* $f(x_1, x_2, \ldots, x_p)$. *All other bands, if any, must be clean.*

We say that f *is* T*-**computable** (the* T *is for Turing) if there exists a machine* M *that computes* f.

Remarks (1) We require that the machine cleans any bands used for calculations before halting. This is not really necessary but it will simplify matters when we wish to construct a machine that computes a function dened by composition or by recursion.

(2) There are many other ways to dene Turing machines: some have tapes with only one band, others allow more symbols, etc. These ways are all equivalent in the sense that they all compute the identical set of partial functions. The particular choice presented here was made because, in our opinion, it permits a less complicated proof of the fundamental theorem of this section, namely that the T-computable functions are precisely the partial recursive functions, while at the same time keeping the necessary codings to a minimum.

We are now going to give some examples of T-computable functions. To describe the corresponding Turing machine, we must, for each, explicitly specify the number of bands, the states, and the transition table. In fact, most often, it is only the values assumed by M on some subset of $S^p \times E$ that are relevant; some of these values are imposed once and for all by the constraints described earlier while others may never intervene. So we will limit ourselves to providing the essential part of M.

Example 5.23 The successor function is T-computable: it can be computed by a machine with a set of two states, $\{e_i, e_f\}$. Here is the relevant part of its table:

$$M(d, d, e_i) = (d, d, e_i, +1);$$
$$M(|, b, e_i) = (|, |, e_i, +1);$$
$$M(b, b, e_i) = (b, |, e_f, 0).$$

Example 5.24 The function $\lambda x.2x$ is T-computable: this machine also has two bands but has four states, $\{e_i, e_f, e_1, e_2\}$. Here is how it operates: it will read the rst band from left to right and each time it encounters a stroke it will write one below it on the second band as well as another stroke at the `end' of the second band. Here is its table:

$M(d, d, e_i) = (d, d, e_i, +1)$	start;
$M(\vert, b, e_i) = (\vert, \vert, e_1, +1)$ $M(b, b, e_i) = (b, b, e_f, 0)$	add a rst stroke to the second band unless, of course, x equals zero;
$M(\vert, b, e_1) = (\vert, b, e_1, +1)$ $M(b, \vert, e_1) = (b, \vert, e_1, +1)$	move to the end of the word;
$M(b, b, e_1) = (b, \vert, e_2, -1)$	add a stroke to the second band;
$M(b, \vert, e_2) = (b, \vert, e_2, -1)$ $M(\vert, b, e_2) = (b, \vert, e_2, -1)$	return to the last stroke that has just been doubled;
$M(\vert, \vert, e_2) = (\vert, \vert, e_i, +1)$ $M(b, \vert, e_i) = (b, , e_f, 0)$	repeat, until all the strokes have been doubled.

In fact, we could have dispensed with the preceding example for we are now going to present a general proof that all partial recursive functions are T-computable. To do this, we need to show that the constant functions, the projections, and the successor function are T-computable and that the set of T-computable functions is closed under composition, denitions by recursion, and the μ-operator. The case of the successor function was already treated above.

- Let us begin with the projection function P_p^i. We can easily describe a machine that will compute this function; it has $p + 1$ bands, two states e_i and e_f, and the following transition table, M:

$$M(d, d, \ldots, d, e_i) = (d, d, \ldots, d, e_i, +1);$$
$$M(s_1, s_2, \ldots, s_p, b, e_i) = (s_1, s_2, \ldots, s_p, |, e_i, +1) \quad \text{if } s_i = |;$$
$$M(s_1, s_2, \ldots, s_p, b, e_i) = (s_1, s_2, \ldots, s_p, b, e_f, 0) \quad \text{if } s_i = b.$$

- The constant functions are also T-computable; here is a description of a machine that will calculate the function of p variables whose constant value is k. This machine has $p + 1$ bands and $k + 2$ states, $e_i, e_f, e_1, e_2, \ldots, e_k$. Here is its table:

$M(d, d, ..., d, e_i) = (d, d, ..., d, e_i, +1)$;	
$M(s_1, s_2, ..., s_p, b, e_n) = (s_1, s_2, ..., s_p, \vert, e_{n+1}, +1)$	for all $s_1, s_2, ..., s_p$ and for all n between 1 and $k - 1$;
$M(s_1, s_2, ..., s_p, b, e_k) = (s_1, s_2, ..., s_p, \vert, e_f, 0)$	for all $s_1, s_2, ..., s_p$.

We must now show that the set of T-computable functions is closed under composition, denitions by recursion, and the μ-operator. Let us begin with composition: assume that $f_1, f_2, \ldots, f_n$ are in $\mathcal{F}_p^*$, $g \in \mathcal{F}_n^*$, and that all these functions are T-computable. We must construct a Turing machine $\mathcal{M}$ that will compute the partial function $h = g(f_1, f_2, \ldots, f_n)$. We know that for i between 1 and n, there is a Turing machine $\mathcal{M}_i$ that computes f_i; we may suppose that this machine has p_i bands ($p_i \geq p + 1$) and that its set of states is E_i. At the cost of renaming the elements of the E_i, we may assume that the sets E_i are pairwise disjoint (so that their initial and nal states are all distinct). The function g is also computable by a machine $\mathcal{N}$; this machine has n' bands and its set of states is E; we will also assume that E is disjoint from all the E_i. The set of states of the machine $\mathcal{M}$ is $E \cup (\bigcup_{1 \leq i \leq n} E_i)$; its initial state is the initial state of $\mathcal{M}_1$ and its nal state is the nal state of $\mathcal{N}$. The number of bands of $\mathcal{M}$ is $p' = p + \sum_{i=1}^{i=n}(p_i - p) + n' - n$. We will restrict ourselves to describing the operation of the machine $\mathcal{M}$ that computes h and we leave it to the reader, if desired, to write down its exact table.

So assume that at the instant $t = 0$, the integers $x_1, x_2, \ldots, x_p$ are represented on the rst p bands and that all other bands are clean. The machine begins by computing $f_1(x_1, x_2, \ldots, x_p)$, behaving as $\mathcal{M}_1$ would, except that it does not use band number $p + 1$; instead, it uses p_1 of the p' bands which it has at its disposal (specically, the rst p bands and $p_1 - p$ of the others) and ignores the remainder. When it nishes this computation (if it ever nishes), the integers $x_1, x_2, \ldots, x_p$ are still represented on the rst p bands and the result, $f(x_1, x_2, \ldots, x_p)$, is represented on another band which we label B_1. The nal state of the rst machine returns the head to the beginning of the tape; when it then reads the sequence of ds, it places $\mathcal{M}$ in the initial state of $\mathcal{M}_2$. Using the rst p bands and $p_2 - p$ new bands, it will now compute $f_2(x_1, x_2, \ldots, x_p)$, which, at the end of this computation (if it does end), will be represented on a band labelled B_2; after this, it will return its head to the beginning of the tape and will calculate $f_3(x_1, x_2, \ldots, x_p)$, and so on. The only precaution it needs to take is to avoid using band number $p + 1$ while computing the intermediate results $f_i(x_1, x_2, \ldots, x_p)$. When this is nished and the head is back at the beginning of the tape, $\mathcal{M}$ moves into the initial state of $\mathcal{N}$ and works, as $\mathcal{N}$ would, using B_1 [which, as we recall, now represents $f_1(x_1, x_2, \ldots, x_p)$] as its rst band, B_2 as its second band, and so on, this time using band $p + 1$ to write the result $h(x_1, x_2, \ldots, x_p)$. In the end, it remains only to erase the contents of bands B_1, B_2, etc.

Now let us see how to compute a function dened by recursion. This involves computing the partial function $f \in \mathcal{F}_{p+1}^*$ whose denition is

$$f(x_1, x_2, \ldots, x_p, 0) = g(x_1, x_2, \ldots, x_p),$$
$$f(x_1, x_2, \ldots, x_p, y + 1) = h(x_1, x_2, \ldots, x_p, y, f(x_1, x_2, \ldots, x_p, y)),$$

where $g \in \mathcal{F}_p^*$ and $h \in \mathcal{F}_{p+2}^*$ are partial functions that are computed by machines $\mathcal{M}$ and $\mathcal{M}'$, respectively. We assume that machines $\mathcal{M}$ and $\mathcal{M}'$ have $p + 1 + k$

and $p+3+k'$ bands, respectively, that their sets of states are E and E', and that E and E' are disjoint. The nal state of $\mathcal{M}$ is e_f and that of $\mathcal{M}'$ is e'_f. The machine $\mathcal{N}$ that computes f has $p+4+k+k'$ bands. Its set of states is

$$E \cup E' \cup \{e_0, e_1, e_2, e_3, e_4, e_5, e_6, e_7\},$$

where the e_i, for i between 0 and 7, are new states that do not already belong to $E \cup E'$; its initial state is the initial state of the machine $\mathcal{M}$. The machine $\mathcal{N}$ must be constructed so that if the integers $x_1, x_2, \ldots, x_p, x_{p+1}$ are represented on bands 1 through $p+1$ at the instant $t = 0$, the integer $f(x_1, x_2, \ldots, x_p, x_{p+1})$ will be represented on band $p+2$ at the end of the computation. Here is how it operates.

To begin, it will behave as $\mathcal{M}$ does using bands $1, 2, \ldots, p$ and $p+4$ as well as k additional bands for the calculations. So at the end of this rst stage, band $p+4$ represents $g(x_1, x_2, \ldots, x_p)$; then, the head returns to the beginning of the tape and $\mathcal{N}$ enters state e_0.

In this way, the machine will compute $f(x_1, x_2, \ldots, x_p, 1)$, $f(x_1, x_2, \ldots, x_p, 2)$, through $f(x_1, x_2, \ldots, x_p, p+1)$ in succession. In the process of computing $f(x_1, x_2, \ldots, x_p, y+1)$, the number y is coded on band $p+2$ and the value $f(x_1, x_2, \ldots, x_p, y)$ is coded on band $p+3$. When it nds itself in state e_0, it transfers the contents of band $p+4$ onto band $p+3$ (while erasing band number $p+4$) and compares the contents of band $p+2$ with those of band $p+1$ (which represents x_{p+1}); if it sees that these two numbers are equal, it erases the contents of band $p+2$ and halts; if not, it returns to the beginning of the tape and enters the initial state of the machine $\mathcal{M}'$. It then operates as this machine would, treating bands $1, 2, \ldots, p, p+2$ and $p+3$ as input, and writes the result on band $p+4$. When this computation is nished, the machine adds a stroke to band $p+2$, brings its head to the beginning of the tape and returns to state e_0. The reader who is amused by this sort of thing can write down the transition table N of the machine $\mathcal{N}$ and assign precise roles to the states e_0 through e_7.

The μ-operator: we now wish to construct a machine $\mathcal{N}$ that will compute

$$g(x_1, x_2, \ldots, x_p) = \mu y\,[f(x_1, x_2, \ldots, x_p, y) = 0],$$

where f is itself a partial function that is computed by machine $\mathcal{M}$. We will suppose that $\mathcal{M}$ has $p+2+k$ bands and that its set of states is E. The machine $\mathcal{N}$ also has $p+2+k$ bands and its set of states is $E \cup \{e_0, e_1, e_2, e_3\}$, where e_0, e_1, e_2, and e_3 are not already in E; the initial state of $\mathcal{N}$ is that of $\mathcal{M}$ and its nal state is e_3.

Once again, we will content ourselves with a description of the behaviour of $\mathcal{N}$. It begins to operate exactly as $\mathcal{M}$ would with the input $x_1, x_2, \ldots, x_p, 0$; so it will write the value $f(x_1, x_2, \ldots, x_p, 0)$ (if this is dened, of course) on band $p+2$; it then returns its head to the beginning of the tape and enters state e_0. It then moves one step to the right and reads the contents of the second cell on band $p+2$; if this is a blank, b, then the computation is nished and the machine passes into state e_3. If not, it passes into state e_1, whose effect is to have the machine replace the rst blank that it nds on band $p + 1$ by a stroke; it then passes to state e_2, whose effect is to bring the head back to the beginning of the tape and return the machine to the initial state of $\mathcal{M}$. The machine will thereby successively compute $f(x_1, x_2, \ldots, x_p, 0)$, $f(x_1, x_2, \ldots, x_p, 1)$, and so on and will halt only if and when the result produced is a 0.

We have thus concluded the proof of the following theorem:

Theorem 5.25 *Every partial recursive function is T-computable.*

Remark It is important to realize that the machines we have described for computing partial functions dened by composition, by recursion, or with the μ-operator do not just compute the values of these functions when they are dened; they also have the property that they do not halt when the function in question is not dened on the initial input.

5.3.3 *T*-computable partial functions are recursive

For this and the next three subsections, we x a Turing machine $\mathcal{M}$ and assume that $\mathcal{M}$ computes a partial function $f \in \mathcal{F}_p^*$. To prove that, under these conditions, f is recursive, we will rst code the status of the machine $\mathcal{M}$ at the instant t by an integer and will show that this code is a primitive recursive function of t and the initial conditions. To do this, we will need to use the functions α_n introduced in the earlier subsection on coding of sequences.

It is clear that actual names of the states is of no importance. At the cost of renaming them, we may suppose that the set of states is $\{0, 1, 2, \ldots, m\}$ and, for convenience, that the initial state is 0 and the terminal state is 1. Also, we will identify the blank symbol b with 0, the symbol d that marks the beginning of a band with 1, and the stroke symbol with 2.

Denition 5.26 *Suppose that $\mathcal{M}$ has n bands. The innite sequence $C(t) = (s_0, s_1, \ldots, s_i, \ldots)$, where for all n, u, and v $(0 \leq v < n)$, s_{u+v} is the symbol written in cell number $u+1$ of band number $v+1$, is called the* ***conguration of $\mathcal{M}$ at the instant*** *t; note that, according to the convention above, s_i is an integer equal to 0 or 1 or 2. The* ***situation of $\mathcal{M}$ at the instant*** *t is the triple $S(t) = (e, k, C(t))$, where e is the state the machine is in at the instant t, $C(t)$ is the conguration of the machine at the instant t, and k is the number of the cells above which the head of the machine is positioned at the instant t.*

To put this another way, $C(t)$ is the sequence obtained by placing end to end the sequences $\sigma_1, \sigma_2, \ldots, \sigma_i, \ldots$ where σ_i is the sequence $(t_i^1, t_i^2, \ldots, t_i^n)$ and t_i^j is the symbol written in the ith cell of the jth band. We have already noted that this innite sequence has only a nite number of non-blank (or non-zero) terms. We will use the following method to code innite sequences of symbols that contain only a nite number of non-zero terms: to the sequence $C = (s_0, s_1, \ldots, s_i, \ldots)$ will correspond the code

$$\Gamma(C) = \sum_{i \geq 0} s_i \cdot 3^i.$$

We will use this same coding for nite sequences: to $C = (s_0, s_1, \ldots, s_q)$ will correspond the code

$$\Gamma(C) = \sum_{0 \leq i \leq q} s_i \cdot 3^i.$$

If we know the code $\Gamma(C)$ of the conguration C, we can easily recover the symbol that is written on any one of the bands. To do this, let $q(x, y)$ and $r(x, y)$ denote the quotient and remainder that result from division of x by y [if $y = 0$, we will arbitrarily set $q(x, y) = r(x, y) = 0$]. The symbol written on cell number u of band number v is

$$r(q(\Gamma(C), 3^{n(u-1)+v-1}), 3).$$

We can just as easily recover the sequence σ of n symbols written on the cells numbered u of the different bands: set

$$\varepsilon(x, y, z) = r(q(x, 3^{z(y-1)}), 3^z).$$

Then the code $\Gamma(\sigma)$ of the sequence σ is

$$\Gamma(\sigma) = \varepsilon(\Gamma(C), u, n).$$

The situation $S = (e, k, C)$ of the machine will be coded by the integer

$$\Gamma(S) = \alpha_3(e, k, \Gamma(C)).$$

The next lemma expresses the fact that we can deduce the situation of the machine at the instant $t + 1$ if we know its situation at the instant t.

Lemma 5.27 *There is a primitive recursive function $g \in \mathcal{F}_1$ such that if x is the code of the situation of the machine at the instant t, then $g(x)$ is the code of the situation of the machine at the instant $t + 1$.*

Proof The function g is dened by cases [precisely $3^n \cdot (m + 1) + 1$ cases]. For each sequence $\sigma = (s_0, s_1, \ldots, s_{n-1})$ of elements of $\{0, 1, 2\}$ and for every $j \in \{0, 1, \ldots, m\}$, we will describe what happens at the instant t if the machine is in state j and reads the sequence σ.

The state the machine is in, the position of the cells being read, and the cong-uration of the bands are, respectively, $\beta_3^1(x)$, $\beta_3^2(x)$, and $\beta_3^3(x)$. The code of the sequence that the head is in the process of reading is $\varepsilon(\beta_3^3(x), \beta_3^2(x), n) = c$.

For every c between 0 and $3^n - 1$ inclusive [these are the possible values that $\Gamma(\sigma)$ can have if σ is a sequence of length n] and for every j between 0 and m:

(a) If $\beta_3^1(x) = j$, if $\varepsilon(\beta_3^3(x), \beta_3^2(x), n) = c$, if $\Gamma(s_0, s_1, \ldots, s_{n-1}) = c$ [in other words, for every i between 0 and $n - 1$ inclusive, set $s_i = r(q(c, 3^i), 3) = \varepsilon(c, i + 1, 1)$], if $M(s_0, s_1, \ldots, s_{n-1}, j) = (t_0, t_1, \ldots, t_{n-1}, h, \omega)$ (so that the t_i are all equal to 0 or 1 or 2, h is between 0 and m inclusive and ω is equal to -1 or 0 or $+1$), and if $\Gamma(t_0, t_1, \ldots, t_{n-1}) = c'$, then
- the new state will be h;
- the new position of the head will be $\beta_3^2(x) + \omega$;
- the new conguration will differ from the old one only in cells numbered $\beta_3^2(x)$, which correspond, in the conguration of $\mathcal{M}$, to those indices between $n(\beta_3^2(x) - 1)$ and $n(\beta_3^2(x) - 1) + n - 1$ inclusive at which the t_i will replace the s_i. Its new code is therefore

$$\beta_3^3(x) + 3^{n \cdot (\beta_3^2(x) - 1)}(c - c').$$

To recapitulate,

$$g(x) = \alpha_3(h, \beta_3^2(x) + \omega, \beta_3^3(x) + 3^{n \cdot (\beta_3^2(x) - 1)}(c - c')).$$

(b) If $\beta_3^1(x) > m$ or if $\varepsilon(\beta_3^3(x), \beta_3^2(x), m)$ is strictly greater than $3^n - 1$ (this case will never happen if x is really the code of a situation), we arbitrarily set $g(x) = 0$.

The function g we have just dened is certainly primitive recursive because it is dened by cases using functions that are primitive recursive and sets that are primitive recursive. The function M that occurs in the denition is totally inoffensive. ∎

Let us now prove the fact, which is intuitively clear, that if we know the initial situation of the machine, we can deduce its situation at any instant. Dene the function $Sit(t, x_1, x_2, \ldots, x_p)$ by induction. Set

$$Sit(0, x_1, x_2, \ldots, x_p) = \alpha_3(0, 1, \Gamma(C)),$$

where C is the conguration in which band 1 represents x_1, band 2 represents x_2, and so on through the pth band representing x_p, all other bands being clean, and set

$$Sit(t + 1, x_1, x_2, \ldots, x_p) = g(Sit(t, x_1, x_2, \ldots, x_p)).$$

Lemma 5.28 *The function $Sit(t, x_1, x_2, \ldots, x_p)$ is primitive recursive. For all $t, x_1, x_2, \ldots, x_p$, $Sit(t, x_1, x_2, \ldots, x_p)$ is equal to the code $\Gamma(S)$ of the situation of the machine at the instant t, assuming that, at $t = 0$, the integers $x_1, x_2, \ldots, x_p$ are represented on bands $1, 2, \ldots, p$ and that all other bands are clean.*

Proof With what we have just seen, it is sufcient to prove that $Sit(0, x_1, x_2, \ldots, x_p)$ is a primitive recursive function of $x_1, x_2, \ldots, x_p$. We have

$$Sit(0, x_1, x_2, \ldots, x_p) = \alpha_3(0, 1, \Gamma(C)),$$

where C is the initial conguration of the bands. So it sufces to show that $\Gamma(C)$ is a primitive recursive function of $x_1, x_2, \ldots, x_p$. Let $\rho(i, x)$ denote the function that is equal to 2 if $i \leq x$ and equal to 0 otherwise. Now, if $C = (s_0, s_1, \ldots, s_i, \ldots)$ is the initial conguration of the machine, then s_i is the symbol written on band number $r(i, n) + 1$ in cell number $q(i, n)$. Thus we have

$$s_i = \begin{cases} 1 & \text{if } 0 \leq i \leq n-1, \\ \rho(q(i, n), x_{r(i,n)+1}) & \text{if } i \geq n; \end{cases}$$

hence the function $\lambda i x_1 x_2 \ldots x_p . s_i$ is a primitive recursive function and so is the function $\Gamma(C)$, which is equal to

$$\sum_{i=0}^{i=(n+1)\cdot\sup(x_1, x_2, \ldots, x_p)} 3^i \cdot s_i. \qquad \blacksquare$$

We may now complete the proof that f, the function computed by the machine $\mathcal{M}$, is a partial recursive function. If it is dened, $f(x_1, x_2, \ldots, x_p)$ is equal to the number of strokes written on band number $p + 1$ when the machine nishes its computation. We begin by nding the **computation time** (the rst instant that the machine is in its nal state):

$$T(x_1, x_2, \ldots, x_p) = \mu t\, [\beta_3^1(Sit(t, x_1, x_2, \ldots, x_p)) = 1],$$

which is dened if and only if $f(x_1, x_2, \ldots, x_p)$ is also dened; and once we know the situation of the machine at this instant $T(x_1, x_2, \ldots, x_p)$, it is not difcult to count the number of strokes written on the $(p + 1)$st band. Let α be the function dened by

$$\alpha(x) = \mu y\, [r(q(\beta_3^3(x), 3^{n\cdot(y+1)+p}), 3) = 0];$$

indeed, if x is the code of the situation of the machine,

$$r(q(\beta_3^3(x), 3^{n\cdot(y+1)+p}), 3) = 0$$

means that the symbol in cell $y + 2$ is the blank. So we see that $\alpha(x)$ is truly the number of consecutive strokes (recall that the rst cell contains a d) at the beginning of band number $p + 1$.

We may now compute f:

$$f(x_1, x_2, \ldots, x_p) = \alpha(Sit(T(x_1, x_2, \ldots, x_p), x_1, x_2, \ldots, x_p)).$$

To summarize, we have

Theorem 5.29 *If f is a partial function that is T-computable, then f is recursive.*

We may observe that the only use of the unbounded μ-operator is for the definition of the function T. Indeed, to dene the function α, it sufces to use the bounded μ-operator; this follows from the fact that if x is the code of the situation of the machine, then there are certainly no more than x strokes written on its bandsóso α is actually primitive recursive. Precisely,

$$\alpha(x) = \mu y \leq x \, [r(q(\beta_3^3(x), 3^{n \cdot (y+1)+p}), 3) = 0].$$

In later subsections, we will exploit more deeply the argument that was just made. For the present, we will just make a few observations that follow from the way that the function f was written above.

- If a function $f \in \mathcal{F}_p$ is computable by a Turing machine in a time $T(x_1, x_2, \ldots, x_p)$ that is a primitive recursive function, then the function f is itself primitive recursive.
- We also see that the set of partial recursive functions is at most equal to the smallest subset $\mathcal{A}$ of $\mathcal{F}_p$ that contains the primitive recursive functions and is closed under composition and the μ-operator (in other words, if we already have all the primitive recursive functions, then denitions by recursion are no longer necessary): for suppose that $\mathcal{M}$ is a machine that computes a partial function $f \in \mathcal{F}_p$; the partial function T we have just dened clearly belongs to $\mathcal{A}$ (the function *Sit* is primitive recursive) as does f since it is obtained using T and functions that are primitive recursive.
- Next, consider the smallest subset $\mathcal{B}$ of $\mathcal{F}$ that contains the primitive recursive functions and is closed under composition and the total μ-operator (i.e. the μ-operator can be used only if the function it denes is total). This set is exactly equal to the set of total recursive functions: for if f is a total recursive function computed by a machine $\mathcal{M}$, the function T that corresponds to it belongs to $\mathcal{B}$, so we see that f does also.

5.3.4 Universal Turing machines

Thus far, we have constructed a Turing machine for each partial recursive function that we wished to compute. We will now see that there is a single Turing machine that is able to compute all the partial recursive functions (of a xed number of variablesóbut this is not a real restriction since we know how to code a partial function in $\mathcal{F}_n^*$ by a partial function in $\mathcal{F}_1^*$ using the function α_n). The idea is to construct a machine whose input consists not only of the values of the variables but also includes the instructions that it should follow. We will begin instead by constructing a universal (in a sense that will be clear by the end of this section) recursive function. To do this, it is essential to establish a coding of Turing machines.

We have already said that a Turing machine is determined by

- the number of its bands;
- the set E of its states, which we continue to assume is of the form $\{0, 1, \ldots, m\}$ with 0 as the initial state and 1 as the nal state;
- its transition table M, which is a map from $S^n \times E$ into $S^n \times E \times \{-1, 0, +1\}$, i.e. from a nite set into a nite set. It is entirely possible, though a bit complicated, to code this mapping by an integer. For each sequence $\sigma = (s_1, s_2, \ldots, s_n, e)$ in $S^n \times E$, let us dene, successively,

$$\begin{aligned} r_1 &= \alpha_2(\Gamma(s_1, s_2, \ldots, s_n), e); \\ r_2 &= \alpha_3(\Gamma(t_1, t_2, \ldots, t_n), e', \varepsilon + 1), \\ &\qquad \text{where } (t_1, t_2, \ldots, t_n, e', \varepsilon) = M(\sigma); \\ n(\sigma) &= [\pi(r_1)]^{r_2}. \end{aligned}$$

[Recall that $\pi(i)$ is the $(i + 1)$st prime number.]
The code of the table M will be the integer u dened by

$$u = \prod_{\rho \in S^n \times E} n(p).$$

It is easy to recover M from its code: if we wish to know

$$(t_1, t_2, \ldots, t_n, e', \varepsilon) = M(s_1, s_2, \ldots, s_n, e),$$

we compute the code $c = \Gamma(s_1, s_2, \ldots, s_n)$ and $r = \alpha_2(c, e)$. We then make use of the function δ introduced in Section 5.1:

$$\delta(r, u) = \alpha_3(c', e', \varepsilon + 1), \quad \text{where } c' = \Gamma(t_1, t_2, \ldots, t_n),$$

and the decoding can be completed without difculty.

Denition 5.30 *The **index** of a machine $\mathcal{M}$ is the integer $a_3(n, m, u)$, where n is the number of bands of $\mathcal{M}$, $m + 1$ is the number of its states, and u is the code of its transition table. Clearly, the condition '. . . is the index of a Turing machine' is very restrictive and that the rst integer that satises it is very large. For every integer p, set*

$$I_p = \{x : x \text{ is the index of a Turing machine that has at least } p + 1 \text{ bands}\}.$$

It would be terribly annoying, although very easy, to verify that these sets are primitive recursive.

Next, let us dene the function $ST^p(i, t, x_1, x_2 \ldots, x_p)$ as follows:

- If $i \in I_p$, $ST^p(i, t, x_1, x_2, \ldots, x_p) = \Gamma(S(t))$ is the code at the instant t of the situation of the machine with index i which began to operate at the instant $t = 0$ with the following conguration: the integers $x_1, x_2, \ldots, x_p$ are represented on bands $1, 2, \ldots, p$ and all other bands are clean (observe that this code is never equal to zero).

- $ST^p(i, t, x_1, x_2, \ldots, x_p) = 0$ otherwise.

Theorem 5.31 *For every integer p, the function $ST^p(i, t, x_1, x_2, \ldots, x_p)$ is primitive recursive.*

Proof The function ST^p is dened by cases according as $i \in I_p$ or not. If $i \notin I_p, ST^p(i, t, x_1, x_2, \ldots, x_p) = 0$. It is the other case that requires slightly more work.

It is not hard to see that there exists a primitive recursive function $h(i, x_1, x_2, \ldots, x_p)$ whose value, if i is the index of a Turing machine $\mathcal{M}$ with at least $p + 1$ bands, is the code of the initial situation of $\mathcal{M}$ when $x_1, x_2, \ldots, x_p$ are represented on bands $1, 2, \ldots, p$, the other bands being clean. It now sufces to repeat the calculations made for Lemma 5.28, replacing n by $\beta_3^1(i)$ (which is the number of bands of the machine whose index is i). We do not have to worry about the values this function will take (it is primitive recursive, so is always dened) if i is not the index of a Turing machine or if $\beta_3^1(x) < p + 1$.

We then have to prove that there exists a primitive recursive function $g(i, x)$ such that, if i is the index of a machine $\mathcal{M}$ and x is the code of the situation of $\mathcal{M}$ at the instant t, then $g(i, x)$ is the code of the situation of $\mathcal{M}$ at the instant $t + 1$. We imitate the proof of Lemma 5.27; note that here we do not even need a denition by cases. The sequence σ of symbols that the head is in the process of reading has $c = \varepsilon(\beta_3^3(x), \beta_3^2(x), \beta_3^1(i))$ as its code. If c' is the code of the sequence that will be written in place of σ, e' is the new state of the machine and ε [$\in \{-1, 0, +1\}$] is the movement of its head, we have

$$\alpha_3(c', e', \varepsilon + 1) = \delta(\alpha_2(c, \beta_3^1(x)), \beta_3^3(i)).$$

To simplify the notation, set $\delta(\alpha_2(c, \beta_3^1(x)), \beta_3^3(i)) = \delta$; then

$$g(i, x) = \alpha_3(e', k', \Gamma(C')),$$

where

$$\begin{aligned} e' &= \beta_3^2(\delta); \\ k' &= \beta_3^2(x) + \beta_3^3(\delta) - 1; \\ \Gamma(C') &= \beta_3^3(x) + 3^{\beta_3^1(i)\cdot(\beta_3^2(x)-1)}(c' - c), \quad \text{where } c' = \beta_3^1(\delta). \end{aligned}$$

■

Finally, we prove by induction (as for Lemma 5.28) that the situation of a Turing machine at the instant t is a primitive recursive function $ST^p(i, t, x_1, x_2, \ldots, x_p)$ of its index, its initial situation, and t. For each strictly positive integer p, let us dene the partial function $\phi^p(i, x_1, x_2, \ldots, x_p)$ as follows:

- if $i \notin I_p, \phi^p(i, x_1, x_2, \ldots, x_p)$ is undened;
- if $i \in I_p$, then we put the machine with index i into operation with $x_1, x_2, \ldots, x_p$ represented on bands $1, 2, \ldots, p$, other bands being clean and

declare that:

ó $\phi^p(i, x_1, x_2, \ldots, x_p)$ is undened if this machine never halts;

ó $\phi^p(i, x_1, x_2, \ldots, x_p)$ is equal to the number of consecutive strokes at the beginning of band number $p + 1$ if this machine does halt.

Theorem 5.32. *(The enumeration theorem) For every integer p, ϕ^p is a partial recursive function. Moreover, if f is a partial recursive function of p variables, there exists an integer i such that*

$$f = \lambda x_1 x_2 \ldots x_p . \phi^p(i, x_1, x_2, \ldots, x_p).$$

Proof Once again, the proof imitates that of Lemma 5.28. We introduce a partial recursive function T^p and primitive recursive predicates B^p and C^p that will be of use later. $T^p(i, x_1, x_2, \ldots, x_p)$ is the computation time of the machine whose index is i on the input $x_1, x_2, \ldots, x_p$ if $i \in I_p$; it is undened if this computation does not halt.

$$T^p(i, x_1, x_2, \ldots, x_p) = \mu t\,[\beta_3^1(ST^p(i, t, x_1, x_2, \ldots, x_p)) = 1].$$

Observe that if $i \notin I_p$, $T^p(i, x_1, x_2, \ldots, x_p)$ is undened. For each integer p, we set

$$B^p = \{(i, t, x_1, x_2, \ldots, x_p) : ST^p(i, t, x_1, x_2, \ldots, x_p) = 1\}, \quad \text{and}$$
$$B^p(i) = \{(t, x_1, x_2, \ldots, x_p) : (i, t, x_1, x_2, \ldots, x_p) \in B^p\}.$$

These sets are primitive recursive and $(t, x_1, x_2, \ldots, x_p) \in B^p(i)$ means (assuming that i is the index of a Turing machine) that this machine, when run with input $x_1, x_2, \ldots, x_p$ on its rst p bands, with all other bands clean, completes its computation at the instant t. Continuing with our denitions, we set

$$\begin{aligned} C^p = \{(i, y, t, x_1, x_2, \ldots, x_p) : {} & i \in I_p, (i, t, x_1, x_2, \ldots, x_p) \in B^p \\ & \text{and the number of strokes at the instant } t \text{ on band } p+1 \text{ of} \\ & \text{the machine with index } i \text{ which started with } x_1, x_2, \ldots, x_p \text{ on} \\ & \text{its rst } p \text{ bands, with all other bands clean, is exactly } \; y\}, \end{aligned}$$

and

$$C^p(i) = \{(y, t, x_1, x_2, \ldots, x_p) : (i, y, t, x_1, x_2, \ldots, x_p) \in C^p\}.$$

Once again, it is easy to see that these sets are primitive recursive. This allows us to dene the partial function ϕ^p as follows:

$$\phi^p(i, x_1, x_2, \ldots, x_p) = \mu y[(i, y, T^p(i, x_1, x_2, \ldots, x_p), x_1, x_2, \ldots, x_p) \in C^p];$$

this shows clearly that ϕ^p is recursive. ■

We could, as before, slightly improve the presentation to emphasize the fact that the only place the unbounded μ-operator is needed is in the denition of T^p;

indeed, the number of strokes on the band of a Turing machine at any instant must be less than the code of the situation of the machine. Setting

$$\psi(i, t, x_1, x_2, \ldots, x_p)$$
$$= \mu y \leq ST^p(i, t, x_1, x_2, \ldots, x_p)[(i, y, t, x_1, x_2, \ldots, x_p) \in C^p],$$

we note that ψ is a primitive recursive function and that

$$\phi^p(i, x_1, x_2, \ldots, x_p) = \psi(i, T(i, x_1, x_2, \ldots, x_p), x_1, x_2, \ldots, x_p).$$

Since the partial function ϕ^p is itself recursive, it is computable by some Turing machine $\mathcal{M}$; this machine can therefore compute all the partial recursive functions of p variables.

For each integer i, let

$$\phi_i^p = \lambda x_1 x_2 \ldots x_p . \phi^p(i, x_1, x_2, \ldots, x_p).$$

We may then observe that the set $\{\phi_i^p : i \in \mathbb{N}\}$ is equal to the set of all partial recursive functions of p variables.

Denition 5.33 *Let $f \in \mathcal{F}_p^*$ be a partial recursive function. We say that $i \in \mathbb{N}$ is an* ***index*** *of f if $f = \phi_i^p$.*

In particular, if i is the index of a Turing machine that computes f, it is also an index of f; but it is clear, for example, that every integer that does not belong to I_p is an index of the partial function in $\mathcal{F}_p^*$ whose domain is empty. It can also happen that j is the index of a machine $\mathcal{M}$ that has at least $p + 1$ bands but that this machine does not compute a partial function in $\mathcal{F}_p^*$ for the simple reason that, when it halts, its bands are not in a conguration required by Denition 5.22; so we again have an example of an integer i that is the index of a function f although the machine whose index is i does not compute f in the strict sense of the term.

5.4 Recursively enumerable sets

5.4.1 Recursive and recursively enumerable sets

Denition 5.34 *Let $A \subseteq \mathbb{N}^p$; we say that A is* ***recursive*** *if its characteristic functions χ_A is (total) recursive. We say that A is* ***recursively enumerable*** *if it is the domain of a partial recursive function.*

The domain of the partial function whose index is x (i.e. ϕ_x^p) will be denoted by W_x^p. It is clear that the set $\{W_x^p : x \in \mathbb{N}\}$ is the set of all recursively enumerable subsets of $\mathbb{N}^p$. If $A = W_x^p$, we will say that x is an **index** of A. In this section, we will prove a few simple facts about recursive and recursively enumerable sets.

Lemma 5.35 *Every recursive set is recursively enumerable.*

Proof The partial functions $f = \mu y\ [y + 1 = x]$ is recursive, undened at 0 and dened for all other values. If χ_A is the characteristic function of a recursive set A, then $f \circ \chi_A$ is a partial recursive function whose domain is A. ■

Lemma 5.36 *For every integer p, the set of recursive subsets of $\mathbb{N}^p$ is closed under Boolean operations.*

Proof The proof is the same as for the set of primitive recursive functions. ■

Lemma 5.37 *The union and the intersection of two recursively enumerable subsets of $\mathbb{N}^p$ are both recursively enumerable.*

Proof Let A_1 and A_2 be recursively enumerable subsets of $\mathbb{N}^p$ and suppose that they are, respectively, the domains of the partial functions f_1 and f_2 computed by machines whose indices are i_1 and i_2.

To begin with, it is clear that $A_1 \cap A_2$ is the domain of the function $f_1 + f_2$. On the other hand, $A_1 \cup A_2$ is the domain of the partial function

$$\mu t\ [(t, x_1, x_2, \ldots, x_p) \in B^p(i_1) \cup B^p(i_2)],$$

which is recursive since the sets $B^p(i)$ are primitive recursive, as we saw in the proof of the enumeration theorem. ■

The three properties that follow are so important that they deserve to be labelled theorems.

Theorem 5.38 *Let $A \subseteq \mathbb{N}^p$; A is recursive if and only if A and $\mathbb{N}^p - A$ are both recursively enumerable.*

Proof In one direction, this is clear: if A is recursive, then so is $\mathbb{N}^p - A$ by Lemma 5.36, so these two sets are recursively enumerable by Lemma 5.35.

Let i be the index of a machine that computes a partial function whose domain is A and let i' be the index of a machine that computes a partial function whose domain is $\mathbb{N}^p - A$. Then

$$h(x_1, x_2, \ldots, x_p) = \mu t\ [(t, x_1, x_2, \ldots, x_p) \in B^p(i) \cup B^p(i')]$$

is a total recursive function and

$$(x_1, x_2, \ldots, x_p) \in A \text{ if and only if } (h(x_1, x_2, \ldots, x_p), x_1, x_2, \ldots, x_p) \in B^p(i).$$

Therefore, if $\chi(t, x_1, x_2, .., x_p)$ is the characteristic function of $B^p(i)$, then the characteristic function of A is

$$\chi(h(x_1, x_2, \ldots, x_p), x_1, x_2, \ldots, x_p),$$

which shows that A is recursive. ■

Theorem 5.39 *The projection of a recursively enumerable set is recursively enumerable.*

This means that if $A \subseteq \mathbb{N}^{p+1}$ *is recursively enumerable, then the set*

$$B = \{(x_1, x_2, \ldots, x_p) : \textit{there exists an } x_0 \textit{ such that } (x_0, x_1, x_2, \ldots, x_p) \in A\}$$

is also recursively enumerable.

Proof Let i be the index of a Turing machine that computes a function whose domain is A. So we see that

$$(x_0, x_1, x_2, \ldots, x_p) \in A$$

if and only if

there exists an integer t such that $(t, x_0, x_1, x_2, \ldots, x_p) \in B^p(i)$.

Also,

$$(x_0, x_1, x_2, \ldots, x_p) \in B$$

if and only if

there exist integers t and x_0 such that $(t, x_0, x_1, x_2, \ldots, x_p) \in B$.

This shows that B is the domain of the partial recursive function

$$g(x_1, x_2, \ldots, x_p) = \mu z\,[(\beta_2^1(z), \beta_2^2(z), x_1, x_2, \ldots, x_p) \in B^p(i)]. \qquad \blacksquare$$

Theorem 5.40 *Every recursively enumerable subset of* $\mathbb{N}^p$ *is the projection of a primitive recursive subset of* $\mathbb{N}^{p+1}$.

Proof This means that if $A \subseteq \mathbb{N}^p$ is recursively enumerable, there exists a primitive recursive subset $B \subseteq \mathbb{N}^{p+1}$ such that

$$(x_0, x_1, x_2, \ldots, x_p) \in A$$

if and only if

there exists an x_0 such that $(x_0, x_1, x_2, \ldots, x_p) \in B$.

It sufces to take for B the set $B^p(i)$, where i is the index of a Turing machine that computes a function whose domain is A. $\blacksquare$

Here are a few corollaries of these theorems.

Corollary 5.41 *The graph of a partial recursive function is a recursively enumerable set.*

Proof Let $f \in \mathcal{F}_p^*$; we have to show that the set

$$G = \{(x_1, x_2, \ldots, x_p, y) : y = f(x_1, x_2, \ldots, x_p)\}$$

is recursively enumerable.

If i is the index of a machine that computes f, we see that $(x_1, x_2, \ldots, x_p, y) \in G$ if and only if there exists an integer t such that $(y, t, x_1, x_2, \ldots, x_p) \in C^p(i)$;

this shows that G is the projection of a primitive recursive set and is therefore recursively enumerable. ■

The converse is true. If the graph G of a partial function f is recursively enumerable, then f is partial recursive: there exists a primitive recursive set A such that $(x_1, x_2, \ldots, x_p, y) \in G$ if and only if $\exists t\ (x_1, x_2, \ldots, x_p, y, t) \in A$; consequently,

$$f(x_1, x_2, \ldots, x_p) = \beta_2^1(\mu t\,[(x_1, x_2, \ldots, x_p, \beta_2^1(t), \beta_2^2(t)) \in A]).$$

The range of f, i.e. the set of values assumed by f, is itself a projection of the graph of f; hence

Corollary 5.42 *The range of a partial recursive function is recursively enumerable.*

The converse is true; in fact, we have more:

Corollary 5.43 *Every non-empty recursively enumerable subset of* $\mathbb{N}$ *is the range of a primitive recursive function in* $\mathcal{F}_1$.

Proof Let A be a non-empty recursively enumerable subset of $\mathbb{N}$; choose an integer $n \in A$ and let i be an index of A. We then have

$$x \in A \quad \text{if and only if there exists a } t \text{ such that } (t, x) \in B^1(i).$$

It is easy to verify that A is the range of the primitive recursive function g dened by

$$g(z) = \begin{cases} \beta_2^2(z) & \text{if } (\beta_2^1(z), \beta_2^2(z)) \in B^1(i); \\ n & \text{if } (\beta_2^1(z), \beta_2^2(z)) \notin B^1(i). \end{cases}$$

■

We proceed next to a rather subtle point that generalizes the principle of denition by cases to the context of partial recursive functions.

Theorem 5.44 *Let* $g(x_1, x_2, \ldots, x_p)$ *and* $g'(x_1, x_2, \ldots, x_p)$ *be two partial recursive functions and let* A *be a recursive set. Then the function* f *dened by*

$$f(x_1, x_2, \ldots, x_p) = \begin{cases} g(x_1, x_2, \ldots, x_p) & \textit{if } (x_1, x_2, \ldots, x_p) \in A, \\ g'(x_1, x_2, \ldots, x_p) & \textit{otherwise} \end{cases}$$

is a partial recursive function.

[It is important to understand the exact meaning of this denition: if $(x_1, x_2, \ldots, x_p) \in A$, then $f(x_1, x_2, \ldots, x_p)$ is dened if and only if $g(x_1, x_2, \ldots, x_p)$ is dened, and if this is the case, they are equal; the same remark applies to g' if $(x_1, x_2, \ldots, x_p) \notin A$. One should be convinced that f is not equal to

$$g \cdot \chi(A) + g' \cdot \chi(\mathbb{N} - A)$$

in general.]

Proof Let i and i' be indices for g and g'. Consider the following subset C of $\mathbb{N}^{p+2}$:

$$\begin{aligned} C = \{(y, t, x_1, x_2, \ldots, x_p) :& \\ &[(y, t, x_1, x_2, \ldots, x_p) \in C^p(i) \text{ and } (x_1, x_2, \ldots, x_p) \in A], \quad \text{or} \\ &[(y, t, x_1, x_2, \ldots, x_p) \in C^p(i') \text{ and } (x_1, x_2, \ldots, x_p) \notin A]\}. \end{aligned}$$

This set is recursive. The meaning of $(y, t, x_1, x_2, \ldots, x_p) \in C$ is that either $(x_1, x_2, \ldots, x_p) \in A$ and the machine whose index is i nished its computation at the instant t with the value y, or else $(x_1, x_2, \ldots, x_p) \notin A$ and the machine whose index is i' nished its computation at the instant t with the value y. So we see that $f(x_1, x_2, \ldots, x_p)$ is equal to the least y such that there exists a t for which $(y, t, x_1, x_2, \ldots, x_p) \in C$. This implies that

$$f(x_1, x_2, \ldots, x_p) = \beta_2^1(\mu z\, [(\beta_2^1(z), \beta_2^2(z), x_1, x_2, \ldots, x_p) \in C]),$$

which shows that f is partial recursive. ■

5.4.2 The halting problem

Thus far, we have carefully avoided a problem that simply must be dealt with: do there exist recursively enumerable sets that are not recursive? With what we already know, this question amounts to asking if there exists a recursively enumerable set whose complement is not recursively enumerable.

The answer is yes. Let us reconsider the function $\phi^1(i, x)$; set $g(x) = \phi^1(x, x)$ and let A be the domain of g. This set is certainly recursively enumerable. But its complement is not. In fact, for every integer x, $x \in A$ if and only if $x \in W_x^1$. To see this, suppose, to the contrary, that there exists an integer n such that $\mathbb{N} - A = W_n^1$, i.e. it is such that, for every integer x,

$$x \notin A \quad \text{if and only if} \quad x \in W_n^1.$$

Setting x equal to n in these two equivalences leads to

$$\begin{aligned} n \in A \quad &\text{if and only if} \quad n \in W_n^1, \quad \text{and} \\ n \notin A \quad &\text{if and only if} \quad n \in W_n^1, \end{aligned}$$

which is manifestly absurd.

This form of reasoning, very popular among logicians, is known as a **diagonal argument**. Let us give a more precise analysis of this argument which will justify the word `diagonal'; below, we have displayed a two-way table of 0s and 1s whose entries are indexed by integers in such a way that the sequence written on the rst row is the characteristic function of W_0^1, on the second row is the characteristic function of W_1^1, and so on.

In this table, $\varepsilon_{p,n}$ is equal to 1 if $n \in W_p^1$ and to 0 if not. We may then observe that the characteristic function of the set A constructed above is the diagonal of this table and that the sequence corresponding to the characteristic function of the complement of A is

$$1 - \varepsilon_{0,0}, \quad 1 - \varepsilon_{1,1}, \quad \ldots, \quad 1 - \varepsilon_{n,n}, \quad \ldots$$

	0	1	2	...	n	...
W_0^1	$\varepsilon_{0,0}$	$\varepsilon_{0,1}$	$\varepsilon_{0,2}$	...	$\varepsilon_{0,n}$	...
W_1^1	$\varepsilon_{1,0}$	$\varepsilon_{1,1}$	$\varepsilon_{1,2}$	...	$\varepsilon_{1,n}$	...
W_2^1	$\varepsilon_{2,0}$	$\varepsilon_{2,1}$	$\varepsilon_{2,2}$	...	$\varepsilon_{2,n}$	...
$\vdots$		$\vdots$		$\vdots$		
W_n^1	$\varepsilon_{n,0}$	$\varepsilon_{n,1}$	$\varepsilon_{n,2}$	...	$\varepsilon_{n,n}$	...

If $\mathbb{N} - A$ were equal, for a certain n, to W_n^1, then the $(n + 1)$st row of this table would be

$$1 - \varepsilon_{0,0}, \quad 1 - \varepsilon_{1,1}, \quad \ldots, \quad 1 - \varepsilon_{n,n}, \quad \ldots$$

But at the intersection of this row with the $(n + 1)$st column (corresponding to the integer n), which is also on the diagonal, one would simultaneously nd $\varepsilon_{n,n}$ and $1 - \varepsilon_{n,n}$, which is absurd.

Corollary 5.45 *The set* $\{(m, x) : \phi^1(m, x)$ *is dened*$\}$ *is not recursive.*

Proof Indeed, if this set were recursive, then the set

$$\{(x, x) : \phi^1(x, x) \text{ is dened } \}$$

would also be recursive and we have just seen that this is not the case. ■

Our intuition is that a subset $A \subseteq \mathbb{N}$ is recursive if there exists an algorithm which allows us to decide whether an integer belongs to A or not. It is recursively enumerable if there is an algorithm $\mathcal{A}$ that enumerates A. If A is recursively enumerable and we ask whether a given integer n belongs to A or not, we can set the algorithm $\mathcal{A}$ to work. If, at some point, the integer n appears in the sequence enumerated by $\mathcal{A}$, then we are certain that $n \in A$. On the other hand, as long as the integer n has not appeared, we may not conclude anything.

The preceding corollary expresses the fact that when we are given the index of a Turing machine (which intuitively represents its instructions) and its initial conguration, we have no effective way to know whether this machine will stop or not. We rephrase this fact by saying that the **halting problem** for Turing machines is undecidable.

For a great number of problems, it is of interest to know whether they are decidable or not. For example, is there an algorithm for deciding whether a given integer p is prime or not? We have long known (at least since Eratosthenes and his sieve) that the answer is yes. When we wish to be formal about this, we say that the set $\{p \in \mathbb{N}: p \text{ is prime}\}$ is recursive. This is an opportune moment to give a precise denition.

Denition 5.46 *Let $B(x_1, x_2, \ldots, x_p)$ be a property that applies to integers $x_1, x_2, \ldots, x_p$. The problem `does the sequence $(x_1, x_2, \ldots, x_p)$ satisfy B?' is called* ***decidable*** *if the set of sequences $(x_1, x_2, \ldots, x_p)$ for which $B(x_1, x_2, \ldots, x_p)$ is true is a recursive set.*

We need not be restricted to properties that apply (directly) to integers; as was the case for Turing machines, we may use the integers to code other things. The next chapter is full of examples of this type. As employed here, the word `decidable' is intended to reect the intuitive notion of decidability, in other words, decidability by mechanical means; it is entirely justied if we accept the validity of Church's thesis.

5.4.3 The *smn* theorem

The bizarre name of this theorem hides an extremely important result whose meaning is the following: if we consider the partial function $f \in \mathcal{F}^*_{n+m}$ of index i and we x values, say $a_1, a_2, \ldots, a_n$, for the rst n variables, we are left with a partial function $g \in \mathcal{F}^*_m$ given by

$$g = \lambda y_1 y_2 \ldots y_m . f(a_1, a_2, \ldots, a_n, y_1, y_2, \ldots, y_m),$$

and that is clearly recursive. The point is that an index for this function g is effectively computable from i and $a_1, a_2, \ldots, a_n$.

Theorem 5.47. *(The smn theorem) For every pair of integers m and n, there exists a primitive recursive function s^m_n of $n+1$ variables such that, for all $i, x_1, x_2, \ldots, x_n, y_1, y_2, \ldots, y_m$, we have*

$$\begin{aligned}&\phi^{n+m}(i, x_1, x_2, \ldots, x_n, y_1, y_2, \ldots, y_m)\\ &\quad = \phi^m(s^m_n(i, x_1, x_2, \ldots, x_n), y_1, y_2, \ldots, y_m).\end{aligned}$$

Proof The value of $s^m_n(i, x_1, x_2, \ldots, x_n)$ is dened by cases according as $i \in I_{n+m}$ (which is the set of indices of Turing machines which have at least $n + m + 1$ bands and which is, as we recall, a primitive recursive set) or not. Let i_0 be an integer that is not the index of a Turing machine (0, for example, will do perfectly well).

(1) If $i \notin I_{n+m}$, we set $s^m_n(i, x_1, x_2, \ldots, x_n) = i_0$; in this case, neither $\phi^{n+m}(i, x_1, x_2, \ldots, x_n, y_1, y_2, \ldots, y_m)$ nor $\phi^m(s^m_n(i, x_1, x_2, \ldots, x_n), y_1, y_2, \ldots, y_m)$ is dened.

(2) The interesting case is when $i \in I_{m+n}$. Let $\mathcal{M}$ be the machine whose index is i and let $a_1, a_2, \ldots, a_n$ be xed integers. It is not difcult to imagine another machine $\mathcal{M}'$ that has the same number of bands as $\mathcal{M}$ and that behaves as follows:

(a) it begins by writing a_1 strokes on band number $m+2$, a_2 strokes on band number $m+3$, etc., and a_n strokes on band number $m+n+1$;

(b) it then behaves as $\mathcal{M}$ would, but with the roles of the different bands permuted: it treats band number $m+2$ (which represents a_1) as its rst band, and, in general for k between 1 and n, treats band number $m+k+1$ (which represents a_k) as its kth band; moreover, for k between 1 and m, the kth band should be considered as band number $n+k$; as for band number $m+1$, it plays the role of band number $n+m+1$ (so it is on this band that the nal result will be written);

(c) nally, it erases the contents of bands number $m+2, m+3, \ldots, m+n$.

Two facts are now more or less clear.

First, the description of $\mathcal{M}'$ is completely explicit and effective based on $\mathcal{M}$ and the given $a_1, a_2, \ldots, a_n$; it would be horribly boring, though very easy, to nd a primitive recursive function of $n+1$ variables, that we will call s_n^m, with the property that $s_n^m(i, a_1, a_2, \ldots, a_n)$ is the index of the machine $\mathcal{M}'$ when $i \in I_{m+n}$.

Second, if we set machines $\mathcal{M}$ and $\mathcal{M}'$ in operation with the following initial conguration:

- for k from 1 to n inclusive, the contents of the kth band of $\mathcal{M}'$ is equal to the contents of band number $n+k$ of $\mathcal{M}$;
- all other bands of $\mathcal{M}'$ are clean;
- for k from 1 to n, a_k is represented on band number k of $\mathcal{M}$;

then, up to permutation of the bands, these two machines will operate exactly the same way; in particular, one will halt if and only if the other does also, and, in this case, the contents of band number $m+1$ of $\mathcal{M}'$ will equal the contents of band number $n+m+1$ of $\mathcal{M}$. Consequently, referring to the denition of the functions ϕ^p, we see that, for all $x_1, x_2, \ldots, x_n, y_1, y_2, \ldots, y_m$,

$$\phi^{n+m}(i, x_1, x_2, \ldots, x_n, y_1, y_2, \ldots, y_m)$$
$$= \phi^m(s_n^m(i, x_1, x_2, \ldots, x_n), y_1, y_2, \ldots, y_m). \qquad \blacksquare$$

We will now present some applications of this theorem.

Example 5.48 There exists a primitive recursive function $pl(i, j)$ such that if $f = \phi_i^1$ and $g = \phi_j^1$, then $pl(i, j)$ is an index for the partial function $f + g$.

Proof Consider the partial function

$$\lambda ijx.(\phi^1(i, x) + \phi^1(j, x)).$$

It is obviously recursive, so there exists an integer k such that this function is equal to ϕ_k^3. Now, for all i, j, and x,

$$\phi_k^3(i, j, x) = \phi^3(k, i, j, x) = \phi^1(s_2^1(k, i, j), x) = \phi^1(i, x) + \phi^1(j, x).$$

So it sufces to take $pl = \lambda ij.s_2^1(k, i, j)$. ■

We could do exactly the same thing for multiplication or for any other partial recursive function.

Example 5.49 Let n and p be integers. There exists a primitive recursive function $\mathsf{Comp}(i_1, i_2, \ldots, i_n, j)$ such that if, for k from 1 to n, $f_k \in \mathcal{F}_p^*$ is the partial function whose index is i_k and if $g \in \mathcal{F}_n^*$ is the partial function whose index is j, then $\mathsf{Comp}(i_1, i_2, \ldots, i_n, j)$ is an index for the partial function $h = g(f_1, f_2, \ldots, f_n)$.

Proof The proof is completely analogous to the one just above. Consider the partial function

$$\lambda i_1 i_2 \ldots i_n j x_1 x_2 \ldots x_p.\phi^n(j, \phi^p(i_1, x_1, x_2, \ldots, x_p),$$
$$\phi^p(i_2, x_1, x_2, \ldots, x_p), \ldots, \phi^p(i_n, x_1, x_2, \ldots, x_p)).$$

It is recursive, so there exists an integer k such that this partial function is equal to ϕ_k^{n+p+1}. So we have

$$\phi_k^{n+p+1}(k, i_1, i_2, \ldots, i_n, j, x_1, x_2, \ldots, x_p)$$
$$= \phi^p(s_{n+1}^p(k, i_1, i_2, \ldots, i_n, j), x_1, x_2, \ldots, x_p),$$

and we may take

$$\mathsf{Comp} = \lambda i_1 i_2 \ldots i_n j.s_{n+1}^p(k, i_1, i_2, \ldots, i_n, j).$$ ■

The theorem that we will prove next is known as Rice's theorem and is another example of an application of the *smn* theorem. It will allow us to show that certain sets of integers are not recursive.

Theorem 5.50. *(Rice's theorem) Let $\mathcal{X}$ be a set of partial recursive functions of one variable that we will assume is not empty and is not equal to the set of all partial recursive functions. Then the set $A = \{x : \phi_x^1 \in \mathcal{X}\}$ is not recursive.*

Proof It is equivalent to show that either A or its complement is not recursive; so by interchanging these two sets, if necessary, and by replacing $\mathcal{X}$ by its complement in the set of partial recursive functions of one variable, we may assume that the partial function θ_0 whose domain is empty is an element of $\mathcal{X}$.

Fix an integer b that does not belong to A and dene the following partial recursive function $\psi \in \mathcal{F}_3^*$:

$$\psi(x, y, z) = \phi^1(b, z) + \phi^1(x, y) - \phi^1(x, y).$$

Also, set

$$\psi_{x,y} = \lambda z.\psi(x, y, z).$$

If $\phi^1(x, y)$ is not dened, the partial function $\psi_{x,y}$ is never dened (so is equal to θ_0), hence it is in $\mathcal{X}$; if not, then $\psi_{x,y}$ is equal to ϕ_b^1, so it is not in $\mathcal{X}$. Thus, $\psi_{x,y}$ belongs to $\mathcal{X}$ if and only if $\phi^1(x, y)$ is undened. Now we apply the *smn* theorem: there exists an integer k such that

$$\psi(x, y, z) = \phi^3(k, x, y, z) = \phi^1(s_2^1(k, x, y), z).$$

The function $h = \lambda xy.s_2^1(k, x, y)$ is primitive recursive and $h(x, y)$ is an index of $\psi_{x,y}$.

We will now use the fact that the set $W = \{(x, y) : \phi^1(x, y) \text{ is undened}\}$ is not recursive (we showed in Corollary 5.45 that its complement is not recursive) and we observe that $(x, y) \in W$ if and only if $h(x, y) \in A$. This shows that A cannot be recursive, otherwise W would also be. ■

Remark The hypothesis `$\mathcal{X}$ is not empty and is not equal to the set of all partial recursive functions' is obviously indispensable, otherwise A is equal to either the empty set or the whole of $\mathbb{N}$ and the conclusion of the theorem would be false. This hypothesis was used when we chose an integer b that is not an element of A.

Here are some corollaries of Rice's theorem.

- If $f \in \mathcal{F}_p^*$ is a partial recursive function, the set of indices of f is not recursive (just take $\mathcal{X} = \{f\}$ in Rice's theorem); in particular, it is not nite.

Intuitively, if a partial function is computable, there are innitely many machines that will compute it. And in fact, we have more: there is no effective description for the set of all the machines that compute f.

- The problem of deciding whether two machines compute the same partial function is undecidable: for every integer p, the set

$$X = \{(i, j) : \phi_i^p = \phi_j^p\}$$

is not recursive.

Indeed, if this set were recursive, then the set

$$\{i : (i, 0) \in X\} = \{i : \phi_i^p = \phi_0^p\}$$

would also be recursive but we have just seen that this is not the case.

- Also, for example, the set $\{n : \phi_n^1 \text{ is total}\}$ is not recursive.

It sufces to take $\mathcal{X}$ to be the set of all total recursive functions.

According to the rst corollary, if a partial function has an index i, it has another index that is greater than i. The next theorem is a more precise statement of this fact.

Theorem 5.51 *For every integer p, there exists a primitive recursive function* α *of two variables such that*

- *for all i and n,* $\phi_i^p = \phi_{\alpha(i,n)}^p$*;*
- *for all i, the function* $\lambda n.\alpha(i, n)$ *is strictly increasing.*

Proof It sufces to construct a primitive recursive function β of one variable such that for all i, $\beta(i) > i$ and $\phi_{\beta(i)}^p = \phi_i^p$; α will then be dened by recursion:

$$\alpha(i, 0) = i;$$
$$\alpha(i, n+1) = \beta(\alpha(i, n)).$$

Without going into details, we will explain how $\beta(i)$ is computed. If i is not the index of a Turing machine, we set $\beta(i)$ equal to an integer that is greater than i and is also not the index of a Turing machine (this can easily be found). If i is the index of a Turing machine $\mathcal{M}$, we arbitrarily produce a more complicated machine $\mathcal{M}'$ (e.g. by adding a new state that will never be used). The index of this new machine, if we have done things carefully, is strictly greater than i and is a primitive recursive function of i; obviously, these two machines will behave exactly the same way and will compute the same function. ∎

Exercise 26 presents a proof of this theorem based on the *smn* theorem and the xed point theorem (see below).

5.4.4 The xed point theorems

These theorems are also very important and are due to S. Kleene. They are sometimes called the **recursion theorems** (this name will be justied by the examples that follow).

Theorem 5.52. *(The xed point theorem, rst version)* *Let p be a positive integer and let* α *be a (total) recursive function of one variable; then there exists an integer i such that*

$$\phi_i^p = \phi_{\alpha(i)}^p.$$

Proof Consider the partial function

$$\lambda y x_1 x_2 \ldots x_p.\phi^p(\alpha(s_1^p(y, y)), x_1, x_2, \ldots, x_p).$$

It is recursive, so it has an index a and we have, for all $x_1, x_2, \ldots, x_p$ and y,

$$\begin{aligned}\phi^{p+1}(a, y, x_1, x_2, \ldots, x_p) &= \phi^p(\alpha(s_1^p(y, y)), x_1, x_2, \ldots, x_p)\\ &= \phi^p(s_1^p(a, y), x_1, x_2, \ldots, x_p).\end{aligned}$$

By setting $y = a$ in the preceding equalities and setting $i = s_1^p(a, a)$, we obtain

$$\phi_i^p = \phi_{\alpha(i)}^p.$$ ∎

Remark 5.53 There is a primitive recursive way to nd the integer i from an index of α. Suppose that $\alpha = \phi^p_j$. We must rst compute an index a of the partial function

$$\lambda y x_1 x_2, \ldots, x_p . \phi^p(\alpha(s^p_1(y, y)), x_1, x_2, \ldots, x_p).$$

This is another application of the *smn* theorem: let b be an index for the partial function

$$\lambda j y x_1 x_2 \ldots x_p . \phi^p(\phi^1(j, s^p_1(y, y)), x_1, x_2, \ldots, x_p);$$

we then have, for all $x_1, x_2, \ldots, x_p$ and y,

$$\begin{aligned}\phi^p(\alpha(s^p_1(y, y)), x_1, x_2, \ldots, x_p) &= \phi^p(\phi^1(j, s^p_1(y, y)), x_1, x_2, \ldots, x_p)\\ &= \phi^{p+2}(b, j, y, x_1, x_2, \ldots, x_p)\\ &= \phi^{p+1}(s^{p+1}_1(b, j), y, x_1, x_2, \ldots, x_p).\end{aligned}$$

So we may take $a = s^{p+1}_1(b, j)$ and we again set $i = s^p_1(a, a)$. This remark constitutes a proof of the following theorem.

Theorem 5.54. *(The xed point theorem, second version)* *For every positive integer p, there is a primitive recursive function h_p of one variable such that, for all j, if $\alpha = \phi^1_j$ is a total function, then*

$$\phi^p_{h_p(j)} = \phi^p_{\alpha(h_p(j))}.$$

Here is one last version of the xed point theorem.

Theorem 5.55. *(The xed point theorem, third version)* *Let α be a total recursive function of $p + 1$ variables and let n and p be integers, with n greater than zero. Then there exists a primitive recursive function h of p variables such that, for all $x_1, x_2, \ldots, x_p$, we have*

$$\phi^n_{\alpha(x_1, x_2, \ldots, x_p, h(x_1, x_2, \ldots, x_p))} = \phi^p_{h(x_1, x_2, \ldots, x_p)}.$$

Proof Let a be an index for the partial function

$$\begin{aligned}&\lambda z x_1 x_2 \ldots x_p y_1 y_2 \ldots y_n .\\ &\quad \phi^n(\alpha(x_1, x_2, \ldots, x_p, s^n_{p+1}(z, z, x_1, x_2, \ldots, x_p)), y_1, y_2, \ldots, y_n).\end{aligned}$$

So for all $x_1, x_2, \ldots, x_p, y_1, y_2, \ldots, y_n$ and z, we have

$$\begin{aligned}&\phi^n(\alpha(x_1, x_2, \ldots, x_p, s^n_{p+1}(z, z, x_1, x_2, \ldots, x_p)), y_1, y_2, \ldots, y_n)\\ &\quad = \phi^{n+p+1}(a, z, x_1, x_2, \ldots, x_p, y_1, y_2, \ldots, y_n)\\ &\quad = \phi^n(s^n_{p+1}(a, z, x_1, x_2, \ldots, x_p), y_1, y_2, \ldots, y_n).\end{aligned}$$

By letting $z = a$, we obtain

$$\begin{aligned}&\phi^n(\alpha(x_1, x_2, \ldots, x_p, s^n_{p+1}(z, z, x_1, x_2, \ldots, x_p)), y_1, y_2, \ldots, y_n)\\&\quad = \phi^n(s^n_{p+1}(a, a, x_1, x_2, \ldots, x_p), y_1, y_2, \ldots, y_n)\end{aligned}$$

and we may take $h(x_1, x_2, \ldots, x_p) = s^n_{p+1}(a, a, x_1, x_2, \ldots, x_p)$. ∎

Remark 5.56 Here too, there is a primitive recursive way to compute an index for h from an index for α.

We will now give some examples of how these theorems can be applied. These examples illustrate how the xed point theorems allow us to generalize the procedure of denition by recursion.

Example 5.57 Consider the partial function f of two variables, to be specic (we would do the same thing if it were a function of $p+1$ variables) that is dened by recursion using

$$\begin{aligned}f(x, 0) &= g(x),\\ f(x, y+1) &= h(x, y, f(x, y)),\end{aligned}$$

where g and h are partial recursive functions. Then we may compute, in a primitive recursive manner, an index for f from an index for g and an index for h.

Proof Consider the mapping from $\mathcal{F}_2^*$ into $\mathcal{F}_2^*$ which assigns to ψ the partial function ψ^* whose denition is

$$\psi^*(x, y) = \begin{cases} g(x) & \text{if } y = 0,\\ h(x, y-1, \psi(x, y-1)) & \text{otherwise.}\end{cases}$$

First, we note that f is the only xed point of this mapping: it is the only partial function that satises $f = f^*$. Moreover, if ψ is recursive, so is ψ^* and we can also compute an index for ψ^* from the respective indices i_1, i_2, and i_3 for g, h, and ψ. This last claim is another application of the *smn* theorem like the ones we have seen previously: we consider the partial recursive function $k(i_1, i_2, i_3, x, y)$ dened by

$$k(i_1, i_2, i_3, x, y) = \begin{cases} \phi^1(i_1, x) & \text{if } y = 0,\\ \phi^3(i_2, x, y-1, \phi^2(i_3, x, y-1)) & \text{otherwise.}\end{cases}$$

The partial function ψ^* is precisely equal to $\lambda xy.k(i_1, i_2, i_3, x, y)$. If a is an index for k, we have

$$k(i_1, i_2, i_3, x, y) = \phi^5(a, i_1, i_2, i_3, x, y) = \phi^2(s^2_3(a, i_1, i_2, i_3), x, y).$$

If we set $\alpha(i_1, i_2, i_3) = s^2_3(a, i_1, i_2, i_3)$, then α is a primitive recursive function that computes an index for ψ^*, as promised:

$$(\phi^2_{i_3})^* = \phi^2_{\alpha(i_1, i_2, i_3)}.$$

Now we apply the third version of the xed point theorem: there exists a primitive recursive function $j \in \mathcal{F}_2$ such that, for all z and t,

$$\phi^2_{\alpha(z,t),j(z,t)} = \phi^2_{j(z,t)},$$

which shows that

$$\left(\phi^2_{j(i_1,i_2)}\right)^* = \phi^2_{j(i_1,i_2)};$$

hence, because of the uniqueness mentioned above, $\phi^2_{j(i_1,i_2)} = f$. ■

Example 5.58 We conclude this chapter by proving that Ackerman's function is recursive. We could have done this earlier by constructing a Turing machine that computes it, but the argument that follows is much more elegant. The denition of this function certainly involved a recursive procedure; but this procedure did not respect the scheme for dening functions by recursion that was described at the very beginning of this chapter. We will see how the xed point theorems allow us to prove that functions dened using this procedure are nonetheless recursive.

Proof Consider the mapping from $\mathcal{F}_2^*$ into $\mathcal{F}_2^*$ which assigns to ψ the partial function ψ^* whose denition is

$$\psi^*(y,x) = \begin{cases} 2^x & \text{if } y = 0; \\ 1 & \text{if } x = 0; \\ \psi(y-1, \psi(y, x-1)) & \text{in all other cases.} \end{cases}$$

When we refer to the denition of Ackerman's function, we realize that it is the unique xed point of this mapping. Thus, if we can prove that there exists a partial recursive function ζ such that $\zeta = \zeta^*$, then ζ is necessarily equal to Ackerman's function, which is consequently recursive. Our argument is similar to the one above. If ψ is a partial recursive function, then so is ψ^* and here is how we compute, in a primitive recursive fashion, an index for ψ^* from an index for ψ: dene $\theta \in \mathcal{F}_3^*$ by

$$\theta(i,y,x) = \begin{cases} 2^x & \text{if } y = 0; \\ 1 & \text{if } x = 0; \\ \phi^2(i, y-1, \phi^2(i, y, x-1)) & \text{in all other cases.} \end{cases}$$

The partial recursive function θ was dened in such a way that $\lambda xy.\theta(i,x,y)$ is equal to ψ^* if $\psi = \phi_i^2$. Let a be an index of θ. Then

$$\theta(i,x,y) = \phi^3(a,i,x,y) = \phi^2(s_1^2(a,i),x,y).$$

Set $\alpha(i) = s_1^2(a,i)$; α is a primitive recursive function which associates an index of ψ^* with an index of ψ. When we then apply the rst version of the xed point theorem, we obtain an integer j such that $\phi_j^2 = \phi^2_{\alpha(j)}$ and, hence, such that $(\phi_j^2)^* = \phi_j^2$. This proves that Ackerman's function is recursive. ■

EXERCISES FOR CHAPTER 5

1. Show that every nite subset of $\mathbb{N}$ is primitive recursive.

2. Show that the function f dened by

$$f(0) = f(1) = 1,$$
$$f(n+2) = f(n) + f(n+1)$$

is primitive recursive.

(This series is known as the **Fibonacci series**).

3. Set $\mathcal{S}^* = \bigcup_{p>0} \mathbb{N}^p$ and dene a map α from $\mathcal{S}^*$ into $\mathbb{N}$ as follows: if σ is a sequence of integers of length p, then $\alpha(\sigma) = \alpha_2(p, \alpha_p(\sigma))$.

(a) Show that the function α is injective and that its range is a primitive recursive set.

(b) Show that there exists a primitive recursive function g such that if $\sigma = (a_1, a_2, \ldots, a_n)$ and if $b = \sup(n, a_1, a_2, \ldots, a_n)$, then $\alpha(s) \leq g(b)$.

(c) Show that the function ϕ dened by

$$\phi(p, i, x) = \begin{cases} \beta_p^i(x) & \text{if } 1 \leq i \leq p, \\ 0 & \text{otherwise,} \end{cases}$$

is primitive recursive.

(d) We now dene another coding: let γ be the function which, with every $(a_0, a_1, a_2, \ldots, a_p) \in \mathcal{S}^*$, associates the integer

$$\gamma((a_0, a_1, \ldots, a_p)) = \pi(0)^{a_0+1} \cdot \pi(1)^{a_1+1} \cdot \ldots \cdot \pi(p)^{a_p+1};$$

it is understood that the value of γ on the empty sequence is 1. Show that γ is an injective map and that its range is a primitive recursive set.

(e) Show that the two codings can be obtained from one another in a primitive recursive way; more precisely, show that there exist two primitive recursive functions f and h of one variable such that

(i) for all x in the range of α, $f(x) = \gamma(\sigma)$, where σ is the non-empty sequence satisfying $\alpha(\sigma) = x$;

(ii) for all x in the range of γ, $h(x) = \alpha(\sigma)$, where σ is the non-empty sequence satisfying $\gamma(\sigma) = x$.

4. Show that the function whose value at n is the nth digit in the decimal expansion of e (the real number that is the basis for natural logarithms) is primitive recursive.

5. (a) Let p be a positive integer. Show that the set

$$E = \{(a_0, a_1, \ldots, a_p) \in \mathbb{N}^{p+1} : \text{the polynomial } a_0 + a_1X + \cdots + a_pX^p \text{ has a root in } \mathbb{Z}\}$$

is primitive recursive.

(b) Repeat question (a) replacing $\mathbb{Z}$ by $\mathbb{Q}$.

(c) Show that the set

$$F = \{\Omega(\sigma) : p \text{ is an integer, } \sigma = (a_0, a_1, \ldots, a_p) \text{ and the polynomial } a_0 + a_1X + \cdots + a_pX^p \text{ has a root in } \mathbb{Z}\}$$

is primitive recursive. (Ω is dened in the section on codings of sequences.)

6. Let L be a language whose only symbol R represents a binary predicate and let F be a closed formula of L. The **spectrum** of F, which we will denote by $Sp(F)$, is dened to be the set

$$\{n \in \mathbb{N} : F \text{ has a model of cardinality } n\}.$$

(See Exercise 10 of Chapter 3.)

Show that $Sp(F)$ is a primitive recursive set.

7. For each of the following functions, construct a Turing machine that computes it: (a) $\lambda x.x^2$, (b) $\lambda xy.xy$, (c) $\lambda x.x \dot{-} 1$, (d) $\lambda xy.x \dot{-} y$

8. Construct a Turing machine that halts if and only if the integer represented on its rst band at the initial instant is even.

9. (a) Show that if a partial function $f \in \mathcal{F}_1^*$ is T-computable, then it is computable by a Turing machine that has exactly three bands.

(b) Consider the set $\mathcal{M}_n$ of Turing machines that have three bands and n states. Set these machines in operation with an initial conguration in which all bands are clean. If machine $\mathcal{M}$ halts, we let $\sigma(\mathcal{M})$ be the number of strokes written on its second band at the instant it halts; otherwise, we set $\sigma(\mathcal{M}) = 0$. Show that the set

$$\{\sigma(\mathcal{M}) : \mathcal{M} \in \mathcal{M}_n\}$$

is bounded. We will denote the upper bound of this set by $\Sigma(n)$.

(c) Let f be a partial function of one variable that is computable by a machine $\mathcal{M}$ in $\mathcal{M}_n$. For every integer p, construct a machine $\mathcal{N}_p$ with three bands which, when started in an initial conguration in which all bands are clean, begins by writing p strokes on its rst band, then returns its head to the beginning of the tape, and continues to behave exactly as $\mathcal{M}$ would.

How many states does $\mathcal{N}_p$ have?

(d) Show that the function Σ is not T-computable.

10. Let $f \in \mathcal{F}_1$. Show that f is recursive if and only if its graph

$$G = \{(x, y) \in \mathbb{N}^2 : y = f(x)\}$$

is recursive.

11. The purpose of this exercise is to provide a direct proof of the fact that Ackerman's function is recursive.

Dene the following binary relation $\ll$ on $\mathbb{N}^3$: $(a, b, c) \ll (a', b', c')$ if and only if

$$\begin{aligned}
&\sup(a, b, c) < \sup(a', b', c'), \quad \text{or} \\
&\sup(a, b, c) = \sup(a', b', c') \text{ and } a < a', \quad \text{or} \\
&\sup(a, b, c) = \sup(a', b', c') \text{ and } a = a' \text{ and } b < b', \quad \text{or} \\
&\sup(a, b, c) = \sup(a', b', c') \text{ and } a = a' \text{ and } b = b' \text{ and } c \leq c'.
\end{aligned}$$

(a) Show that $\ll$ is a total ordering.

If α and β belong to $\mathbb{N}^3$, we will say that α is **less than or equal** (respectively, **greater than or equal**) to β if $\alpha \ll \beta$ (respectively, $\beta \ll \alpha$). We will say that α is **strictly less** (respectively, **strictly greater**) than β if, in addition, $\alpha \neq \beta$.

Show that for all $(a, b, c) \in \mathbb{N}^3$, the set

$$\{(x, y, z) \in \mathbb{N}^3 : (x, y, z) \ll (a, b, c)\}$$

has at most $(\sup(a, b, c)+1)^3$ elements. Show that every element $(a, b, c) \in \mathbb{N}^3$ has an immediate successor [i.e. there exists an element that is strictly greater than (a, b, c) and is less than or equal to all elements that are strictly greater than (a, b, c)]. We will explicitly describe this immediate successor.

(b) Show that there exist three primitive recursive functions γ_1, γ_2, and γ_3 from $\mathbb{N}$ into $\mathbb{N}$ such that

(i) the function Γ from $\mathbb{N}$ into $\mathbb{N}^3$ dened by

$$\Gamma(n) = (\gamma_1(n), \gamma_2(n), \gamma_3(n))$$

is a bijection;

(ii) for all integers n and m, $n \leq m$ if and only if $\Gamma(n) \ll \Gamma(m)$.

(c) Let H be the subset of $\mathbb{N}$ dened recursively by the following condition: $n \in H$ if and only if

$$\begin{aligned}
&\gamma_2(n) = 0 \quad \text{and} \quad \gamma_1(n) = 2^{\gamma_3(n)}; \quad \text{or} \\
&\gamma_3(n) = 0 \quad \text{and} \quad \gamma_1(n) = 1; \quad \text{or}
\end{aligned}$$

$$\gamma_2(n) \neq 0 \quad \text{and} \quad \gamma_3(n) \neq 0$$

and there exist integers p and q strictly less than n such that $p \in H, q \in H$, $\gamma_2(p) = \gamma_2(n), \gamma_3(p) = \gamma_3(n) - 1, \gamma_2(q) = \gamma_2(n) - 1, \gamma_3(q) = \gamma_1(p)$ and $\gamma_1(n) = \gamma_1(q)$.

Show that H is primitive recursive.

As in the body of Chapter 5, let ζ denote Ackerman's function. Show that, for every integer n, $n \in H$ if and only if $\gamma_1(n) = \zeta(\gamma_2(n), \gamma_3(n))$.

(d) Show that the graph

$$G = \{(y, x, z) : z = \zeta(y, x)\}$$

of Ackerman's function is primitive recursive. Show that Ackerman's function is recursive.

12. Show that if f is a function of one variable that is recursive and increasing, then its range is a recursive set. Conversely, show that every innite recursive set is the range of a strictly increasing recursive function.

13. Suppose that $f \in \mathcal{F}_1$ is a recursive function and assume that its image is innite. Show that there exists a recursive function $g \in \mathcal{F}_1$ that is recursive and injective and satises $\mathsf{Im}(f) = \mathsf{Im}(g)$. Conclude from this that there exists an injective recursive function whose image is not recursive.

14. Show that every innite recursively enumerable set includes an innite recursive set.

15. Let α be a recursive function that is injective. We set

$$A = \mathsf{Ran}(\alpha);$$
$$B = \{x : \text{there exists } y > x \text{ such that } \alpha(y) < \alpha(x)\}.$$

(a) Show that B is recursively enumerable and that its complement is innite.

(b) Assume that there exists an innite recursively enumerable subset $C \subset \mathbb{N}$ that is disjoint from B. Show that A is recursive.

(c) Show that there exists a recursively enumerable set which has (1) a non-empty intersection with every innite recursively enumerable set, and (2) an innite complement.

16. (a) Show that the set of recursive bijections from $\mathbb{N}$ onto $\mathbb{N}$ is a subgroup of the group of permutations of $\mathbb{N}$.

The remainder of this exercise is devoted to showing that this assertion is false if we replace recursive by primitive recursive.

(b) Let ϕ be a (total) function of one variable that is recursive but not primitive recursive; let e be an index for a machine $\mathcal{M}$ that computes ϕ. Consider the function T which associates, with x, the time required by $\mathcal{M}$ to compute $\phi(x)$; more precisely, $T(x) = \mu t\,[(e, t, x) \in B^1)]$.

Show that if f is a function that satises $f(x) \geq T(x)$ for all $x \in \mathbb{N}$, then f is not primitive recursive; also show that the graph G of T is primitive recursive.

(c) We set

$$g(x) = \sup\{T(y) : y \leq x\} + 2x.$$

Show that g is a strictly increasing recursive function and that it is not primitive recursive. Show that the graph G_1 and the range I of g are primitive recursive sets.

(d) Show that there is a unique strictly increasing primitive recursive function g' whose range is the complement of I.

(e) Dene the function h by

$$h(2x) = g(x),$$
$$h(2x+1) = g'(x),$$

where g and g' are the functions dened in (c) and (d) above. Show that h is a bijective recursive function that is not primitive recursive. Show that its inverse, h^{-1}, is primitive recursive.

17. Exhibit a recursive set $A \subseteq \mathbb{N}^2$ such that the set

$$B = \{x : \text{for all } y \in \mathbb{N},\ (x, y) \in A\}$$

is not recursively enumerable.

18. Show that there exists a primitive recursive function α of one variable that has the following property: for every integer x, if ϕ_x^1 is a bijection from $\mathbb{N}$ onto $\mathbb{N}$, then $\alpha(x)$ is an index for the inverse bijection.

19. Let g, α, and h be partial recursive functions with g and α in $\mathcal{F}_1^*$ and $h \in \mathcal{F}_3^*$. Show that there exists one and only one function $f \in \mathcal{F}_2^*$ such that, for all x and y,

$$f(0, y) = g(y),$$
$$f(x+1, y) = h(f(x, \alpha(y)), y, x),$$

and f is partial recursive.

20. Let $A \subseteq \mathbb{N}$ be a recursively enumerable set that is not recursive; let f be a partial recursive function whose domain is A and let i be an index for a Turing machine that computes f. Show that the function $\lambda x.T^1(i, x)$ cannot be extended to a total recursive function [here, T^1 is the function dened in the proof of the enumeration theorem whose value is the time required to compute $f(x)$].

21. The purpose of this exercise is to prove the following fact:

(∗) *There exists a (total) recursive function $\psi(x, y)$ such that if we set $\psi_x = \lambda y.\psi(x, y)$, then the set $\{\psi_x : x \in \mathbb{N}\}$ is precisely the set of all primitive recursive functions of one variable.*

(a) Show that if $f \in \mathcal{F}_p$, then the following two conditions are equivalent:
   (i) f is primitive recursive;
   (ii) there exists an index i and a primitive recursive function $g \in \mathcal{F}_p$ such that the machine whose index is i computes f and the computation time $T(i, x_1, x_2, \ldots, x_p)$ is less than or equal to $g(x_1, x_2, \ldots, x_p)$.

(b) We will make use of Ackerman's function ζ and of the functions $\zeta_n = \lambda x.\zeta(n, x)$. Show that if f is a primitive recursive function of one variable, then there exist two integers n and A such that, for all x, we have

$$f(x) \leq \sup(A, \zeta_n(x)).$$

(c) Let g be the function of four variables dened by

$$g(i, A, n, x) = \mu y \leq \sup(A, \zeta(n, x))$$
$$[\exists t \leq \sup(A, \zeta(n, x))\, (i, t, x, y) \in C^1];$$

[recall that $(i, t, x, y) \in C^1$ means that when the machine whose index is i is set in operation with x on its rst band, it will halt at the instant t with output y]. Show that, for all i, A, and n, the function $\lambda x.g(i, A, n, x)$ is primitive recursive and that, conversely, if f is any primitive recursive function of one variable, then there exist integers i, A, and n such that $f = \lambda x.g(i, A, n, x)$.

(d) Use these results to prove $(*)$.

(e) Show that there exists a recursive set that is not primitive recursive.

22. Let $\mathcal{T}$ be a set of partial recursive functions of one variable. We say that $\mathcal{T}$ has a **recursive listing** if there exists a partial recursive function F of two variables such that, if we set $F_x = \lambda y.F(x, y)$, then

$$\mathcal{T} = \{F_x : x \in \mathbb{N}\}.$$

Exercise 21 showed that the set of primitive recursive functions has a recursive listing.

(a) Show that the set of total recursive functions does not have a recursive listing.

(b) Show that the set of strictly increasing primitive recursive functions has a recursive listing.

(c) Show that the set of injective primitive recursive functions has a recursive listing.

(d) Let $F \in \mathcal{F}_2$ be a recursive function and assume that, for all $x \in \mathbb{N}$, the set

$$A_x = \{F(x, y) : y \in \mathbb{N}\}$$

is innite. Show that there exists an innite recursive set, B, which is distinct from all the sets A_x. Conclude from this that the set of strictly increasing recursive functions does not have a recursive listing, nor does the set of injective recursive functions.

23. Let A and B be two subsets of $\mathbb{N}$. We say that A is **reducible to** B and write $A \leq B$ if there exists a (total) recursive function f such that

$$x \in A \quad \text{if and only if } f(x) \in B.$$

(a) Show that the relation $\leq$ is reexive and transitive.

(b) Assume that A is reducible to B. Show that if B is recursively enumerable, then A is recursively enumerable; show also that if B is recursive, then so is A.

Set

$$X = \{x : \phi^1(x, x) \text{ is dened }\};$$
$$Y = \{\alpha_2(x, y) : \phi^1(x, y) \text{ is dened }\}.$$

(c) Show that a set $A \subseteq \mathbb{N}$ is recursively enumerable if and only if $A \leq Y$.

(d) Let A and B be two subsets of $\mathbb{N}$. Let

$$C = \{2n : n \in A\} \cup \{2n + 1 : n \in B\}.$$

Show that A and B are reducible to C and that if D is a subset of $\mathbb{N}$ such that A and B are both reducible to D, then C is reducible to D.

(e) We will say that A is **self-dual** if $A \leq \mathbb{N} - A$. Show that for every $B \subseteq \mathbb{N}$, there exists a $C \subseteq \mathbb{N}$ that is self-dual and is such that $B \leq C$.

(f) Let $\mathcal{T}$ be a set of partial recursive functions of one variable that is not empty and is not equal to the set of all partial recursive functions of one variable. Set

$$A = \{x : \phi_x^1 \in \mathcal{T}\}.$$

(i) Show that if the partial function whose domain is empty belongs to $\mathcal{T}$, then $X \leq \mathbb{N} - A$.

(ii) Show that, in the opposite case, $X \leq A$.

(iii) Show that A is not self-dual.

(g) Show that $Y \leq X$.

24. The goal of this exercise is to show that the precautions which we took in dening the unbounded μ-operator (see Denition 5.19) are necessary.

Show that the partial function $\psi(x, y)$ dened by

$$\psi(x, y) = \begin{cases} \phi^1(x, y) - \phi^1(x, y) & \text{if } y = 0, \\ 0 & \text{otherwise;} \end{cases}$$

is partial recursive.

Dene the function g by

$$g(x) = \text{ the least integer } y \text{ such that } \psi(x, y) = 0.$$

Show that g is a total function that is not recursive.

25. Consider the following sets:

$$A = \{x : \phi_x^1(0) \text{ is dened }\};$$
$$B = \{x : \phi_x^1 \text{ is a total function}\}.$$

(a) Show that the complement of A is not recursively enumerable.

(b) Show that there exists a primitive recursive function $f \in \mathcal{F}_1$ such that for all i, $i \in A$ if and only if $\alpha(i) \in B$. Show that the complement of B is not recursively enumerable.

(c) Let F be the following partial function:

$$F(x, y) = \begin{cases} 1 & \text{if for all } z < y,\ \neg B^1(e, z, x), \\ \text{undened} & \text{otherwise,} \end{cases}$$

where B^1 is the predicate dened in the proof of the enumeration theorem and e is the index of a partial function whose domain is A.

Show that the partial function $\lambda y.F(x, y)$ is total if and only if $x \notin A$. Conclude from this that B is not recursively enumerable.

(d) By generalizing the results from (b) and (c), prove the following:

Proposition *Let f be a partial recursive function of one variable whose domain is innite; then neither the set $\{x : \phi_x^1 = f\}$ nor its complement is recursively enumerable.*

26. In this exercise, we will give an alternate proof of the fact that there exists a primitive recursive function β of one variable such that, for all i,

$$\phi_i^1 = \phi_{\beta(i)}^1 \quad \text{and} \quad \beta(i) > i.$$

(See Theorem 5.51.) This proof is based only on the xed point theorems and no longer involves Turing machines.

(a) Show that there exists a primitive recursive function δ such that, for all n, $\phi_{\delta(n)}^1$ is the constant function equal to n.

(b) Dene the function $\gamma(n, t, z)$ by

$$\gamma(n, t, z) = \begin{cases} \delta(n) & \text{if } z < t; \\ t & \text{otherwise.} \end{cases}$$

By applying the third version of the xed point theorem (see Theorem 5.55) to this function, show that there exists a primitive recursive function $h(n, t)$

such that

$$\phi^1_{h(n,t)} = \begin{cases} \phi^1_{\delta(n)} & \text{if } h(n,t) \le t, \\ \phi^1_t & \text{otherwise.} \end{cases}$$

(c) Show that, for all t, the set $A_t = \{n : h(n,t) \le t\}$ has at most $t+1$ elements. Use this to conclude that the desired function β exists.

27. When we constructed the functions ϕ^p, we used a certain number of codings and, for this purpose, we had to make some completely arbitrary choices. In this exercise, our concern is to know what sort of functions would have been obtained instead of the ϕ^p if our choices had been different. The only assumption we will make is that these choices are reasonable and sufcient for proving the enumeration theorem and the xed point theorems.

Let $\Psi = \{\psi^p : p \ge 1\}$ be a family of partial recursive functions such that, for all p, $\psi^p \in \mathcal{F}^*_{p+1}$. We set

$$\psi^p_x = \lambda y_1 y_2 \ldots y_p . \psi^p(x, y_1, y_2, \ldots, y_p).$$

Consider the following conditions on the family Ψ:

- (enu) For every $p > 0$, the set $\{\psi^p_i : i \in \mathbb{N}\}$ is equal to the set of all partial recursive functions of p variables.
- (smn) For every pair of integers m and n, there exists a total recursive function σ^m_n of $n+1$ variables such that for all $i, x_1, x_2, \ldots, x_n, y_1, y_2, \ldots, y_m$, we have

$$\begin{aligned} &\psi^{n+m}(i, x_1, x_2, \ldots, x_n, y_1, y_2 \ldots, y_m) \\ &\quad = \psi^m(\sigma^m_n(i, x_1, x_2, \ldots, x_n), y_1, y_2, \ldots, y_m). \end{aligned}$$

(a) Let θ be a partial recursive function of two variables. For every integer x, we set $\theta_x = \lambda y.\theta(x, y)$. Show that the following two conditions are equivalent:

(i) there exists a family $\Psi = \{\psi^p : p \ge 1\}$ that satises conditions (enu) and (smn) and is such that $\psi^1 = \theta$;

(ii) there exists a recursive function β such that, for all x, $\phi^1_x = \theta_{\beta(x)}$.

(b) Assume once more that the family Ψ satises the conditions (enu) and (smn). Show that the xed point theorems are valid for the family Ψ.

(c) Assume that the function θ satises conditions (i) or (ii) from (a). Show that there exist two injective recursive functions α and β such that, for all x,

$$\phi^1_x = \theta_{\beta(x)} \quad \text{and} \quad \theta_x = \phi^1_{\alpha(x)}.$$

(d) (Difcult!) Under these same hypotheses, show that there exists a recursive function ε that is total and bijective and is such that, for all x, $\phi^1_x = \theta_{\varepsilon(x)}$.

6 Formalization of arithmetic, Gödel's theorems

Of all the branches of mathematics that we could choose to formalize, arithmetic is no doubt the most natural choice. This is what we undertake in the present chapter.

In Section 6.1, we describe the language of arithmetic and present the set of its axioms, commonly known as Peano's axioms, which we denote by $\mathcal{P}$. The purpose of some of these axioms (A_1 through A_7) is to force addition and multiplication to behave correctly; the others (the axiom scheme IS) are to sanction the well-known proofs by induction. Supercially, these are very simple axioms and we could even ask ourselves whether they are not too simple. Also, a question which comes immediately to mind is whether we have forgotten to include anything that mathematicians commonly use. The answer is no, but we will not attempt to convince the reader about this. We will be satised to derive some easy consequences of these axioms, for example, the commutativity and associativity of addition and multiplication. Nothing stops the reader from deriving, from Peano's axioms alone, theorems such as those of Gauss or Bezout. Even far more complicated theorems, such as those concerning the distribution of primes, can not only be expressed as rst-order formulas but can also be proved from these axioms.

There are then two natural questions that arise. The rst concerns the completeness of $\mathcal{P}$: is it true that every closed formula of the language of arithmetic is either provable or refutable (i.e. its negation is provable) in $\mathcal{P}$? The second concerns its decidability: is there an algorithm that allows us to determine whether a closed formula of the language of arithmetic is derivable from $\mathcal{P}$? The answer to both these questions is negative and the concluding part of this chapter is devoted to a proof of these facts, the famous theorems of Godel.

To answer the second of these questions requires a coding of formulas by integers. This dirty work is done in Section 6.3; in Section 6.2, we proceed in a different direction. We will show that recursive functions can be represented, in a very strong sense, by rst-order formulas. To answer the questions we have raised, we will use a `diagonal' argument of the type we used in Chapter 5 for showing that there exist recursively enumerable sets that are not recursive. As it applies to answering the second question, this argument reminds us of the famous paradox of Epimenides, the Cretan, who claimed that all Cretans are liars (see Exercise 15).

In our situation, this amounts to constructing a formula which asserts that it is itself unprovable. We will see that this formula is true in $\mathbb{N}$, *that it is unprovable in* $\mathcal{P}$, *and that it is equivalent (modulo* $\mathcal{P}$*) to a formula asserting that* $\mathcal{P}$ *is a consistent theory.*

In this chapter, we deal simultaneously with the set $\mathbb{N}$ *of 'true integers' and with arbitrary models of Peano's axioms. As pointed out in the introduction, we must adopt two different attitudes: we will not hesitate to use all the known properties of* $\mathbb{N}$*; but those properties that are true in all other models of* $\mathcal{P}$ *must be derived, sometimes laboriously, directly from* $\mathcal{P}$, *at least in principle.*

This chapter contains some rather indigestible codings. The reader who is convinced that such codings are possible and do in fact permit us to obtain the expected results may, of course, pass up a detailed reading.

6.1 Peano's axioms

6.1.1 The axioms

The language $\mathcal{L}_0$ that will allow us to describe arithmetic is a nite language with four symbols:

- a constant symbol: $\underline{0}$;
- a unary function symbol: $\underline{S}$;
- two binary function symbols: $\underline{+}$ and $\underline{\times}$.

(*Caution*! The symbol $\underline{+}$ is an underlined plus symbol, to distinguish it from the operation $+$. It has nothing to do with the sign meaning 'plus or minus'.)

We will agree to break the rules for writing terms of the language $\mathcal{L}_0$ so as to recover the more familiar notations, $v_0 \underline{+} v_1$ and $v_0 \underline{\times} v_1$ instead of $\underline{+} v_0 v_1$ and $\underline{\times} v_0 v_1$, respectively. This clearly necessitates the use of parentheses (as explained in Chapter 3) for writing terms. If a problem ever arises relating to the syntax of formulas of arithmetic, it is always possible to insist on the standard way of writing formulas, which is the only legitimate way.

From now on, when we speak of $\mathbb{N}$, we mean the $\mathcal{L}_0$-structure whose base set is the set of natural numbers and in which $\underline{0}$ is interpreted by the integer 0, $\underline{S}$ by the successor function $S = \lambda n.n + 1$, $\underline{+}$ by addition and $\underline{\times}$ by multiplication.

Denition 6.1 *The set* $\mathcal{P}$ *of* ***Peano's axioms*** *consists of the seven axioms* A_1 *through* A_7 *below, together with an innite number of axioms which we will call the* ***induction scheme*** *and denote by IS.*

$$
\begin{aligned}
A_1 &: \ \forall v_0 \neg \underline{S} v_0 \simeq \underline{0} \\
A_2 &: \ \forall v_0 \exists v_1 (\neg v_0 \simeq \underline{0} \Rightarrow \underline{S} v_1 \simeq v_0) \\
A_3 &: \ \forall v_0 \forall v_1 (\underline{S} v_0 \simeq \underline{S} v_1 \Rightarrow v_0 \simeq v_1) \\
A_4 &: \ \forall v_0 v_0 \underline{+} \underline{0} \simeq v_0
\end{aligned}
$$

$$A_5: \ \forall v_0 \forall v_1 v_0 \underline{+} \underline{S} v_1 \simeq \underline{S}(v_0 \underline{+} v_1)$$
$$A_6: \ \forall v_0 v_0 \underline{\times} \underline{0} \simeq \underline{0}$$
$$A_7: \ \forall v_0 \forall v_1 v_0 \underline{\times} \underline{S} v_1 \simeq (v_0 \underline{\times} v_1) \underline{+} v_0.$$

Finally, the induction scheme *IS* is the set of all formulas of $\mathcal{L}_0$ which are of the form

$$\forall v_1 \ldots \forall v_n ((F[\underline{0}, v_1, \ldots, v_n] \wedge \forall v_0 (F[v_0, v_1, \ldots, v_n] \Rightarrow F[\underline{S} v_0, v_1, \ldots, v_n])) \Rightarrow \forall v_0 F[v_0, v_1, \ldots, v_n])$$

where n is an integer and $F[v_0, v_1, \ldots, v_n]$ is any formula of $\mathcal{L}_0$ whose only free variables are $v_0, v_1, \ldots, v_n$.

Remark When we wish to prove, using *IS*, that a formula $F[v_0, v_1, \ldots, v_n]$ is provable from $\mathcal{P}$, it will sufce to establish the following two facts:

- $\mathcal{P} \vdash \forall v_0 F[\underline{0}, v_1, v_2, \ldots, v_n]$;
  (This is called the **basis step** or **initial step** of the induction, which consists in `letting $v_0 = 0$' in F.)
- $\mathcal{P} \vdash \forall v_0 \forall v_1 \ldots \forall v_n \ (F[v_0, v_1, v_2, \ldots, v_n] \Rightarrow F[\underline{S} v_0, v_1, v_2, \ldots, v_n])$;
  (This is called the **induction step**.)

It is clear that $\mathbb{N}$, viewed as an $\mathcal{L}_0$-structure, is a model of $\mathcal{P}$. We usually call this the **standard model** of $\mathcal{P}$. We will immediately show that it is not the only one.

Theorem 6.2 *There exist models of $\mathcal{P}$ that are not isomorphic to $\mathbb{N}$.*

Proof For each integer n, we let $\underline{n}$ denote the term $\underline{S}\,\underline{S} \ldots \underline{S}\,\underline{0}$ consisting of n occurrences of the symbol $\underline{S}$ followed by the symbol $\underline{0}$. Thus, $\underline{n} = \underline{S}\,\underline{S} \ldots \underline{S}\,\underline{0}$. Let us say that an element of an $\mathcal{L}_0$-structure is **standard** if it is the interpretation of a term of the form $\underline{n}$, where $n \in \mathbb{N}$. We see that in the standard model (and in any model that is isomorphic to it), every element is standard. Now consider a new language $\mathcal{L}$ which is obtained by adding a new constant symbol c to $\mathcal{L}_0$ and let T be the following theory:

$$T = \{\neg c \simeq \underline{n} : \ n \in \mathbb{N}\} \cup \mathcal{P}.$$

Every nite subset of T has a model; indeed, if T_0 is such a set, it is included in a set of the form

$$\{\neg c \simeq \underline{n} : \ n \in I\} \cup \mathcal{P},$$

where I is a nite subset of $\mathbb{N}$; we can obtain a model of T_0 by taking $\mathbb{N}$ as the base set for an $\mathcal{L}$-structure and interpreting c by any integer that does not belong to I. Then, by applying the compactness theorem (see Theorem 3.78), we conclude that there exists a model $\mathcal{M}$ of T. The model $\mathcal{M}$ is also, obviously, a model of $\mathcal{P}$ and it contains a point, namely, the interpretation of c, that is not standard. The reduct of

$\mathcal{M}$ to $\mathcal{L}_0$ (which, we recall, is the $\mathcal{L}_0$-structure obtained in a natural way from $\mathcal{M}$ by ignoring the interpretation of c) is therefore a non-standard model of $\mathcal{P}$. ■

Nearly all the theorems of arithmetic that are expressible as rst-order formulas of $\mathcal{L}_0$ can, in fact, be proved from $\mathcal{P}$ (despite the fact that their `classical' proofs may use notions that do not belong to arithmetic). To illustrate how these axioms operate, the induction scheme in particular, we will show that in models of $\mathcal{P}$, addition and multiplication are associative and commutative, as well as other properties of a similar nature. We will observe that, in $\mathcal{P}$, the derivation of these simple facts can be rather lengthy.

Theorem 6.3 *In every model $\mathcal{M}$ of $\mathcal{P}$, addition and multiplication are associative and commutative, and multiplication distributes over addition; moreover, we have*

- *the cancellation law for addition:*

$$\mathcal{M} \models \forall v_0 \forall v_1 \forall v_2 ((v_0 \underline{+} v_1 \simeq v_0 \underline{+} v_2) \Rightarrow v_1 \simeq v_2);$$

- *the cancellation law for multiplication:*

$$\mathcal{M} \models \forall v_0 \forall v_1 \forall v_2 ((\neg v_0 \simeq 0 \wedge v_0 \underline{\times} v_1 \simeq v_0 \underline{\times} v_2) \Rightarrow v_1 \simeq v_2);$$

- *the formula $\exists v_2 (v_2 \underline{+} v_0 \simeq v_1)$ denes a total ordering on $\mathcal{M}$ and this ordering is compatible with addition and multiplication.*

Proof This theorem is a consequence of the following twenty-four (!) facts.

(1) $\mathcal{P} \vdash \forall v_0 (\underline{0} \underline{+} v_0 \simeq v_0)$.

Using A_4 and A_5 we see that

$$\mathcal{P} \vdash \underline{0} \underline{+} \underline{0} \simeq \underline{0} \wedge \forall v_0 (\underline{0} \underline{+} v_0 \simeq v_0 \Rightarrow \underline{0} \underline{+} \underline{S} v_0 \simeq \underline{S} v_0).$$

If we then use the particular case of the *IS* where the formula F is the formula $\underline{0} \underline{+} v_0 \simeq v_0$, we may conclude that

$$\mathcal{P} \vdash \forall v_0 (\underline{0} \underline{+} v_0 \simeq v_0).$$

(2) $\mathcal{P} \vdash \forall v_0 \forall v_1 \underline{S}(v_1 \underline{+} v_0) \simeq \underline{S} v_1 \underline{+} v_0$.

First of all, using A_4 (twice), we have

$$\mathcal{P} \vdash \underline{S}(v_1 \underline{+} \underline{0}) \simeq \underline{S} v_1 \underline{+} \underline{0}.$$

On the other hand, using A_5,

$$\mathcal{P} \vdash \underline{S}(v_1 \underline{+} \underline{S} v_0) \simeq \underline{S}\, \underline{S}(v_1 \underline{+} v_0) \wedge \underline{S} v_1 \underline{+} \underline{S} v_0 \simeq \underline{S}\,(\underline{S} v_1 \underline{+} v_0),$$

and hence

$$\mathcal{P} \vdash \underline{S}(v_1 \mathbin{\underline{+}} v_0) \simeq \underline{S}v_1 \mathbin{\underline{+}} v_0 \Rightarrow \underline{S}(v_1 \mathbin{\underline{+}} \underline{S}v_0) \simeq (\underline{S}v_1 \mathbin{\underline{+}} \underline{S}v_0),$$

and the conclusion now follows from *IS*.

(3) $\mathcal{P} \vdash \forall v_0(\underline{1} \mathbin{\underline{+}} v_0 \simeq \underline{S}v_0)$.

(Recall that $\underline{1}$ is an abbreviation for the term $\underline{S}\,\underline{0}$.)

It is an obvious consequence of (2) that

$$\mathcal{P} \vdash \underline{S}(\underline{0} \mathbin{\underline{+}} v_0) \simeq \underline{1} \mathbin{\underline{+}} v_0;$$

the conclusion then follows using (1).

(4) $\mathcal{P} \vdash \forall v_0 \forall v_1 v_0 \mathbin{\underline{+}} v_1 \simeq v_1 \mathbin{\underline{+}} v_0$.

For $v_1 = \underline{0}$, this is true by A_4 and (1). On the other hand, using A_5,

$$\mathcal{P} \vdash v_0 \mathbin{\underline{+}} \underline{S}v_1 \simeq \underline{S}(v_0 \mathbin{\underline{+}} v_1),$$

and from (2), we have

$$\mathcal{P} \vdash \underline{S}v_1 \mathbin{\underline{+}} v_0 \simeq \underline{S}(v_1 \mathbin{\underline{+}} v_0).$$

So it sufces to invoke *IS*.

(5) $\mathcal{P} \vdash \forall v_0 \forall v_1 \forall v_2 v_0 \mathbin{\underline{+}} (v_1 \mathbin{\underline{+}} v_2) \simeq (v_0 \mathbin{\underline{+}} v_1) \mathbin{\underline{+}} v_2$.

Once again, it is the induction scheme that will provide our proof. For the initial step ($v_2 = \underline{0}$), this equality is easily obtained from A_4. We also have, from A_5, that

$$\mathcal{P} \vdash v_0 \mathbin{\underline{+}} (v_1 \mathbin{\underline{+}} \underline{S}v_2) \simeq v_0 \mathbin{\underline{+}} \underline{S}(v_1 \mathbin{\underline{+}} v_2) \simeq \underline{S}(v_0 \mathbin{\underline{+}} (v_1 \mathbin{\underline{+}} v_2))$$

and

$$\mathcal{P} \vdash (v_0 \mathbin{\underline{+}} v_1) \mathbin{\underline{+}} \underline{S}v_2 \simeq \underline{S}((v_0 \mathbin{\underline{+}} v_1) \mathbin{\underline{+}} v_2).$$

Let us now turn to multiplication.

(6) $\mathcal{P} \vdash \forall v_0(\underline{0} \mathbin{\underline{\times}} v_0 \simeq \underline{0})$.

Indeed, using A_6,

$$\mathcal{P} \vdash \underline{0} \mathbin{\underline{\times}} \underline{0} \simeq \underline{0}$$

and, by A_7,

$$\mathcal{P} \vdash \underline{0} \mathbin{\underline{\times}} \underline{S}v_0 \simeq (\underline{0} \mathbin{\underline{\times}} v_0) \mathbin{\underline{+}} \underline{0};$$

so (6) follows, once again using A_4 and *IS*.

(7) $\mathcal{P} \vdash \forall v_0(v_0 \mathbin{\underline{\times}} \underline{1} \simeq v_0)$.

Invoke A_7, A_6, and (1).

(8) $\forall v_0(\underline{1} \underline{\times} v_0 \simeq v_0)$.

By *IS*: the case '$v_0 = \underline{0}$' comes from A_6; also, using A_7, A_5, and A_4, we see that

$$\mathcal{P} \vdash \underline{1} \underline{\times} \underline{S} v_0 \simeq \underline{S}(\underline{1} \underline{\times} v_0).$$

(9) $\mathcal{P} \vdash \forall v_0 \forall v_1 \forall v_2 v_0 \underline{\times} (v_1 \underline{+} v_2) \simeq (v_0 \underline{\times} v_1) \underline{+} (v_0 \underline{\times} v_2)$.

Use *IS* once again: the initial step with $v_2 \simeq \underline{0}$ is a consequence of A_6 and A_4; on the other hand, from A_5 and A_7, we have

$$\mathcal{P} \vdash v_0 \underline{\times} (v_1 \underline{+} \underline{S} v_2) \simeq (v_0 \underline{\times} (v_1 \underline{+} v_2)) \underline{+} v_0;$$

and if

$$\mathcal{P} \vdash (v_0 \underline{\times} (v_1 \underline{+} v_2)) \underline{+} v_0 \simeq ((v_0 \underline{\times} v_1) \underline{+} (v_0 \underline{\times} v_2)) \underline{+} v_0$$

(the induction hypothesis), then, by (5),

$$\mathcal{P} \vdash ((v_0 \underline{\times} v_1) \underline{+} (v_0 \underline{\times} v_2)) \underline{+} v_0 \simeq (v_0 \underline{\times} v_1) \underline{+} ((v_0 \underline{\times} v_2) \underline{+} v_0);$$

and nally, by A_7,

$$\mathcal{P} \vdash (v_0 \underline{\times} v_1) \underline{+} ((v_0 \underline{\times} v_2) \underline{+} v_0) \simeq (v_0 \underline{\times} v_1) \underline{+} (v_0 \underline{\times} \underline{S} v_2).$$

(10) $\mathcal{P} \vdash \forall v_0 \forall v_1 \forall v_2 ((v_0 \underline{\times} v_1) \underline{\times} v_2 \simeq v_0 \underline{\times} (v_1 \underline{\times} v_2))$.

Use *IS* once again: for the initial step with $v_2 \simeq \underline{0}$, invoke A_6; then, using A_7, we have

$$\mathcal{P} \vdash (v_0 \underline{\times} v_1) \underline{\times} \underline{S} v_2 \simeq ((v_0 \underline{\times} v_1) \underline{\times} v_2) \underline{+} (v_0 \underline{\times} v_1),$$

and, using A_7 and (9),

$$\begin{aligned} \mathcal{P} \vdash v_0 \underline{\times} (v_1 \underline{\times} \underline{S} v_2) &\simeq v_0 \underline{\times} ((v_1 \underline{\times} v_2) \underline{+} v_1) \\ &\simeq (v_0 \underline{\times} (v_1 \underline{\times} v_2)) \underline{+} (v_0 \underline{\times} v_1); \end{aligned}$$

the conclusion now follows from (4).

(11) $\mathcal{P} \vdash \forall v_0 \forall v_1 (v_0 \underline{\times} v_1 \simeq v_1 \underline{\times} v_0)$.

Begin, using the same type of proof as above with *IS*, by showing that

$$\mathcal{P} \vdash \forall v_0 \forall v_1 (\underline{S} v_0 \underline{\times} v_1 \simeq (v_1 \underline{\times} v_0) \underline{+} v_1);$$

then, use *IS* once again.

(12) $\mathcal{P} \vdash \forall v_0 \forall v_1 \forall v_2 (v_0 \underline{+} v_2 \simeq v_1 \underline{+} v_2 \Rightarrow v_0 \simeq v_1)$.

We use *IS*: A_4 for the case $v_2 \simeq \underline{0}$; then A_5 and A_3.

(13) $\mathcal{P} \vdash \forall v_0 \forall v_1 (\neg v_1 \simeq \underline{0} \Rightarrow \neg v_0 \underline{+} v_1 \simeq \underline{0})$.

Indeed, from A_2 and A_5, we obtain

$$\mathcal{P} \vdash \neg v_1 \simeq \underline{0} \Rightarrow \exists v_2 (v_1 \simeq \underline{S} v_2 \wedge v_0 \underline{+} v_1 \simeq \underline{S}(v_0 \underline{+} v_2)),$$

and, from A_1,

$$\mathcal{P} \vdash \neg \underline{S}(v_0 \underline{+} v_2) \simeq \underline{0}.$$

(14) $\mathcal{P} \vdash \forall v_0 \forall v_1 (v_0 \underline{+} v_1 \simeq \underline{0} \Rightarrow (v_0 \simeq \underline{0} \wedge v_1 \simeq \underline{0}))$.

Use (13) and A_4.

(15) $\mathcal{P} \vdash \forall v_0 \forall v_1 (v_0 \underline{+} v_1 \simeq v_0 \Rightarrow v_1 \simeq \underline{0})$.

Use (12), A_4, and (4).

To be continued in the next issue ...

6.1.2 The ordering on the integers

Notation Henceforth, $v_0 \leq v_1$ will be an abbreviation for the formula

$$\exists v_2 (v_2 \underline{+} v_0 \simeq v_1)$$

and $v_0 < v_1$ an abbreviation for $(v_0 \leq v_1 \wedge \neg v_0 \simeq v_1)$; the expressions $v_0 \geq v_1$ and $v_0 > v_1$ will be synonyms of $v_1 \leq v_0$ and $v_1 < v_0$, respectively.

We will show that, in every model of $\mathcal{P}$, the relation $\leq$ is a total order relation and, moreover, that it is compatible with addition and multiplication. Obviously, in the standard model, $\leq$ is the natural ordering of the integers. Note that we are abusing language here by using the same symbol $\leq$ to denote both the abbreviation in the language $\mathcal{L}_0$ and the binary relation dened in a given model by the formula $v_0 \leq v_1$, i.e. the set of pairs of elements of the model that satisfy this formula. We will use, thanks to (4), the fact that

$$\mathcal{P} \vdash \forall v_0 \forall v_1 (v_0 \leq v_1 \Leftrightarrow \exists v_2 (v_0 \underline{+} v_2 \simeq v_1)).$$

(16) $\mathcal{P} \vdash \forall v_0 (v_0 \leq v_0)$.

Because $\mathcal{P} \vdash \underline{0} \underline{+} v_0 \simeq v_0$.

(17) $\mathcal{P} \vdash \forall v_0 \forall v_1 \forall v_2 ((v_0 \leq v_1 \wedge v_1 \leq v_2) \Rightarrow v_0 \leq v_2)$.

By (5).

(18) $\mathcal{P} \vdash \forall v_0 \forall v_1 ((v_0 \leq v_1 \wedge v_1 \leq v_0) \Rightarrow v_0 \simeq v_1)$.

This follows from (5), (15) and (4), and (14).

(19) $\mathcal{P} \vdash \forall v_0 \forall v_1 \forall v_2 (v_0 \underline{+} v_2 \leq v_1 \underline{+} v_2 \Leftrightarrow v_0 \leq v_1)$.

By (5) and (12).

(20) $\mathcal{P} \vdash \forall v_0 \forall v_1 (v_0 \leq v_1 \vee v_1 \leq v_0)$.

Here, we will have to use *IS* again. The result is clear for $v_0 \simeq \underline{0}$, by (1). On the other hand, we have, in succession,

$$\mathcal{P} \vdash \forall v_0 (v_0 \leq \underline{S} v_0) \quad \text{(by } A_5\text{, taking } v_1 = \underline{0}, \text{ and } A_4\text{);}$$

$$\mathcal{P} \vdash \forall v_0 \forall v_1 (v_1 \leq v_0 \Rightarrow v_1 \leq \underline{S} v_0) \quad \text{(by (17));} \qquad (*)$$

$$\mathcal{P} \vdash \forall v_0 \forall v_1 ((v_0 \leq v_1 \wedge \neg v_1 \simeq v_0)$$
$$\Rightarrow \exists v_2 (\neg v_2 \simeq \underline{0} \wedge v_1 \simeq v_0 \underline{+} v_2)) \quad \text{(by } A_4\text{);}$$

$$\mathcal{P} \vdash \forall v_0 \forall v_1 ((v_0 \leq v_1 \wedge \neg v_1 \simeq v_0) \Rightarrow \exists v_3\, v_1 \simeq v_0 \underline{+}\, \underline{S} v_3) \quad \text{(by } A_2\text{);}$$

$$\mathcal{P} \vdash \forall v_0 \forall v_1 ((v_0 \leq v_1 \wedge \neg v_1 \simeq v_0)$$
$$\Rightarrow \exists v_3\, v_1 \simeq \underline{S} v_0 \underline{+} v_3) \quad \text{(by } A_5 \text{ and (2));}$$

$$\mathcal{P} \vdash \forall v_0 \forall v_1 ((v_0 \leq v_1 \wedge \neg v_1 \simeq v_0) \Rightarrow \underline{S} v_0 \leq v_1). \qquad (**)$$

We may now deduce, from $(*)$ and $(**)$, that

$$\mathcal{P} \vdash \forall v_0 (\forall v_1 (v_0 \leq v_1 \vee v_1 \leq v_0) \Rightarrow \forall v_1 (\underline{S} v_0 \leq v_1 \vee v_1 \leq \underline{S} v_0));$$

this completes the induction step.

(21) $\mathcal{P} \vdash \forall v_0 \forall v_1 \forall v_2 (v_0 \leq v_1 \Rightarrow v_0 \underline{\times} v_2 \leq v_1 \underline{\times} v_2)$.

By (9) and (11).

(22) $\mathcal{P} \vdash \forall v_0 \forall v_1 (\neg v_1 \simeq \underline{0} \Rightarrow v_0 \underline{\times}\, v_1 \geq v_0)$.

Apply A_2 and A_7.

(23) $\mathcal{P} \vdash \forall v_0 \forall v_1 ((\neg v_0 \simeq \underline{0} \wedge \neg v_1 \simeq \underline{0}) \Rightarrow \neg v_0 \underline{\times} v_1 \simeq \underline{0})$.

After observing that

$$\mathcal{P} \vdash \forall v_2 \forall v_3 \underline{S} v_2 \,\underline{\times}\, \underline{S} v_3 \simeq \underline{S}((\underline{S} v_2 \,\underline{\times}\, v_3) \,\underline{+}\, v_2) \quad \text{(by } A_7 \text{ and } A_5\text{),}$$

we may apply A_2 and A_1.

(24) $\mathcal{P} \vdash \forall v_0 \forall v_1 \forall v_2 (v_0 \underline{\times} v_2 \simeq v_1 \underline{\times} v_2 \Rightarrow (v_0 \simeq v_1 \vee v_2 \simeq \underline{0}))$.

Let $\mathcal{M} = \langle M, 0, S, +, \cdot \rangle$ be a model of $\mathcal{P}$ and let a, b, and c be elements of M such that $a \cdot c = b \cdot c$. According to (20), we have $a \leq b$ or $b \leq a$; in the rst case, for example, there exists a d such that $d + a = b$; hence [by (11) and (9)], $b \cdot c = (d \cdot c) + (a \cdot c)$. By (4) and (15), $d \cdot c = 0$, so the conclusion follows from (23).

This concludes the proof of Theorem 6.3. ■

Notation We will let $\mathcal{P}_0$ denote the theory consisting of axioms A_1 through A_7. We will observe that this theory is extremely weak; we cannot even prove from these axioms that addition is commutative (see Exercise 1). Nonetheless, we will show that every model of $\mathcal{P}_0$ (and hence every model of $\mathcal{P}$ also) `begins' with a structure that is isomorphic to $\mathbb{N}$. First, let us specify what we mean by `begin'.

Denition 6.4 *Let $\mathcal{M}$ and $\mathcal{N}$ be two models of $\mathcal{P}_0$ and assume that $\mathcal{N}$ is a substructure of $\mathcal{M}$. We say that $\mathcal{N}$ is an* ***initial segment of*** *$\mathcal{M}$, or, equivalently, that $\mathcal{M}$ is an* ***end-extension of*** *$\mathcal{N}$, if for every a belonging to $\mathcal{N}$ and every b belonging to $\mathcal{M}$,*

(1) *if $\mathcal{M} \models b \leq a$, then b belongs to $\mathcal{N}$;*

(2) *if $b \notin \mathcal{N}$, then $\mathcal{M} \models a \leq b$.*

We must use caution, because $\mathcal{P}_0$ does not prove that the relation $\leq$ is an order relation (see Exercise 1). However:

Theorem 6.5 *Let $\mathcal{M}$ be a model of $\mathcal{P}_0$; then the following subset of $\mathcal{M}$,*

$$\{a : \text{there exists an integer } n \text{ such that } a \text{ is the interpretation of } \underline{n} \text{ in } \mathcal{M}\},$$

is a substructure of $\mathcal{M}$ that is an initial segment of $\mathcal{M}$ and is isomorphic to $\mathbb{N}$.

Proof Facts (25)ñ(29) which follow show that the map ϕ, from $\mathbb{N}$ into $\mathcal{M}$, that sends an integer $n \in \mathbb{N}$ into the interpretation in $\mathcal{M}$ of the term $\underline{n}$ is an injective homomorphism. Properties (30) and (31) show that the image of this homomorphism is an initial segment of $\mathcal{M}$. Before starting the proof, we offer a brief remark: these statements involve the integers (the `true' integers!), and the fact that *IS* is not part of $\mathcal{P}_0$ does not prevent us in any way from using proofs by induction on these integers.

(25) *For every integer n, we have*

$$\mathcal{P}_0 \vdash \underline{n} \,\underline{+}\, \underline{1} \simeq \underline{S}\,\underline{n}.$$

In fact, there is nothing to prove: $\underline{n} \,\underline{+}\, \underline{1}$ and $\underline{S}\,\underline{n}$ represent the same term, consisting of $n + 1$ occurrences of the symbol $\underline{S}$, followed by a single occurrence of the symbol $\underline{0}$.

(26) *For all integers m and n, we have*

$$\mathcal{P}_0 \vdash \underline{m} \,\underline{+}\, \underline{n} \simeq \underline{m + n}.$$

This is proved by induction on n. For $n = 0$, we certainly have

$$\mathcal{P}_0 \vdash \underline{m} \,\underline{+}\, \underline{0} \simeq \underline{m} \quad \text{(by } A_4\text{)}.$$

For $n + 1$, under the assumption that $\mathcal{P}_0 \vdash \underline{m} \,\underline{+}\, \underline{n} \simeq \underline{m + n}$, we have

$$\mathcal{P}_0 \vdash \underline{n} \,\underline{+}\, \underline{1} \simeq \underline{S}\,\underline{n} \quad \text{and} \quad \mathcal{P}_0 \vdash \underline{m + n + 1} \simeq \underline{S}\,\underline{m + n} \qquad \text{(by (25))},$$

and

$$\mathcal{P}_0 \vdash \underline{m} \,\underline{+}\, \underline{S}\,\underline{n} \simeq \underline{S}(\underline{m} \,\underline{+}\, \underline{n}) \quad \text{(by } A_5\text{)};$$

putting all this together yields

$$\mathcal{P}_0 \vdash \underline{m} \,\underline{+}\, \underline{n+1} \simeq \underline{m+n+1}.$$

(27) *For all integers m and n, we have*

$$\mathcal{P}_0 \vdash \underline{m} \,\underline{\times}\, \underline{n} \simeq \underline{m \cdot n}.$$

The argument is again by induction on n. For $n = 0$, this is A_6. On the other hand, by the induction hypothesis, we have

$$\mathcal{P}_0 \vdash \underline{m} \,\underline{\times}\, \underline{n} \simeq \underline{m \cdot n},$$

and, by (26),

$$\mathcal{P}_0 \vdash \underline{m \cdot n} \,\underline{+}\, \underline{m} \simeq \underline{m \cdot (n+1)}.$$

(28) *For every non-zero integer n, we have*

$$\mathcal{P}_0 \vdash \neg(\underline{n} \simeq \underline{0}).$$

Let $m = n - 1$. From (25), we obtain

$$\mathcal{P}_0 \vdash \underline{n} \simeq \underline{S}\,\underline{m};$$

the conclusion now follows from A_1.

(29) *For all distinct integers m and n, we have*

$$\mathcal{P}_0 \vdash \neg(\underline{m} \simeq \underline{n}).$$

By induction on $\inf(m, n)$. If one of the integers m or n is zero, the preceding fact applies. If not, then, by (25),

$$\mathcal{P}_0 \vdash \underline{m} \simeq \underline{n} \Rightarrow \underline{S}\,\underline{m-1} \simeq \underline{S}\,\underline{n-1},$$

and so, using A_3, we have

$$\mathcal{P}_0 \vdash \underline{m} \simeq \underline{n} \Rightarrow \underline{m-1} \simeq \underline{n-1};$$

the result now follows from the induction hypothesis.

(30) *For every integer n, we have*

$$\mathcal{P}_0 \vdash \forall v_0(v_0 \leq \underline{n} \Rightarrow (v_0 \simeq \underline{0} \vee v_0 \simeq \underline{1} \vee \cdots \vee v_0 \simeq \underline{n})).$$

By induction on n. Let us rst deal with $n = 0$. We must show that

$$\mathcal{P}_0 \vdash \forall v_0 \forall v_1 (v_1 \,\underline{+}\, v_0 \simeq \underline{0} \Rightarrow v_0 \simeq \underline{0}).$$

We may invoke (14) [whose proof, incidentally, did not use *IS*, nor did the proof of (13); so the replacement of $\mathcal{P}$ by $\mathcal{P}_0$ is legitimate]. Consequently,

$$\mathcal{P}_0 \vdash \forall v_0 \forall v_1 (v_1 \,\underline{+}\, v_0 \simeq \underline{0} \Rightarrow v_0 \simeq \underline{0} \wedge v_1 \simeq \underline{0}).$$

Next, assuming the property is true for n, we will prove it for $n+1$: so let $\mathcal{M}$ be a model of $\mathcal{P}_0$ and a be a point of $\mathcal{M}$ such that $\mathcal{M} \models a \leq \underline{n+1}$. It sufces to show that there exists a $p \in \mathbb{N}$ such that $p \leq n+1$ and $\mathcal{M} \models a = \underline{p}$.

There exists a point b of $\mathcal{M}$ such that $\mathcal{M} \models b \underline{+} a \simeq \underline{S}\,\underline{n}$. If $a = \underline{0}$, we are done; if not, by A_2, there exists a point c of $\mathcal{M}$ such that $\mathcal{M} \models a = \underline{S}c$; by A_5 and A_3, we see that $\mathcal{M} \models b \underline{+} c \simeq \underline{n}$, hence $\mathcal{M} \models c \leq \underline{n}$ and we may use the induction hypothesis to conclude that there exists $m \leq n$ such that $\mathcal{M} \models c = \underline{m}$; thus $\mathcal{M} \models \underline{S}c \simeq \underline{S}\underline{m}$, i.e. $\mathcal{M} \models a = \underline{m+1}$.

(31) *For every integer n, we have*

$$\mathcal{P}_0 \vdash \forall v_0 (v_0 \leq \underline{n} \vee \underline{n} \leq v_0).$$

By induction on n. For $n = 0$, this is obvious from A_4 and the denition of $\leq$. Suppose the property is true for n. Consider a model $\mathcal{M}$ of $\mathcal{P}_0$ and a point a of $\mathcal{M}$. We have to show that $\mathcal{M} \models a \leq \underline{n+1}$ or $\mathcal{M} \models \underline{n+1} \leq a$. If $a = \underline{0}$, this is obvious. If not, there exists a $b \in \mathcal{M}$ such that $\mathcal{M} \models a = \underline{S}b$; so it follows by the induction hypothesis that either $\mathcal{M} \models b \leq \underline{n}$ or $\mathcal{M} \models \underline{n} \leq b$. In the rst case, there exists a $c \in \mathcal{M}$ such that $\mathcal{M} \models c \underline{+} b \simeq \underline{n}$, so by A_5 and (25), $\mathcal{M} \models c \underline{+} a = \underline{n+1}$ and so $\mathcal{M} \models a \leq \underline{n+1}$. In the second case, there exists a $d \in \mathcal{M}$ such that $\mathcal{M} \models d \underline{+} \underline{n} \simeq b$, so $\mathcal{M} \models d \underline{+} \underline{n+1} \simeq a$ and so $\mathcal{M} \models \underline{n+1} \leq a$. ∎

Some additional properties of models of $\mathcal{P}$ will be found in Exercise 2.

6.2 Representable functions

Recall that $\mathcal{F}_p$ denotes the set of total functions from $\mathbb{N}^p$ into $\mathbb{N}$.

Denition 6.6 *Let $f \in \mathcal{F}_p$ and let $F[v_0, v_1, \ldots, v_p]$ be a formula of $\mathcal{L}_0$ with no free variables other than $v_0, v_1, \ldots, v_p$. We say that $F[v_0, v_1, \ldots, v_p]$* ***represents*** *f if, for every p-tuple of integers $(n_1, n_2, \ldots, n_p)$, we have*

$$\mathcal{P}_0 \vdash \forall v_0 (F[v_0, \underline{n}_1, \underline{n}_2, \ldots, \underline{n}_p] \Leftrightarrow v_0 \simeq \underline{f(n_1, n_2, \ldots, n_p)}).$$

The function f is said to be ***representable*** *if there exists a formula that represents it.*

Therefore, to say that a formula F represents f means that, for every model $\mathcal{M}$ of $\mathcal{P}_0$ and for every sequence of integers $(n_1, n_2, \ldots, n_p)$, there exists one and only one element x of $\mathcal{M}$ satisfying $F[x, \underline{n}_1, \underline{n}_2, \ldots, \underline{n}_p]$ and this element is the (standard) element of $\mathcal{M}$ that interprets the term $\underline{f(n_1, n_2, \ldots, n_p)}$ which, we recall, consists of the symbol $\underline{S}$ repeated $f(n_1, n_2, \ldots, n_p)$ times, followed by $\underline{0}$.

This denition can be adapted to subsets:

Denition 6.7 *Let $A \subseteq \mathbb{N}^p$ and let $F[v_0, v_1, \ldots, v_p]$ be a formula of $\mathcal{L}_0$ with no free variables other than $v_0, v_1, \ldots, v_p$. We say that $F[v_0, v_1, \ldots, v_p]$* ***represents*** *A*

if, for every p-tuple of integers $(n_1, n_2, \ldots, n_p)$*, we have*

- *if* $(n_1, n_2, \ldots, n_p) \in A$ *then* $\mathcal{P}_0 \vdash F[\underline{n}_1, \underline{n}_2, \ldots, \underline{n}_p]$*;*
- *if* $(n_1, n_2, \ldots, n_p) \notin A$ *then* $\mathcal{P}_0 \vdash \neg F[\underline{n}_1, \underline{n}_2, \ldots, \underline{n}_p]$.

We say that the set A is ***representable*** *if there exists a formula that represents it.*

Remark A subset $A \subseteq \mathbb{N}^p$ is representable if and only if its characteristic function is representable: it is easy to verify that, if F represents A, then the formula

$$(F[v_1, \ldots, v_p] \wedge v_0 \simeq \underline{1}) \vee (\neg F[v_1, \ldots, v_p] \wedge v_0 \simeq \underline{0})$$

represents the characteristic function of A; conversely, if $G[v_0, v_1, \ldots, v_p]$ represents the characteristic function of A, then $G[\underline{1}, v_1, \ldots, v_p]$ represents A.

Let us give a few examples of representable functions with their corresponding formulas:

- The successor function is represented by the formula $v_0 \simeq \underline{S} v_1$ [see item (25) of the previous subsection].
- Addition $\lambda xy.x + y$ is represented by the formula $v_0 \simeq v_1 \underline{+} v_2$ [item (26)].
- Multiplication $\lambda xy.x \cdot y$ is represented by the formula $v_0 \simeq v_1 \underline{\times} v_2$ [item (27)].
- The projection functions are also representable: the function P^i_p is represented by the formula $v_0 \simeq v_i$.
- The constant function equal to n is represented by the formula $v_0 \simeq \underline{n}$.

In fact, every recursive function is representable.

Theorem 6.8. *(The representation theorem) Every (total) recursive function is representable.*

Proof With what we have seen, it sufces to show that the set of representable functions is closed under composition, the (total) μ-operator and recursion (see the remark that followed Theorem 5.29). This is the purpose of the following lemmas.

Lemma 6.9 *The set of representable functions is closed under composition.*

Proof Let $f_1, f_2, \ldots, f_n \in \mathcal{F}_p$ and $g \in \mathcal{F}_n$ and suppose that, for every i from 1 to n inclusive, f_i is represented by $F_i[v_0, v_1, \ldots, v_p]$ and that g is represented by $G[v_0, v_1, \ldots, v_n]$. It is immediate to verify that $g(f_1, f_2, \ldots, f_n)$ is represented by

$$\exists w_1 \exists w_2 \ldots \exists w_n \left(G[v_0, w_1, w_2, \ldots, w_n] \wedge \bigwedge_{1 \leq i \leq n} F_i[w_i, v_1, v_2, \ldots, v_p] \right).$$

∎

Lemma 6.10 *Let $A \subseteq \mathbb{N}^{p+1}$ be a representable set such that the function*

$$f(x_1, x_2, \ldots, x_p) = \mu y(y, x_1, x_2, \ldots, x_p \in A)$$

is total; then f is representable.

Proof Let $F[v_0, v_1, \ldots, v_p]$ be a formula that represents A. We will show that the formula

$$G = F[v_0, v_1, \ldots, v_p] \wedge \forall w < v_0 \neg F[w, v_1, \ldots, v_p]$$

represents f. Indeed, let $\mathcal{M}$ be a model of $\mathcal{P}_0$ and let $n_1, n_2, \ldots, n_p$ be integers. We must show that the interpretation, b, of $\underline{f(n_1, n_2, \ldots, n_p)}$ in $\mathcal{M}$ is the only element of $\mathcal{M}$ that satises the formula $G[v_0, \underline{n}_1, \underline{n}_2, \ldots, \underline{n}_p]$. First of all, since F represents A, we have

$$\mathcal{P}_0 \vdash F[b, \underline{n_1}, \underline{n_2}, \ldots, \underline{n_p}],$$

and, since $\mathcal{M}$ is a model of $\mathcal{P}_0$, b satises $F[v_0, \underline{n_1}, \underline{n_2}, \ldots, \underline{n_p}]$ in $\mathcal{M}$. Also, if c is an element of $\mathcal{M}$ that is less than b, then, according to Theorem 6.5, c is a standard element; so, by denition of f, it does not satisfy $F[v_0, \underline{n_1}, \underline{n_2}, \ldots, \underline{n_p}]$. Hence, b satises $G[v_0, \underline{n_1}, \underline{n_2}, \ldots, \underline{n_p}]$. Moreover, suppose d is an element of $\mathcal{M}$ that satises $G[v_0, \underline{n_1}, \underline{n_2}, \ldots, \underline{n_p}]$; neither $d < b$ nor $b < d$ can hold in $\mathcal{M}$; but, since b is standard, Theorem 6.5 guarantees that $d \leq b$ or $b \leq d$. The conclusion is that $b = d$. ■

We have now arrived at the most delicate point in the argument: it involves denition by recursion. To deal with this, we have to introduce a clever function, **Godel's function**, β, whose role is to code the nite sequences of integers.

Lemma 6.11 *There exists a function β of three variables which is recursive and representable and which has the property that, for all $p \in \mathbb{N}$ and for every sequence $(n_1, n_2, \ldots, n_p) \in \mathbb{N}^p$, there exist integers a and b such that, for all i between 1 and p inclusive, we have $\beta(i, a, b) = n_i$.*

Before we prove this lemma, let us use it to conclude the proof of the representation theorem. The next lemma supplies what remains to be proved.

Lemma 6.12 *Let $g \in \mathcal{F}_p$ and $h \in \mathcal{F}_{p+2}$ be two representable functions. Then the function $f \in \mathcal{F}_{p+1}$ dened by recursion from g and h by*

$$f(x_1, x_2, \ldots, x_p, 0) = g(x_1, x_2, \ldots, x_p)$$
$$f(x_1, x_2, \ldots, x_p, x_{p+1} + 1) = h(x_1, x_2, \ldots, x_p, x_{p+1}, f(x_1, x_2, \ldots, x_p, x_{p+1}))$$

is also representable.

Proof To express that $y = f(x_1, x_2, \ldots, x_p, x_{p+1})$, we state that there exists a nite sequence of integers $(z(0), z(1), \ldots, z(x_{p+1}))$ such that

$$z(0) = g(x_1, x_2, \ldots, x_p), \qquad z(x_{p+1}) = y,$$

and, for all i between 0 and $x_{p+1} - 1$ inclusive,

$$z(i+1) = h(x_1, x_2, \ldots, x_p, i, z(i)).$$

Obviously, to express `there exists a sequence ... ', we say that there exist two integers that code this sequence by means of the function β.

Let g and h be represented by the formulas

$$G[v_0, v_1, \ldots, v_p] \quad \text{and} \quad H[v_0, v_1, \ldots, v_{p+2}],$$

respectively. For the function β, we must be slightly more cautious: let $B[v_0, v_1, v_2, v_3]$ be a formula that represents β. This function will also be represented by the following formula:

$$B'[v_0, v_1, v_2, v_3] = B[v_0, v_1, v_2, v_3] \wedge \forall v_4 < v_0 \neg B[v_4, v_1, v_2, v_3].$$

The advantage of B' over B is that if $\mathcal{M}$ is any model of $\mathcal{P}_0$, if x is a standard element of $\mathcal{M}$ (the interpretation in $\mathcal{M}$ of $\underline{n}$ for some intuitive integer n), and if a, b, and c are three elements of $\mathcal{M}$ such that

$$\mathcal{M} \models B'[x, a, b, c],$$

then there is no other point in $\mathcal{M}$, standard or not, that satises $B'[v_0, a, b, c]$.

We are going to verify that the formula $F[v_0, v_1, v_2, \ldots, v_p, v_{p+1}]$ that follows represents the function f:

$$\begin{aligned}
&\exists w_1 \exists w_2 (\exists w_0 (B'[w_0, \underline{1}, w_1, w_2] \wedge G[w_0, v_1, v_2, \ldots, v_p]) \\
&\quad \wedge B'[v_0, v_{p+1} \underline{+} \underline{1}, w_1, w_2] \\
&\quad \wedge \forall w_3 < v_{p+1} \exists w_4 \exists w_5 (B'[w_4, \underline{S} w_3, w_1, w_2] \wedge B'[w_5, \underline{S}\,\underline{S} w_3, w_1, w_2] \\
&\quad \wedge H[w_5, v_1, v_2, \ldots, v_p, w_3, w_4])).
\end{aligned}$$

[First, a note that explains how to read this formula: the variables w_1 and w_2 represent integers a and b such that, for all i from 0 to x_{p+1} inclusive, $f(x_1, x_2, \ldots, x_p, i)$ is equal to $\beta(i+1, a, b)$, w_0 should assume the value $g(x_1, x_2, \ldots, x_p)$, and if $0 \leq w_3 < n_{p+1}$, then w_4 should be equal to $f(x_1, x_2, \ldots, x_p, w_3)$ and w_5 to $f(x_1, x_2, \ldots, x_p, w_3 + 1)$.]

So, let $n_1, n_2, \ldots, n_{p+1}$ be integers, $\mathcal{M}$ a model of $\mathcal{P}_0$, and c a point in $\mathcal{M}$. First, it is clear that if

$$\mathcal{M} \models c \simeq \underline{f(n_1, n_2, \ldots, n_{p+1})},$$

then

$$\mathcal{M} \models F[c, \underline{n}_1, \underline{n}_2, \ldots, \underline{n}_{p+1}].$$

The values that must be assigned to the variables w_1 and w_2 to witness that this formula is true are precisely the interpretations of $\underline{a}$ and $\underline{b}$, where a and b are integers that code the sequence

$$(f(n_1, n_2, \ldots, n_p, 0), f(n_1, n_2, \ldots, n_p, 1), \ldots, f(n_1, n_2, \ldots, n_p, n_{p+1}))$$

by means of the function β and whose existence is guaranteed by Lemma 6.11.

Conversely, suppose that

$$\mathcal{M} \models F[c, \underline{n_1}, \underline{n_2}, \ldots, \underline{n_{p+1}}];$$

we have to show that c is the standard element that is the interpretation of $\underline{f(n_1, n_2, \ldots, n_{p+1})}$.

Because $F[c, \underline{n_1}, \underline{n_2}, \ldots, \underline{n_{p+1}}]$ is true in $\mathcal{M}$, we know that there exist elements a, b, and d in $\mathcal{M}$ such that

$$\mathcal{M} \models B'[d, \underline{1}, a, b] \wedge G[d, \underline{n_1}, \underline{n_2}, \ldots, \underline{n_p}] \vee B'[c, \underline{n_{p+1}}, a, b]$$

and, for every integer i satisfying $0 \leq i < n_{p+1}$, there are elements r_i and s_i in $\mathcal{M}$ such that

$$\mathcal{M} \models B'[r_i, \underline{Si}, a, b] \wedge B'[s_i, \underline{S\,Si}, a, b] \wedge H[s_i, \underline{n_1}, \underline{n_2}, \ldots, \underline{n_p}, \underline{i}, r_i].$$

Because G represents g, $\mathcal{M} \models d \simeq \underline{g(n_1, n_2, \ldots, n_p)}$. Since the formula B' was chosen so that, for all x, y, and z in $\mathcal{M}$, there is at most one point satisfying $B'[v_0, x, y, z]$, we conclude that $d = r_0$, that $c = s_{n_{p+1}-1}$ and that, for all i with $0 \leq i < n_{p+1}$, $r_{i+1} = s_i$. By using the denition of H, we can then conclude, by induction on $i < n_{p+1}$, that

$$\mathcal{M} \models r_i \simeq \underline{f(n_1, n_2, \ldots, n_p, i)}$$

and hence that $\mathcal{M} \models c \simeq \underline{f(n_1, n_2, \ldots, n_{p+1})}$. ■

We should note that, in Lemma 6.11, it is the representability of β that is dif-cult to guarantee; otherwise, the function δ that was introduced in Chapter 5 would do perfectly well. This function is primitive recursive and we could conclude, in the end, that it is representable; but at the moment, we cannot make this assertion.

Let us now return to the proof of Lemma 6.11.

Proof To dene this function, β, we must use some elementary facts of arithmetic, in particular, the following classical result that is known as the **Chinese remainder theorem.**

Theorem 6.13 *Let $(b_0, b_1, \ldots, b_n)$ be a sequence of elements of $\mathbb{N}$ that are pairwise relatively prime and let $(\alpha_0, \alpha_1, \ldots, \alpha_n)$ be a sequence of the same length of elements of $\mathbb{N}$. Then there exists $a \in \mathbb{N}$ such that, for all i from 0 to n inclusive,*

$$a \text{ is congruent to } \alpha_i \text{ modulo } b_i.$$

(The proof of this theorem is given in Exercise 3.)

By denition, $\beta(i, a, b)$ is the remainder of the (Euclidean) division of b by $a(i + 1) + 1$. First of all, it is easy to see that β is represented by the formula

$$B[v_0, v_1, v_2, v_3] = \exists v_4(v_3 \simeq (v_4 \underline{\times}\, \underline{S}(v_2 \underline{\times}\, \underline{S}v_1))\underline{+}v_0) \\ \wedge\, v_0 < \underline{S}(v_2 \underline{\times}\, \underline{S}v_1).$$

Also, it has the desired property. To see this, let $(\alpha_0, \alpha_1, \ldots, \alpha_n)$ be a sequence of integers. Choose an integer m greater than $n + 1$ such that, if we set $a = m!$, then a is greater than or equal to all the α_i. For i from 0 to n inclusive, the integers $a(i + 1) + 1$ are pairwise relatively prime: for suppose that $0 \leq i < j \leq n$ and that c is a common prime divisor of $a(i + 1) + 1$ and $a(j + 1) + 1$; then c must also divide the difference $a(i - j) = m!(i - j)$ and is hence less than or equal to m; but this is impossible since it must also divide $m!(i + 1) + 1$.

So by the Chinese remainder theorem, there exists an integer b such that, for all i from 0 to n inclusive, we have

$$b \text{ is congruent to } \alpha_i \text{ modulo } a(i + 1) + 1,$$

and, since $\alpha_i \leq a < a(i + 1) + 1$, we do have $\beta(i, a, b) = \alpha_i$. ∎

This concludes the proof of the representation theorem. ∎

It clearly follows from this that every recursive set is representable. Moreover, let $\mathcal{P}'$ be any theory that includes $\mathcal{P}_0$ ($\mathcal{P}$, for example). It is clear that if $f \in \mathcal{F}_p$ is represented by the formula F, and if $(n_1, n_2, \ldots, n_p)$ is a sequence of integers, then $\mathcal{P}' \vdash \forall v_0(F[v_0, n_1, n_2, \ldots, n_p] \Leftrightarrow v_0 \simeq \underline{f(n_1, n_2, \ldots, n_p)})$.

6.3 Arithmetization of syntax

6.3.1 The coding of formulas

In this section, we will code terms and formulas of a nite language by integers. We could do this for any nite language, and even for certain innite languages; but to avoid overly complicated notation, we will be content to treat $\mathcal{L}_0$. Our goal, above all, is to prove that the set of universally valid formulas of this language is recursively enumerable. The coding will make use of the functions α_i and β_i^j introduced in Chapter 5 in the subsection on coding of sequences. We will also need the following little lemma.

Lemma 6.14 *Assume that p and n are integers, that $k_1, k_2, \ldots, k_n \in \mathcal{F}_1$, $g \in \mathcal{F}_p$, $h \in \mathcal{F}_{n+p+1}$, and that, for all $y > 0$ and i from 1 to n inclusive, $k_i(y) < y$. Then the unique function determined by the following conditions*

$$\begin{aligned} f(0, x_1, \ldots, x_p) &= g(x_1, \ldots, x_p); \\ f(y, x_1, \ldots, x_p) &= h(y, f(k_1(y), x_1, \ldots, x_p), f(k_2(y), x_1, \ldots, x_p), \ldots, \\ &\qquad f(k_n(y), x_1, \ldots, x_p), x_1, \ldots, x_p) \quad \textit{if } y > 0; \end{aligned}$$

is primitive recursive.

Proof We have here a denition by recursion that does not quite t the framework of Denition 5.1. To justify it, we will make use of

- the function Ω from Denition 5.5, which will serve to code the sequence of values of $f(i, x_1, x_2, \ldots, x_p)$ for i from 0 to y;
- the function π [$\pi(n)$ is the $(n+1)$st prime number]; and
- the function δ from Denition 5.5 (which allows us to decode Ω).

Dene the function ϕ by

$$\begin{aligned} \phi(0, x_1, x_2, \ldots, x_p) &= 2^{g(x_1, x_2, \ldots, x_p)}; \\ \phi(y+1, x_1, x_2, \ldots, x_p) &= \phi(y, x_1, x_2, \ldots, x_p) \cdot \pi(y+1)^{\gamma}, \end{aligned}$$

where

$$\begin{aligned} \gamma = h(y&+1, \\ &\delta(k_1(y+1), \phi(y, x_1, x_2, \ldots, x_p)), \\ &\delta(k_2(y+1), \phi(y, x_1, x_2, \ldots, x_p)), \ldots, \\ &\delta(k_n(y+1), \phi(y, x_1, x_2, \ldots, x_p)), x_1, x_2, \ldots, x_p). \end{aligned}$$

So the function ϕ is primitive recursive and, as a consequence, so is f since

$$f(y, x_1, x_2, \ldots, x_p) = \delta(y, \phi(y, x_1, x_2, \ldots, x_p)).$$ ■

We can now proceed to the coding of terms. The idea is to code a term t by a triple of integers (a, b, c) whose third coordinate, c, will distinguish whether t is an elementary term, or a term of the form $\underline{S}t_1$, or of the form $t_1 \underline{+} t_2$ or of the form $t_1 \underline{\times} t_2$. Depending on the situation, coordinates a and b will code the elementary term that is equal to t or to the terms t_1 and t_2 from which t is built.

Obviously, the triple (a, b, c) can be viewed as a single integer with the help of the function α_3.

Denition 6.15 *By induction on the term t, we will dene an integer, denoted by #t, that we will call the* ***Gödel number of*** *t.*

$$\begin{array}{ll} \textit{If } t = \underline{0}, & \textit{then } \#t = \alpha_3(0,0,0); \\ \textit{if } t = v_n, & \textit{then } \#t = \alpha_3(n+1,0,0); \\ \textit{if } t = \underline{S}t_1, & \textit{then } \#t = \alpha_3(\#t_1,0,1); \\ \textit{if } t = t_1 \underline{+} t_2, & \textit{then } \#t = \alpha_3(\#t_1,\#t_2,2); \\ \textit{if } t = t_1 \underline{\times} t_2, & \textit{then } \#t = \alpha_3(\#t_1,\#t_2,3). \end{array}$$

Lemma 6.16 *The set* $\mathsf{Term} = \{\#t : t \textit{ is a term of } \mathcal{L}_0\}$ *is primitive recursive.*

Proof Indeed, the characteristic function g of the set Term can be dened in the following way:

$$\begin{array}{ll} g(0) = 1; & g(1) = 1; \\ \text{and for } x > 1, & \\ \text{if } \beta_3^3(x) = 0 \text{ and } \beta_3^2(x) = 0, & \text{then } g(x) = 1; \\ \text{if } \beta_3^3(x) = 0 \text{ and } \beta_3^2(x) \neq 0, & \text{then } g(x) = 0; \\ \text{if } \beta_3^3(x) = 1 \text{ and } \beta_3^2(x) \neq 0 & \text{then } g(x) = 0; \\ \text{if } \beta_3^3(x) = 1 \text{ and } \beta_3^2(x) = 0, & \text{then } g(x) = g(\beta_3^1(x)); \\ \text{if } \beta_3^3(x) = 2, & \text{then } g(x) = g(\beta_3^1(x)) \cdot g(\beta_3^2(x)); \\ \text{if } \beta_3^3(x) = 3, & \text{then } g(x) = g(\beta_3^1(x)) \cdot g(\beta_3^2(x)); \\ \text{if } \beta_3^3(x) > 3, & \text{then } g(x) = 0. \end{array}$$

When we refer to the denitions in the subsection from Chapter 5 on codings of sequences, we see that, if $x > 1$, then $\beta_3^1(x)$, $\beta_3^2(x)$, and $\beta_3^3(x)$ are strictly less than x; so Lemma 6.14 can be applied. ■

This coding is injective: for if $\#t = \#t'$, then $t = t'$. The reader who is not convinced can prove this by induction on the term t.

Next, we proceed to the coding of formulas. We employ the same principle. Atomic formulas will be recognized by having the third coordinate equal to 0; for a negation, the third coordinate will be equal to 1, for a conjunction to 2, and so on. The code of a formula F will, by analogy with the above, be denoted by $\#F$ and will be called the **Godel number of** F.

$$\begin{array}{ll} \text{If } F = t_1 \simeq t_2, & \text{then } \#F = \alpha_3(\#t_1,\#t_2,0); \\ \text{if } F = \neg F_1, & \text{then } \#F = \alpha_3(\#F_1,0,1); \end{array}$$

$$\begin{array}{ll}
\text{if } F = (F_1 \wedge F_2), & \text{then } \#F = \alpha_3(\#F_1, \#F_2, 2); \\
\text{if } F = (F_1 \vee F_2), & \text{then } \#F = \alpha_3(\#F_1, \#F_2, 3); \\
\text{if } F = (F_1 \Rightarrow F_2), & \text{then } \#F = \alpha_3(\#F_1, \#F_2, 4); \\
\text{if } F = (F_1 \Leftrightarrow F_2), & \text{then } \#F = \alpha_3(\#F_1, \#F_2, 5); \\
\text{if } F = \forall v_n F_1, & \text{then } \#F = \alpha_3(\#F_1, n, 6); \\
\text{if } F = \exists v_n F_1, & \text{then } \#F = \alpha_3(\#F_1, n, 7).
\end{array}$$

We have the analogous lemma.

Lemma 6.17 *The set* Form $= \{\#F\colon F$ *is a formula of* $\mathcal{L}_0\}$ *is primitive recursive.*

Proof The same type of proof is always involved. If g is the characteristic function of Term, then the characteristic function h of Form can be dened as follows:

$$\begin{array}{ll}
\text{if } \beta_3^3(x) = 0, & \text{then } h(x) = g(\beta_3^1(x)) \cdot g(\beta_3^2(x)); \\
\text{if } \beta_3^3(x) = 1 \text{ and } \beta_3^2(x) \neq 0, & \text{then } h(x) = 0; \\
\text{if } \beta_3^3(x) = 1 \text{ and } \beta_3^2(x) = 0, & \text{then } h(x) = h(\beta_3^1(x)); \\
\text{if } \beta_3^3(x) = 2, 3, 4 \text{ or } 5, & \text{then } h(x) = h(\beta_3^1(x)) \cdot h(\beta_3^2(x)); \\
\text{if } \beta_3^3(x) = 6 \text{ or } 7, & \text{then } h(x) = h(\beta_3^1(x)); \\
\text{if } \beta_3^3(x) > 7, & \text{then } h(x) = 0.
\end{array}$$
■

We observe, as for terms, that the coding is injective.

We must also prove that the operations that are performed on formulas (e.g. substitutions, the recognition of free or bound variables, etc.) can be coded by primitive recursive functions of Godel numbers.

Lemma 6.18 *The following sets are all primitive recursive:*

$\Theta_0 = \{(\#t, n) :$ *t is a term in which* v_n *does not occur*$\}$;

$\Theta_1 = \{(\#t, n) :$ *t is a term in which* v_n *does occur*$\}$;

$\Phi_0 = \{(\#F, n) :$ *F is a formula in which* v_n *does not occur*$\}$;

$\Phi_1 = \{(\#F, n) :$ *F is a formula with no free occurrence of* $v_n\}$;

$\Phi_2 = \{(\#F, n) :$ *F is a formula with no bound occurrence of* $v_n\}$;

$\Phi_3 = \{\#F :$ *F is a closed formula*$\}$;

$\Phi_4 = \{(\#F, n) :$ *F is a formula that contains a free occurrence of* $v_n\}$;

$\Phi_5 = \{(\#F, n) :$ *F is a formula that contains a bound occurrence of* $v_n\}$.

Proof We will be content to treat Θ_0 and Φ_1. We will again denote the characteristic functions of Term and Form by g and h, respectively. The characteristic function of Θ_0, which we will denote by g_0, can be dened by the following conditions:

if $\beta_3^3(x) = 0$, then $g_0(x, y) = 1$
if and only if $\beta_3^2(x) = 0$ and $\beta_3^1(x) \neq y + 1$;

if $\beta_3^3(x) = 1$ and $\beta_3^2(x) \neq 0$, then $g_0(x, y) = 0$;

if $\beta_3^3(x) = 1$ and $\beta_3^2(x) = 0$, then $g_0(x, y) = g_0(\beta_3^1(x), y)$;

if $\beta_3^3(x) = 2$ or 3, then $g_0(x, y) = g_0(\beta_3^1(x), y) \cdot g_0(\beta_3^2(x), y)$;

if $\beta_3^3(x) > 3$, then $g_0(x, y) = 0$.

Now let h be the characteristic function of Φ_1. Then

if $\beta_3^3(x) = 0$, then $h_1(x, y) = g_0(\beta_3^1(x), y) \cdot g_0(\beta_3^2(x), y)$;

if $\beta_3^3(x) = 1$ and $\beta_3^2(x) \neq 0$, then $h_1(x, y) = 0$;

if $\beta_3^3(x) = 1$ and $\beta_3^2(x) = 0$, then $h_1(x, y) = h_1(\beta_3^1(x), y)$;

if $\beta_3^3(x) = 2, 3, 4$, or 5, then $h_1(x, y) = h_1(\beta_3^1(x), y) \cdot h_1(\beta_3^2(x), y)$;

if $\beta_3^3(x) = 6$ or 7 and $\beta_3^2(x) \neq y$, then $h_1(x, y) = h_1(\beta_3^1(x), y)$;

if $\beta_3^3(x) = 6$ or 7 and $\beta_3^2(x) = y$, then $h_1(x, y) = h_1(\beta_3^1(x))$;

if $\beta_3^3(x) > 7$, then $h_1(x, y) = 0$.

It is obviously Lemma 6.14 which allows us to conclude that these sets are primitive recursive.

Let us now move on to substitutions. It is not surprising that we have

Lemma 6.19 *There exist two primitive recursive functions $Subs_t$ and $Subs_f$ of three variables such that if t and u are terms and if F is a formula, then for every integer n,*

$$Subs_t(n, \#t, \#u) = \#u_{t/v_n};$$
$$Subs_f(n, \#t, \#F) = \#F_{t/v_n}.$$

(The notations u_{t/v_n} and F_{t/v_n} were dened in Chapter 3.)

Proof Once again, we will use Lemma 6.14. First, we will dene $Subs_t$ by the following conditions:

if $\beta_3^3(x) = 0$, then

$$Subs_t(n, y, x) = \begin{cases} x & \text{if } x \neq \alpha_3(n+1, 0, 0), \\ y & \text{if not;} \end{cases}$$

if $\beta_3^3(x) = 1$ and $\beta_3^2(x) = 0$, then

$$Subs_t(n, y, x) = \alpha_3(Subs_t(n, y, \beta_3^1(x)), 0, 1);$$

if $\beta_3^3(x) = 2$ or 3, then

$Subs_t(n, y, x) = \alpha_3(Subs_t(n, y, \beta_3^1(x)), Subs_t(n, y, \beta_3^2(x)), \beta_3^3(x))$;

in all other cases, we arbitrarily set $Subs_t(n, y, x) = x$.

Next, for the function $Subs_f$, the situation is slightly more complicated because the substitution can only take place at free occurrences of the variable:

if $\beta_3^3(x) = 0$, then

$Subs_f(n, y, x) = \alpha_3(Subs_t(n, y, \beta_3^1(x)), Subs_t(n, y, \beta_3^2(x)), 0)$;

if $\beta_3^3(x) = 1$ and $\beta_3^2(x) = 0$, then

$Subs_f(n, y, x) = \alpha_3(Subs_f(n, y, \beta_3^1(x)), 0, 1)$;

if $\beta_3^3(x) = 2, 3, 4,$ or 5, then

$Subs_f(n, y, x) = \alpha_3(Subs_f(n, y, \beta_3^1(x)), Subs_f(n, y, \beta_3^2(x)), \beta_3^3(x))$;

if $\beta_3^3(x) = 6$ or 7, then

$$Subs_f(n, y, x) = \begin{cases} x & \text{if } \beta_3^2(x) = n, \\ \alpha_3(Subs_f(n, y, \beta_3^1(x)), \beta_3^2(x), \beta_3^3(x)) & \text{otherwise;} \end{cases}$$

in all other cases, we arbitrarily set $Subs_f(n, y, x) = x$.

6.3.2 The coding of proofs

We must, at this point, deal with the slightly more difcult question of the decidability of propositional calculus. We will return to this calculus momentarily. Thus we have, in addition to the propositional connectives, an innite set of propositional variables $A_1, A_2, \ldots$. We begin by establishing a coding of propositional formulas analogous to the ones that have preceded. Corresponding to a formula P, we will have its **Godel number**, $\#P$, dened as follows:

$$\begin{array}{ll} \text{if } P = A_n, & \text{then } \#P = \alpha_3(n, 0, 0); \\ \text{if } P = \neg P_1, & \text{then } \#P = \alpha_3(\#P_1, 0, 1); \\ \text{if } P = (P_1 \wedge P_2), & \text{then } \#P = \alpha_3(\#P_1, \#P_2, 2); \\ \text{if } P = (P_1 \vee P_2), & \text{then } \#P = \alpha_3(\#P_1, \#P_2, 3); \\ \text{if } P = (P_1 \Rightarrow P_2), & \text{then } \#P = \alpha_3(\#P_1, \#P_2, 4); \\ \text{if } P = (P_1 \Leftrightarrow P_2), & \text{then } \#P = \alpha_3(\#P_1, \#P_2, 5). \end{array}$$

As is now our habit, we observe that the set

$$\mathsf{Prop} = \{\#P : P \text{ is a proposition}\}$$

is primitive recursive.

Theorem 6.20. *(The decidability of propositional calculus) The set*

$$\mathcal{T} = \{\#P : P \text{ is a tautology}\}$$

is primitive recursive.

Proof With each integer k, we associate the assignment of truth values λ_k, which is dened by

$$\lambda_k(A_n) = \begin{cases} 1 & \text{if } \pi(n) \text{ [the } (n+1)\text{st prime number] divides } k; \\ 0 & \text{otherwise.} \end{cases}$$

Now let c be an integer and let λ be any assignment of truth values. We can easily nd an integer k such that, for all $i \leq c$, $\lambda_k(A_i) = \lambda(A_i)$. It sufces to take

$$k = \prod_{0 \leq i \leq c} \pi(i)^{\lambda(A_i)},$$

and we observe that k can be chosen to be less than or equal to $\pi(c)!$.

Let P be a propositional formula. We wish to determine whether P is a tautology or not. First of all, it is clear that if A_n is a propositional variable that occurs in P, then $n \leq \#P$. From all that we have said, it follows that P is a tautology if and only if, for every integer $k \leq \pi(\#P)!$, $\lambda_k(P) = 1$. So we will begin by proving that

Lemma 6.21 *The function E dened by*

$$E(k, x) = 0 \quad \text{if } x \text{ is not the Godel number of a proposition};$$
$$E(k, x) = \lambda_k(P) \quad \text{if } x \text{ is the Godel number of a proposition};$$

is primitive recursive.

Proof Again, it is Lemma 6.14 that comes to the rescue. Indeed, E can be dened in the following way:

if $x \notin \mathsf{Prop}$, then $E(k, x) = 0$;
if $x \in \mathsf{Prop}$, then
  if $\beta_3^3(x) = 0$, then
    if $\pi(\beta_3^1(x))$ divides k, $E(k, x) = 1$,
    if $\pi(\beta_3^1(x))$ does not divide k, $E(k, x) = 0$;
  if $\beta_3^3(x) = 1$, then $E(k, x) = 1 - E(k, \beta_3^1(x))$;
  if $\beta_3^3(x) = 2$, then $E(k, x) = E(k, \beta_3^1(x)) \cdot E(k, \beta_3^2(x))$;
  if $\beta_3^3(x) = 3$, then $E(k, x) = sg(E(k, \beta_3^1(x)) + E(k, \beta_3^2(x)))$;
  if $\beta_3^3(x) = 4$, then $E(k, x) = sg(E(k, \beta_3^2(x)) + 1 - E(k, \beta_3^1(x)))$;
  if $\beta_3^3(x) = 5$, then $E(k, x) = \begin{cases} 1 & \text{if } E(k, \beta_3^1(x)) = E(k, \beta_3^2(x)); \\ 0 & \text{otherwise.} \end{cases}$ ∎

So to conclude the proof of the theorem, it is sufcient to observe that

$$x \in \mathcal{T} \quad \text{if and only if} \quad \forall k \leq \pi(x)! E(k, x) = 1.$$ ■

Theorem 6.22 *The set*

$$\mathsf{Taut} = \{\#F : F \textit{ is a formula and is a tautology of the predicate calculus}\}$$

is primitive recursive.

Proof With each formula F, we will associate a proposition P_F that is obtained as follows: write F in the form $P[F_1, F_2, \ldots, F_k]$, where P is a proposition whose propositional variables are $A_1, A_2, \ldots, A_k$ and where the formulas $F_1, F_2, \ldots, F_k$ cannot be further decomposed using propositional connectives; in other words, each formula F_i is either an atomic formula or a formula that begins with a quantier. (See Chapter 3 for a denition of the formula $P[F_1, F_2, \ldots, F_k]$.) For each i, set $\#F_i = c(i)$ and

$$P_F = P[A_{c(1)}, A_{c(2)}, \ldots, A_{c(k)}].$$

Then F is a tautology of the predicate calculus if and only if P_F is a tautology: in one direction, from left to right, this is just Lemma 3.49. In the other direction, from right to left, suppose that $F = J[G_1, G_2, \ldots, G_m]$ where the G_j are formulas of the language $\mathcal{L}_0$ and where $J[B_1, B_2, \ldots, B_m]$ is a propositional formula that is a tautology; it is then important to note that there is an obvious relation between the propositional formula J and P_F: precisely, P_F is obtained from J by substituting, for the propositional variables $B_1, B_2, \ldots, B_m$, appropriate propositional formulas constructed from the variables $A_{c(1)}, A_{c(2)}, \ldots, A_{c(k)}$. Without providing a real proof of this assertion, we are satised to note that the formula P_F represents, in a certain way, the maximal decomposition of F into propositions, that this maximal decomposition is unique up to the names of the propositional variables, and that the formula J represents an intermediate stage of the decomposition. It is then a consequence of Corollary 1.23 that if J is a tautology of the propositional calculus, then so is P_F. As a result, it sufces to construct a primitive recursive function γ such that, for every formula F, $\gamma(\#F) = \#P_F$. We will then have

$$x \in \mathsf{Taut} \quad \text{if and only if} \quad x \in \mathsf{Form} \text{ and } \gamma(x) \in \mathcal{T}.$$

As usual, we invoke Lemma 6.14 and dene γ as follows:

$$\begin{array}{ll}
\text{if } \beta_3^3(x) = 0, 6, \text{ or } 7, & \text{then } \gamma(x) = \alpha_3(x, 0, 0); \\
\text{if } \beta_3^3(x) = 1, & \text{then } \gamma(x) = \alpha_3(\gamma(\beta_3^1(x)), 0, 1); \\
\text{if } \beta_3^3(x) = 2, 3, 4, \text{ or } 5, & \text{then } \gamma(x) = \alpha_3(\gamma(\beta_3^1(x)), \gamma(\beta_3^2(x)), \beta_3^3(x)); \\
\text{if } \beta_3^3(x) > 7, & \text{then we arbitrarily set } \gamma(x) = 0.
\end{array}$$ ■

We have now accumulated all we need to prove that the set of logical axioms is a primitive recursive set.

Theorem 6.23 *The set* $\mathsf{Ax} = \{\#F : F$ *is a logical axiom*$\}$ *is primitive recursive.*

Proof (a) *The set*

$$\mathsf{Ax}_1 = \{\#(\exists v F \Leftrightarrow \neg\forall v \neg F) : F \textit{ is a formula and } v \textit{ is a variable}\}$$

is primitive recursive.

To see this, perform the easy calculation which shows that

$$\begin{aligned}&\#(\exists v F \Leftrightarrow \neg\forall v \neg F)\\ &\quad = \alpha_3(\alpha_3(\#F, n, 7), \alpha_3(\alpha_3(\alpha_3(\#F, 0, 1), n, 6), 0, 1), 5).\end{aligned}$$

Hence $x \in \mathsf{Ax}_1$ if and only if there exists $y < x$ and $n < x$ such that $y \in \mathsf{Form}$ and

$$x = \alpha_3(\alpha_3(y, n, 7), \alpha_3(\alpha_3(\alpha_3(y, 0, 1), n, 6), 0, 1), 5).$$

(b) *The set*

$$\begin{aligned}\mathsf{Ax}_2 = \{\#(\forall v(F \Rightarrow G) \Rightarrow (F \Rightarrow \forall v G)) &: F \textit{ and } G \textit{ are formulas}\\ &\textit{and } v \textit{ is a variable that has no free occurrence in } F\}\end{aligned}$$

is primitive recursive.

The argument is the same: $x \in \mathsf{Ax}_2$ if and only if there exists y, z, and n less than x such that $(y, n) \in \Phi_1$, $z \in \mathsf{Form}$ and

$$x = \alpha_3(\alpha_3(\alpha_3(y, z, 4), n, 6), \alpha_3(y, \alpha_3(z, n, 6), 4), 4).$$

(c) *The set*

$$\begin{aligned}\mathsf{Ax}_3 = \{\#\,(\forall v F \Rightarrow F_{t/v}) &: v \textit{ is a variable, } F \textit{ is a formula, } t \textit{ is a term}\\ &\textit{and any free occurrence of } v \textit{ in } F \textit{ does not lie within}\\ &\textit{the scope of a quantier that binds a variable of } t\}\end{aligned}$$

is primitive recursive.

One must rst be persuaded that the set

$$\begin{aligned}B = \{(\#F, n, m) &: \textit{ any free occurrence of } v_m \textit{ in } F \textit{ does not}\\ &\textit{lie in the scope of a quantier } \forall v_n \textit{ or } \exists v_n\}\end{aligned}$$

is primitive recursive. As usual, this is done using Lemma 6.14. The characteristic function g of B can be dened as follows:

if $x \notin \mathsf{Form}$, then $g(x, n, m) = 0$;

if $x \in \mathsf{Form}$, then

 if $\beta_3^3(x) = 0$, then $g(x, n, m) = 1$;

 if $\beta_3^3(x) = 1$, then $g(x, n, m) = g(\beta_3^1(x), n, m)$;

if $\beta_3^3(x) \in \{2, 3, 4, 5\}$, then
$g(x, n, m) = g(\beta_3^1(x), n, m) \cdot g(\beta_3^2(x), n, m)$;
if $\beta_3^3(x) \in \{6, 7\}$, then
if $\beta_3^1(x) = n$ and $(\beta_3^1(x), m) \in \Phi_4$, then $g(x, n, m) = 0$;
otherwise, $g(x, n, m) = g(\beta_3^1(x), n, m)$.

To conclude, observe that

$x \in \mathsf{Ax}_3$ if and only if there exist y, z, and m less than x such that
$y \in \mathsf{Form}$, $z \in \mathsf{Term}$, for all $n < z$, $((z, n) \in \Theta_0$ or $(y, n, m) \in B)$
and $x = \alpha_3(\alpha_3(y, m, 6), Subs_f(m, z, y), 4)$. ■

Denition 6.24 (1) *Let T be a theory; we say that T is* ***recursive*** *if the set*

$$\#T = \{\#F : F \in T\}$$

is recursive.

(2) *Let* $\mathsf{Th}(T) = \{\#F : F$ *is a closed formula and* $T \vdash F\}$
[$\mathsf{Th}(T)$ *is the set of Godel numbers of theorems of* T].

(3) *We say that T is* ***decidable*** *if* $\mathsf{Th}(T)$ *is recursive. An* ***undecidable*** *theory is a theory that is not decidable.*

Remark To be recursive is a reasonable requirement for a theory; we might even say that non-recursive theories are articial: how could we hope to deal with derivations if we do not have effective knowledge of the axioms? By contrast, we will see many natural and interesting theories that are undecidable.

Example The empty theory is recursive; the set of its theorems is simply the set of closed valid formulas. Finite theories such as $\mathcal{P}_0$ are also recursive. It is not difcult to see that $\mathcal{P}$ is recursive.

Notation Let $d = (F_0, F_1, \ldots, F_k)$ be a sequence of formulas of the language $\mathcal{L}_0$; $\#\#d$ is the integer dened by

$$\#\#d = \Omega((\#F_0, \#F_1, \ldots, \#F_k)),$$

where Ω is the function introduced in Denition 5.5. We will again refer to $\#\#d$ as the **Godel number of** d.

Proposition 6.25 *Let T be a recursive theory; then the set*

$$\mathsf{Drv}(T) = \{(n, m) : n = \#F,\ m = \#\#d,\ F \text{ is a formula and } d \text{ is a derivation of } F \text{ in } T\}$$

is primitive recursive.

Proof It sufces to refer to the denition of a derivation (see Denition 4.3) and to realize that the procedure for recognizing whether a sequence of formulas is a derivation is an effective one;

$(n, m) \in \mathsf{Drv}(T)$ if and only if the following three conditions are satised [below, $\mathsf{lg}(m)$ denotes the length of the word coded by m]:

(1) for all $i < \mathsf{lg}(m)$, $\delta(i, m) \in \mathsf{Form}$;

(2) $\delta(\mathsf{lg}(m) - 1, m) = n$;

(3) for all $i < \mathsf{lg}(m)$, $\delta(i, m) \in \mathsf{Ax} \cup \#T$, or there exist $j < i$ and $p < m$ such that $\delta(i, m) = \alpha_3(\delta(j, m), p, 6)$, or there exist $j < i$ and $k < i$ such that $\delta(j, m) = \alpha_3(\delta(k, m), \delta(i, m), 4)$.

Clause (3) expresses that each formula in the derivation is either an axiom [if $\delta(i, m) \in \mathsf{Ax} \cup \#T$], or a formula derived by generalization from a formula already proved [if there exist $j < i$ and $p < m$ such that $\delta(i, m) = \alpha_3(\delta(j, m), p, 6)$], or a formula derived by modus ponens from two formulas that are already proved [if there exist $j < i$ and $k < i$ such that $\delta(j, m) = \alpha_3(\delta(k, m), \delta(i, m), 4)$]. ■

Corollary 6.26 *Let T be a recursive theory; then $\mathsf{Th}(T)$ is recursively enumerable. In particular, the following sets are recursively enumerable:*

$$\{\#F : F \text{ is a valid closed formula}\};$$
$$\{\#F : F \text{ is a theorem of } \mathcal{P}_0\};$$
$$\{\#F : F \text{ is a theorem of } \mathcal{P}\}.$$

Proof Indeed, $n \in \mathsf{Th}(T)$ if and only if $n \in \Phi_3$ and there exists an integer m such that $(n, m) \in \mathsf{Drv}(T)$; thus $\mathsf{Th}(T)$ is the intersection of a recursive set with the projection of a recursive set, so it is recursively enumerable (see Theorem 5.39). ■

We end this section with another corollary.

Corollary 6.27 *If the theory T is complete and recursive, then it is decidable.*

Proof We already know that $\mathsf{Th}(T)$ is recursively enumerable; if we show that its complement is also recursively enumerable, the result follows from Theorem 5.38. Because T is complete, if F is a closed formula that is not a theorem of T, then $\neg F$ is a theorem of T; this fact is equivalent to

$$m \notin \mathsf{Th}(T) \quad \text{if and only if} \quad m \notin \Phi_3 \;\text{ or }\; \alpha_3(m, 0, 1) \in \mathsf{Th}(T). \qquad ■$$

6.4 Incompleteness and undecidability theorems

6.4.1 Undecidability of arithmetic and predicate calculus

The last corollaries of the preceding section leave the following questions wide open: Is the empty theory decidable? Is $\mathcal{P}_0$ decidable? What about $\mathcal{P}$? In this section, we will answer no to each of these three questions and prove the most famous theorems of mathematical logic.

In this entire section, the theories under consideration are expressed in a nite language that includes $\mathcal{L}_0$ (see the introductory remarks at the beginning of the previous section).

Theorem 6.28 *Let T be a consistent theory that extends $\mathcal{P}_0$; then T is undecidable.*

Proof We will assume that T is a decidable theory that includes $\mathcal{P}_0$ and construct a closed formula F of $\mathcal{L}_0$ such that $T \vdash F$ and $T \vdash \neg F$. To do this, we will employ the results from Section 6.2 on the representation of recursive functions.

Consider the set

$$\Theta = \{(m, n) : m \text{ is the Godel number of a formula } F[v_0] \\ \text{whose only free variable, if any, is } v_0 \text{ and } T \vdash F[\underline{n}]\}.$$

It is clear, rst of all, that because T is decidable, Θ is recursive: indeed, the set A of Godel numbers of formulas whose only free variable, if any, is v_0 is recursive since $m \in A$ if and only if, for all p from 1 to n inclusive, $(m, p) \in \Theta_4$ (see Lemma 6.18). The function $\lambda n.\#\underline{n}$ is also recursive; it can be dened by recursion:

$$\#\underline{0} = \alpha_3(0, 0, 0);$$
$$\#\underline{(n+1)} = \alpha_3(\#\underline{n}, 0, 1).$$

So we see that

$$(m, n) \in \Theta \quad \text{if and only if} \quad m \in A \text{ and } Subs_f(0, \#\underline{n}, m) \in \mathsf{Th}(T).$$

It follows that the set

$$B = \{n \in \mathbb{N} : (n, n) \notin \Theta\}$$

is also recursive, so by the representation theorem (Theorem 6.8), there exists a formula $G[v_0]$ which represents it. Consequently, we have that, for all n,

$$n \in B \text{ implies } \mathcal{P}_0 \vdash G[\underline{n}], \quad \text{and hence } T \vdash G[\underline{n}]; \qquad (*)$$
$$n \notin B \text{ implies } \mathcal{P}_0 \vdash \neg G[\underline{n}], \quad \text{and hence } T \vdash \neg G[\underline{n}]. \qquad (**)$$

Moreover, $\#G[v_0]$ is an integer that belongs to A; we will call it a. Observe rst that a cannot belong to B: by denition of B, this would imply that $(a, a) \notin \Theta$ and, by denition of Θ, that $T \vdash G[\underline{a}]$ is false; this contradicts assertion $(*)$ above. So we must conclude that $a \notin B$ and that $(a, a) \in \Theta$. On the one hand, we have $T \vdash G[\underline{a}]$ by denition of Θ; on the other hand, $(**)$ implies that $T \vdash \neg G[\underline{a}]$. So T is inconsistent. ∎

The following corollary is **Church's theorem**. It establishes the **undecidability of predicate calculus**.

Corollary 6.29 *The set*

$$T_0 = \{F : F \text{ is a universally valid closed formula of } \mathcal{L}\}$$

is not recursive.

Proof Let G be the conjunction of all the axioms of $\mathcal{P}_0$ (this is where we congratulate ourselves for having worked with a nite theory!). It is then clear that, for every closed formula F of $\mathcal{L}_0$,

$$\mathcal{P}_0 \vdash F \quad \text{if and only if} \quad (G \Rightarrow F) \in T_0.$$

So if T_0 were recursive, $\mathcal{P}_0$ would be decidable; but this is false according to the previous theorem. ∎

Remarks The statement of this last corollary makes sense only if we have arithmetized the syntax of $\mathcal{L}$; but its proof is independent of this arithmetization as long as it extends the one we have provided for $\mathcal{L}_0$.

The undecidability of predicate calculus has been proved here only for languages that include the language of arithmetic; we will see in Exercise 11 that it sufces to assume that the language contains at least one binary predicate symbol. But the theorem is false for very weak languages that only contain unary predicate symbols.

6.4.2 Godel's incompleteness theorems

Here is the **rst incompleteness theorem** (of **GodelñRosser**).

Theorem 6.30 *Let T be a consistent recursive theory that includes $\mathcal{P}_0$; then T is not complete. In particular, $\mathcal{P}$ is incomplete.*

Proof With what we have so far (Theorem 6.28), we need only recall that a complete recursive theory is decidable (Corollary 6.27). ∎

There are, therefore, closed formulas of $\mathcal{L}_0$ that are neither derivable nor refutable from Peano's axioms. By following the proof of the incompleteness theorem, we could, if we wished, succeed in constructing such a formula. But this would not tell us whether this formula has a meaning, or, if it does, what this meaning is.

Godel's second incompleteness theorem provides a striking response to this question: it presents a formula which expresses that the axioms of Peano are consistent. This formula is true in the standard model, but, since it is not derivable, there are, by the completeness theorem of Chapter 4, models of Peano's axioms in which it is false.

The statement itself of the second incompleteness theorem requires some notation and a bit of work. Let T be a recursive theory that includes $\mathcal{P}$; consider the following two recursive sets, Drv and Drv_0, dened by:

$$\mathsf{Drv} = \{(a,b) : a \text{ is the Godel number of a closed formula } F \text{ and } b \text{ is the Godel number of a derivation of } F \text{ in } T\};$$

$$\mathsf{Drv}_0 = \{(a,b) : a \text{ is the Godel number of a closed formula } F \text{ and } b \text{ is the Godel number of a derivation of } F \text{ in } \mathcal{P}_0\}.$$

According to the representation theorem, there are two formulas, each with two free variables, of the language of arithmetic that represent these sets. We will choose two such formulas and denote them by *Drv* and Drv_0, respectively (we will return later to the question of how this choice is to be made). The consistency of the theory T then becomes expressible by a formula of $\mathcal{L}_0$. It sufces to say that it is impossible to derive both a formula and its negation. For this purpose, we will dene the primitive recursive function ng from $\mathbb{N}$ into $\mathbb{N}$ as follows:

- if n is the Godel number of a closed formula F, then $ng(n)$ is the Godel number of $\neg F$; in other words, $ng(n) = \alpha_3(n, 0, 1)$;
- $ng(n) = 0$, otherwise.

Now, let $Neg[v_0, v_1]$ be a formula that represents this function. The formula $Con(T)$ is then, by denition, equal to the formula

$$\neg\exists v_0 \exists v_1 \exists v_2 \exists v_3 (Drv[v_0, v_2] \wedge Drv[v_1, v_3] \wedge Neg[v_0, v_1]).$$

The formula $Con(T)$ well deserves its name: suppose that a theory T is inconsistent; then there exists a closed formula F and two derivations from T, say d_0 and d_1, of F and of $\neg F$, respectively. If n_0, n_1, m_0, and m_1 are the Godel numbers of F, $\neg F$, d_0, and d_1, respectively, then we see that

$$\mathbb{N} \models Drv[n_0, m_0] \wedge Drv[n_1, m_1] \wedge Neg[n_0, n_1];$$

thus

$$\mathbb{N} \models \neg Con(T).$$

Conversely, if

$$\mathbb{N} \models \neg Con(T),$$

then we can nd integers n_0, n_1, m_0, and m_1 such that

$$\mathbb{N} \models Drv[n_0, m_0] \wedge Drv[n_1, m_1] \wedge Neg[n_0, n_1],$$

and hence $(n_0, m_0) \in Drv$, $(n_1, m_1) \in Drv$ and $ng(n_0) = n_1$: i.e. n_0 is the Godel number of a formula such that both it and its negation are derivable.

We will see that, nonetheless, it is possible to have a model $\mathcal{M}$ of $\mathcal{P}$ in which the formula $\neg Con(T)$ is satised despite the fact that the theory T is consistent.

This simply means that there exist elements a_0, a_1, a_2, and a_3 in $\mathcal{M}$ such that

$$\mathcal{M} \models Drv[a_0, a_2] \wedge Drv[a_1, a_3] \wedge Neg[a_0, a_1];$$

the fact that a_0, a_1, a_2, and a_3 need not be standard integers prevents us from going further and concluding, as was the case for $\mathbb{N}$, that T is inconsistent.

However, in every model of $\mathcal{P}$, the formulas *Drv* and *Neg* do continue to have certain properties to which we are accustomed. Here, for example, is a fact that is a formal consequence of the way in which $Con(T)$ was dened (a fact which we will exploit later on): suppose that $b \in \mathbb{N}$ is the Godel number of a closed formula F and that c is the Godel number of $\neg F$ [in other words, $c = \alpha_3(b, 0, 1)$]; then

$$\mathcal{P}_0 \vdash (\exists v_0 Drv[\underline{b}, v_0] \wedge \exists v_1 Drv[\underline{c}, v_1]) \Rightarrow \neg Con(T).$$

We are now in a position to state **Godel's second incompleteness theorem**:

Theorem 6.31 *Let T be a consistent, recursive theory that extends $\mathcal{P}$. Then $Con(T)$ is not derivable in T.*

We must, at the same time, be a bit prudent. Indeed, although the sets Drv and Drv_0 are perfectly well-dened, we have already observed that this is not the case for the formulas Drv and Drv_0 or, consequently, for the formula $Con(T)$. The only thing we know, *a priori,* about these formulas is that they represent the sets Drv and Drv_0; we know exactly which integers satisfy them but we do not know very much about their behaviour as it relates to non-standard elements; as a matter of fact, we can nd (see Exercise 8) two formulas $D[v_0, v_1]$ and $D'[v_0, v_1]$, both of which represent Drv_0, but which are not equivalent in the sense that the formula

$$\forall v_0 \forall v_1 (D[v_0, v_1] \Leftrightarrow D'[v_0, v_1])$$

is not derivable in $\mathcal{P}$.

Since we require the formula *Drv* in the statement of the second incompleteness theorem [to write $Con(T)$], we need to know which one we are dealing with.

To summarize, the incompleteness theorem stated above is only true if we have made the right choice for the formula *Drv* and it seems that we may have to resign ourselves to writing it down effectively. We could actually do this, constructing it step by step, following the proof of the representation theorem and the proof that the set Drv is recursive. In this way, we would obtain a formula *Drv* that is more or less canonical in the following approximate sense: if two people were to conscientiously make this construction, they would surely, in the end, arrive at two formulas that are equivalent modulo $\mathcal{P}$. But this is not what we will do, for this would require a process of writing and verifying, which is far too lengthy and boring. It is easier to isolate those properties that these formulas must satisfy (there are not very many, in fact) in order to allow a proof of the second incompleteness theorem. Then, later, we will see how to manage with these properties.

To do this, we require a denition.

Denition 6.32 *The set Σ is the smallest set of formulas of the language $\mathcal{L}_0$ that*

(i) *contains all formulas without quantiers;*

(ii) *is closed under conjunction and disjunction (if F and G are in Σ, then so are $F \wedge G$ and $F \vee G$);*

(iii) *is closed under existential quantication;*

(iv) *is closed under bounded universal quantication [if F is in Σ, then so is $\forall v_0(v_0 < v_1 \Rightarrow F)$, which we will rewrite as $(\forall v_0 < v_1)F$].*

*We will say that F is a Σ **formula** (and will write `F is Σ') if F belongs to Σ.*

Remark This set belongs to a well-known family of sets and is generally denoted by Σ_1^0 (read as `sigma zero one'). However, since this is the only member of the family under consideration here, we will not encumber ourselves with indices.

For example, it is not difcult to see that the relations `$n$ divides $m$' and `n is prime' are expressible by Σ formulas. We must beware, however, of the fact that the set Σ is not closed under negation.

Here are the properties which we will require of the formulas Drv and Drv_0:

(P_1) $\vdash \forall v_0 \forall v_1 (Drv_0[v_0, v_1] \Rightarrow Drv[v_0, v_1])$;

(P_2) Drv and Drv_0 are Σ formulas;

(P_3) if F is a closed Σ formula, then $\mathcal{P} \vdash F \Rightarrow \exists v_1 Drv_0[\underline{\#F}, v_1]$.

The rst of these is not difcult to justify. It could hardly be more natural since T includes $\mathcal{P}_0$. In any case, if property (P_1) were not satised, we could replace $Drv[v_0, v_1]$ by $Drv[v_0, v_1] \vee Drv_0[v_0, v_1]$.

For the second property, if we re-examine the proof of the representation theorem, we realize that, in fact, this same proof establishes a **second representation theorem**:

Theorem 6.33 *Every total recursive function can be represented by a Σ formula.*

As a consequence, we will assume that the formulas Drv and Drv_0 and all the formulas that we may need to represent a recursive set or a recursive function are Σ formulas.

Here is another result that illustrates the importance of Σ formulas and begins to justify (P_3).

Proposition 6.34 *If F is a closed Σ formula of $\mathcal{L}_0$, then*

$$\mathbb{N} \models F \Rightarrow \exists v_1 Drv_0[\underline{\#F}, v_1].$$

(In other words, if F is a closed Σ formula, then $\mathbb{N} \models F$ if and only if $\mathcal{P}_0 \vdash F$.)

Proof If F is false in $\mathbb{N}$, then the formula $F \Rightarrow \exists v_1 Drv_0[\underline{\#F}, v_1]$ is obviously true in $\mathbb{N}$. If F is true in $\mathbb{N}$, we will show that it is derivable in $\mathcal{P}_0$ and, to do this, we will invoke the completeness theorem: it sufces to prove that F is true in every model of $\mathcal{P}_0$. The next lemma concludes the proof once we recall that every model of $\mathcal{P}_0$ is an end-extension of $\mathbb{N}$ (see Theorem 6.5).

Lemma 6.35 *Let $\mathcal{N}$ be an $\mathcal{L}_0$-structure and $\mathcal{M}$ an end-extension of $\mathcal{N}$, let $F[v_1, v_2, \ldots, v_p]$ be a Σ formula and let $a_1, a_2, \ldots, a_p$ be elements of $\mathcal{N}$. Then*

$$\mathcal{N} \models F[a_1, a_2, \ldots, a_p] \textit{ implies } \mathcal{M} \models F[a_1, a_2, \ldots, a_p].$$

Proof We argue by induction. Consider the set of those formulas G such that, for all $a_1, a_2, \ldots, a_p$ in $\mathcal{N}$ (where p is the number of free variables in G),

$$\mathcal{N} \models G[a_1, a_2, \ldots, a_p] \text{ implies } \mathcal{M} \models G[a_1, a_2, \ldots, a_p].$$

We easily see that this set contains all formulas without quantiers and is closed under conjunction and disjunction; it is also closed under existential quantication since $\mathcal{N}$ is a substructure of $\mathcal{M}$; nally, it is closed under bounded quantication since $\mathcal{M}$ is an end-extension of $\mathcal{N}$. So this set contains all Σ formulas. ■

Let us set

$$\mathcal{P}_1 = \mathcal{P}_0 \cup \{F \Rightarrow \exists v_1 Drv_0[\underline{\#F}, v_1] : F \text{ is a closed } \Sigma \text{ formula}\}.$$

We have just seen that $\mathbb{N}$ is a model of $\mathcal{P}_1$. It is also an easy consequence of all the lemmas in Section 6.3 that $\mathcal{P}_1$ is a recursive theory. The second incompleteness theorem is therefore a consequence of the next two lemmas.

Lemma 6.36 *Every formula of $\mathcal{P}_1$ is derivable from $\mathcal{P}$.*

Lemma 6.37 *Let T be a consistent recursive theory in which all the formulas of $\mathcal{P}_1$ are derivable. Then $Con(T)$ is not derivable in T.*

Note that the second of these lemmas immediately implies that a consistent recursive theory that includes $\mathcal{P} \cup \mathcal{P}_1$ cannot prove its own consistency; this is already a good approximation to the second incompleteness theorem. (It can be shown that $\mathcal{P}_1$, which is much simpler syntactically than $\mathcal{P}$, is in fact much weaker than $\mathcal{P}$; consequently, this lemma, together with this last fact, provides a strong version of the second incompleteness theorem.) The proof of the second lemma, as we shall see, is not too difcult. The proof of the rst lemma is also not very difcult, but it is long and annoying; it requires a great number of fussy verications. So we will leave the choice to the reader: if he wishes to have a complete argument, he will have to prove the rst lemma on his own; we will restrict ourselves here to presenting some hints concerning this proof. If not, either because the reader accepts the rst lemma as is, or else is content to have a slightly weakened form of the theorem, we offer to rejoin the reader some lines further down for the proof of the second lemma.

Hints for the proof of Lemma 6.36

Proof A word of caution to begin with: here is a precise statement of Lemma 6.36. There exists a formula $Drv_0[v_0, v_1]$ of $\mathcal{L}_0$ which is a Σ formula, which represents the set Drv_0 and is such that, for every Σ formula F of $\mathcal{L}_0$, we have

$$\mathcal{P} \vdash F \Rightarrow \exists v_1 Drv_0[\underline{\#F}, v_1].$$

The idea which will guide us is simple: we take the argument which allowed us to assert that, for every closed Σ formula F,

$$\mathbb{N} \models F \Rightarrow \exists v_1 Drv_0[\underline{\#F}, v_1],$$

and we formalize it in $\mathcal{P}$.

Before setting down this path, we recall some facts and make some comments. Let $n \in \mathbb{N}$ and let $\mathcal{M}$ be a model of $\mathcal{P}$. Let M be the underlying set of $\mathcal{M}$. A subset X of M^n is denable if there exists a formula $F[v_0, v_1, \ldots, v_{n-1}]$ of L such that

$$X = \{(a_0, a_1, \ldots, a_{n-1}) \in M^n : \mathcal{M} \models F[a_0, a_1, \ldots, a_{n-1}]\}.$$

A map from M^n into M is denable if its graph is denable. An element a is denable if $\{a\}$ is denable.

If $F[v_0, v_1, \ldots, v_n]$ is a formula of $\mathcal{L}_0$, the fact that the set dened by F is, in every model of $\mathcal{P}$, the graph of a map from M^n into M can be expressed by

$$\mathcal{P} \vdash \forall v_1 \forall v_2 \ldots \forall v_n \exists! v_0 F[v_0, v_1, \ldots, v_n].$$

The formula $\forall v_1 \forall v_2 \ldots \forall v_n \exists! v_0 F[v_0, v_1, \ldots, v_n]$ will be designated by writing `$F$ `defines a map from` $M^n$ `into` M'. One can also do the same for other properties that can be expressed by formulas of $\mathcal{L}_0$: to assert the property expressed within quotation marks will stand for the (or a) formula that expresses it.

Arguments by induction are allowed since the scheme *IS* is included in $\mathcal{P}$. One can also dene mappings by recursion. To be precise:

Let F and G be formulas of $\mathcal{L}_0$, *let n be an integer, and suppose that*

$$\mathcal{P} \vdash \text{`}F \texttt{ defines a map from } M^n \texttt{ into } M\text{'}$$
$$\wedge \text{`}G \texttt{ defines a map from } M^{n+2} \texttt{ into } M\text{'}.$$

Then for every model $\mathcal{M}$ *of* $\mathcal{P}$, *if we let* f *and* g *denote the functions dened in* $\mathcal{M}$ *by F and G, respectively, there exists one and only one denable function h from* M^{n+1} *into M such that*

- *for all elements* $a_0, a_1, \ldots, a_n$ *of* M, $h(0, a_1, a_2, \ldots, a_n) = f(a_1, a_2, \ldots, a_n)$;
- *for all elements* $a_0, a_1, \ldots, a_n$ *of* M,

$$h(a_0 + 1, a_1, a_2, \ldots, a_n) = g(a_0, a_1, \ldots, a_n, h(a_0, a_1, a_2, \ldots, a_n)).$$

This function is dened by a formula H that depends recursively on F and G (but not on $\mathcal{M}$). Moreover, if F and G are Σ formulas, then H can also be chosen to be a Σ formula.

To prove this result, one begins by proving, in $\mathcal{P}$, certain simple facts of arithmetic in order to generalize Lemma 6.12.

One is then able to construct Σ formulas such as: \`$v_0$ is the code of a closed formula of $\mathcal{L}_0$', \`v_0 is the code of a formula of $\mathcal{L}_0$ that has only one free variable', \`v_0 is the code of a closed formula of $\mathcal{L}_0$ and v_1 is the code of a derivation of this formula', and so on. The advantage of these formulas over the ones that we would obtain by applying the representation theorem is that certain facts pertaining to the property enclosed in quotation marks become theorems of $\mathcal{P}$, e.g. the deduction theorem (Theorem 4.18) and Proposition 4.25.

What remains is the most difcult: the completeness theorem. Let us restrict ourselves to the language $\mathcal{L}_0$. Given a sequence of ve formulas

$$\mathcal{H} = (H_0[v_0], H_1[v_0], H_2[v_0, v_1], H_3[v_0, v_1, v_2], H_4[v_0, v_1, v_2]),$$

we can easily nd a closed formula G (which depends recursively on $\mathcal{H}$) which expresses that, in every model of $\mathcal{P}$, the set X_0 dened by H_0 is not empty, the set dened by H_1 consists of a single element a that belongs to X_0, H_2 denes a function from X_0 into itself, and H_3 and H_4 dene functions from $X_0 \times X_0$ into X_0. If $\mathcal{H}$ satises these conditions and if $\mathcal{M}$ is a model of $\mathcal{P}$, then we will let $\mathcal{M}(\mathcal{H})$ denote the $\mathcal{L}_0$-structure whose base set is $\{a \in M : \mathcal{M} \models H_0[a]\}$, where the interpretation of $\underline{0}$ is the unique element of M that satises H_1, and where the interpretations of $\underline{S}$, $\underline{+}$, and $\underline{\times}$ are the functions dened in $\mathcal{M}$ by H_2, H_3, and H_4. Given a sequence $\mathcal{H}$ as above and a formula $F[v_0, v_1, \ldots, v_k]$ of $\mathcal{L}_0$, it is possible, by ordinary induction on the height of H (also see Exercise 11), to construct a formula of $\mathcal{L}_0$ that we will denote by \`$(v_0, v_1, \ldots, v_k)$ satisfies the formula F in $\mathcal{M}(\mathcal{H})$'. This formula will be such that, for every model $\mathcal{M}$ of $\mathcal{P}$ and for all elements $a_0, a_1, \ldots, a_k$ of M that satisfy H_0, we have

$$\mathcal{M}(\mathcal{H}) \models F[a_0, a_1, \ldots, a_k]$$

if and only if

$$\mathcal{M} \models \text{`}(a_0, a_1, \ldots, a_k)\ \texttt{satisfies the formula}\ F\ \texttt{in}\ \mathcal{M}(\mathcal{H})\text{'}.$$

The formula \`$(v_0, v_1, \ldots, v_k)$ satisfies the formula F in $\mathcal{M}(\mathcal{H})$' depends recursively on F and on $\mathcal{H}$.

We can now state the version of **the completeness theorem in Peano arithmetic**:

Theorem 6.38 *For every closed formula F, there exists a sequence of ve formulas*

$$\mathcal{H} = (H_0[v_0], H_1[v_0], H_2[v_0, v_1], H_3[v_0, v_1, v_2], H_4[v_0, v_1, v_2])$$

that depends recursively on F and is such that

$$\mathcal{P} \vdash Con(F) \Rightarrow \text{`}F\ \texttt{is satisfied in}\ \mathcal{M}(\mathcal{H})\text{'}.$$

Here, $Con(F)$ is the formula

$$\neg\exists v_0(\text{`}v_0\ \texttt{is the code of a derivation of}\ \neg F\text{'})$$

and `F is satised in $\mathcal{M}(\mathcal{H})$' is the formula

$$\text{`}\texttt{the empty sequence satisfies the formula}\ F\ \texttt{in}\ \mathcal{M}(\mathcal{H})\text{'}.$$

The proof of this theorem imitates the proof of the completeness theorem (see Theorem 4.29). We will not insist on this point.

We can now complete the proof of Lemma 6.36. The formula $Drv_0[v_0, v_1]$ is the formula `v_0 is the code of a closed formula of $\mathcal{L}_0$ and v_1 is the code of a derivation of this formula in $\mathcal{P}_0$'. To prove the fact that if F is a closed Σ formula, then $\mathcal{P} \vdash F \Rightarrow \exists v_1 Drv_0[\underline{\#F}, v_1]$, we make use of the ordinary completeness theorem; we show that the formula

$$F \Rightarrow \exists v_1 Drv_0[\underline{\#F}, v_1]$$

is true in every model of $\mathcal{P}$. Consider a model $\mathcal{M} = \langle M, 0, S, +, \times\rangle$ of $\mathcal{P}$. If $\exists v_1 Drv_0[\underline{\#F}, v_1]$ is true in $\mathcal{M}$, then so is the formula $F \Rightarrow \exists v_1 Drv_0[\underline{\#F}, v_1]$. If it is false in $\mathcal{M}$, let G denote the conjunction of the formulas of $\mathcal{P}_0$ and $\neg F$. The formula $\neg\exists v_1 Drv_0[\underline{\#F}, v_1]$ is equivalent to $Con(G)$, and hence

$$\mathcal{M} \models Con(G).$$

One then applies the completeness theorem in Peano arithmetic to conclude that there exists a sequence $\mathcal{H}$ of ve formulas as above that dene an $\mathcal{L}_0$-structure

$$\mathcal{M}(\mathcal{H}) = \langle X, 0', S', +', \times'\rangle$$

such that

$$\mathcal{M} \models \text{`}G\ \texttt{is satisfied in}\ \mathcal{M}(\mathcal{H})\text{'}$$

and hence

$$\mathcal{M}(\mathcal{H}) \models G.$$

In $\mathcal{M}$, one can dene by induction a denable map k from M into X by setting

$$k(0) = 0' \text{ and for all } a \in M, \quad k(S(a)) = S'(k(a)).$$

One shows that k is a monomorphism from $\mathcal{M}$ into $\mathcal{M}(\mathcal{H})$ and that the image of k is an initial segment of $\mathcal{M}(\mathcal{H})$ [to do this, one needs to use the induction scheme in $\mathcal{M}$ and the fact that $\mathcal{M}(\mathcal{H})$ is a model of $\mathcal{P}_0$]. So the structure $\mathcal{M}(\mathcal{H})$ is an end-extension of a structure that is isomorphic to $\mathcal{M}$. Now $\mathcal{M}(\mathcal{H})$ does not satisfy F and, according to Lemma 6.35, since F is a Σ formula, $\mathcal{M}$ does not satisfy F; so

$$\mathcal{M} \models F \Rightarrow \exists v_1 Drv_0[\underline{\#F}, v_1]. \qquad \blacksquare$$

Remark This proof uses the completeness theorem (here, we are speaking of the real completeness theorem, from Chapter 4, not the one that was proved in $\mathcal{P}$) which required the notion of innite set and even the axiom of choice. Syntactic proofs of this lemma (there are some) have the advantage of appealing only to nite notions (integers, nite sequences, etc.).

Proof of Lemma 6.37

Proof Consider the function g from $\mathbb{N}$ into $\mathbb{N}$ dened by the following conditions:

- if n is the Godel number of a formula $F[v_0]$ with one free variable, then $g(n)$ is the Godel number of the formula $F[\underline{n}]$;
- otherwise, $g(n) = 0$.

This function is clearly primitive recursive; let $G[v_0, v_1]$ be a formula that represents it. So for every integer n, we have

$$\mathcal{P}_0 \vdash \forall v_0(G[v_0, \underline{n}] \Leftrightarrow v_0 \simeq \underline{g(n)}). \qquad (*)$$

Let $\varepsilon[v_0]$ denote the formula

$$\exists v_1 \exists v_2(Drv[v_2, v_1] \wedge G[v_2, v_0]).$$

Observe that if n is the Godel number of a formula $F[v_0]$ of one free variable, then

$$\mathbb{N} \models \varepsilon[\underline{n}] \quad \text{if and only if} \quad F[\underline{n}] \text{ is derivable.}$$

Let a be the Godel number of the formula $\neg\varepsilon[v_0]$ and let $b = g(a)$ be the Godel number of $\neg\varepsilon[\underline{a}]$. From the denition of ε and from $(*)$, we conclude that

$$\mathcal{P}_0 \vdash \varepsilon[\underline{a}] \Leftrightarrow \exists v_1 Drv[\underline{b}, v_1]. \qquad (**)$$

First we will show that $\neg\varepsilon[\underline{a}]$ is not derivable in T. We will assume the contrary and deduce that T is inconsistent. So there exists an integer c which is the code of a derivation of $\neg\varepsilon[\underline{a}]$ in T, and hence

$$\mathcal{P}_0 \vdash Drv[\underline{b}, \underline{c}],$$

which, together with $(**)$, shows that $\mathcal{P}_0 \vdash \varepsilon[\underline{a}]$. As T includes $\mathcal{P}_0$, $T \vdash \varepsilon[\underline{a}]$; thus T is inconsistent.

Next, we show that $T \vdash Con(T) \Rightarrow \neg\varepsilon[\underline{a}]$. In fact, we will do better by showing that $\mathcal{P}_1 \vdash \varepsilon[\underline{a}] \Rightarrow \neg Con(T)$. Set $T_1 = \mathcal{P}_1 \cup \{\varepsilon[\underline{a}]\}$. It follows from $(**)$ that

$$T_1 \vdash \exists v_1 Drv[\underline{b}, v_1].$$

But $\varepsilon[\underline{a}]$ is a closed Σ formula. Let d be the Godel number of $\varepsilon[\underline{a}]$. Then

$$\varepsilon[\underline{a}] \Rightarrow \exists v_2 Drv_0[\underline{d}, v_2]$$

is an element of $\mathcal{P}_1$ and

$$T_1 \vdash \exists v_2 Drv_0[\underline{b}, v_2].$$

But, we had assumed that $\vdash \forall v_0 \forall v_1 (Drv_0[v_0, v_1] \Rightarrow Drv[v_0, v_1])$, and hence

$$T_1 \vdash \exists v_1 Drv[\underline{b}, v_1] \wedge \exists v_2 Drv[\underline{d}, v_2],$$

which, when we refer to the denition of $Con(T)$, shows that

$$T_1 \vdash \neg Con(T).$$

Finally, thanks to the deduction theorem,

$$\mathcal{P}_1 \vdash \varepsilon[\underline{a}] \Rightarrow \neg Con(T).$$ ∎

Remark 1 By assuming that the formula Drv satises certain properties that are wholly innocent and natural, essentially that

$$\mathcal{P} \vdash (\exists v_0 Drv[\underline{\#F}, v_0] \wedge \exists v_1 Drv[\underline{\#(F \Rightarrow G)}, v_1]) \Rightarrow \exists v_2 Drv[\underline{\#G}, v_2],$$

we see that the formula $Con(T)$ is equivalent to $\neg\exists v_0 Drv[\#(\underline{0} \simeq \underline{1}), v_0]$.

Remark 2 The formula $\varepsilon[\underline{a}]$ asserts that its negation is derivable. It is obviously false in $\mathbb{N}$.

Remark 3 Godel's theorem asserts that a theory which is consistent and recursive cannot prove its own consistency; by contrast, it may very well prove its own inconsistency, as is the case for the theory $\mathcal{P} \cup \{\neg Con(\mathcal{P})\}$, for example. However, this is not the case for the theory $\mathcal{P}$ itself: there is a model of $\mathcal{P} \cup \{Con(\mathcal{P})\}$, namely $\mathbb{N}$. This generalizes to any recursive theory that has $\mathbb{N}$ as a model.

EXERCISES FOR CHAPTER 6

1. Let X be a non-empty set and let f be a function from $X \times X$ into X. Consider the $\mathcal{L}_0$-structure $\mathcal{M}$ whose base set is $M = \mathbb{N} \cup (X \times \mathbb{Z})$ and in which the symbols $\underline{S}$, $\underline{+}$, and $\underline{\times}$ are interpreted by the functions S, $+$, and $\times$ that are dened by the following conditions:

- $\mathcal{M}$ is an extension of $\mathbb{N}$;
- if $a = (x, n) \in M - \mathbb{N}$, then $S(a) = (x, n + 1)$;
- if $a = (x, n) \in M - \mathbb{N}$ and $m \in \mathbb{N}$, then $a + m = m + a = (x, n + m)$;
- if $a = (x, n)$ and $b = (y, m)$ are elements of $M - \mathbb{N}$, then $(x, n) + (y, m) = (x, n + m)$;
- if $a = (x, n) \in M - \mathbb{N}$ and $m \in \mathbb{N}$, then $(x, n) \times m = (x, n \times m)$ if $m \neq 0$ and $(x, n) \times 0 = 0$;
- if $a = (x, n) \in M - \mathbb{N}$ and $m \in \mathbb{N}$, then $m \times (x, n) = (x, m \times n)$;
- if $a = (x, n)$ and $b = (y, m)$ are elements of $M - \mathbb{N}$, then $(x, n) \times (y, m) = (f(x, y), n \times m)$.

(a) Show that $\mathcal{M}$ is a model of $\mathcal{P}_0$.

(b) Show that none of the following formulas is a consequence of $\mathcal{P}_0$:

(i) $\forall v_0 \forall v_1\, v_0 \underline{+} v_1 \simeq v_1 \underline{+} v_0$;
(ii) $\forall v_0 \forall v_1 \forall v_2\, v_0 \underline{\times} (v_1 \underline{\times} v_2) \simeq (v_0 \underline{\times} v_1) \underline{\times} v_2$;
(iii) $\forall v_0 \forall v_1 ((v_0 \leq v_1 \wedge v_1 \leq v_0) \Rightarrow v_0 \simeq v_1)$;
(iv) $\forall v_0 \underline{0} \underline{\times} v_0 \simeq \underline{0}$.

(c) Construct a model of $\mathcal{P}_0$ in which addition is not associative.

2. Let $\mathcal{M}$ be a model of $\mathcal{P}$ and suppose that $\mathbb{N}$ is a proper substructure of $\mathcal{M}$. On the base set, M, of $\mathcal{M}$, dene the following relation $\approx$: $x \approx y$ if and only if there exist two elements n and m of $\mathbb{N}$ such that

$$\mathcal{M} \models x \underline{+} \underline{n} \simeq y \underline{+} \underline{m}.$$

(a) Show that the relation $\approx$ is an equivalence relation.

(b) Let a, a', b, and b' be elements of M that satisfy $a \approx a'$ and $b \approx b'$. Show that $a + b \approx a' + b'$.

(c) Let E be the set of equivalence classes of M relative to the relation $\approx$. Dene the relation R on E as follows:

if x and y are in E, then xRy if and only if there exist $a \in x$ and $b \in y$ such that $\mathcal{M} \models a \leq b$.

Show that the relation R is a total ordering. Show that E, with respect to this ordering, has a least element but does not have a greatest element. Show that R is a dense ordering of E.

3. Prove the Chinese remainder theorem (Theorem 6.13).

4. Prove the converse of the representation theorem (Theorem 6.8): if a function from $\mathbb{N}^p$ into $\mathbb{N}$ is representable, then it is recursive.

5. Let T be a theory in a nite language. Assume that T is recursively enumerable, i.e. that the set

$$\{\#F : F \in T\}$$

is recursively enumerable. Show that there exists a recursive theory T' that is equivalent to T in the sense that, for every formula G, G is derivable in T if and only if G is derivable in T'.

6. Show that if Fermat's last theorem,

$$\neg(\exists x > 0)(\exists y > 0)(\exists z > 0)(\exists t > 2)(x^t + y^t \simeq z^t),$$

is not refutable in $\mathcal{P}_0$, then it is true in $\mathbb{N}$.

(At the time this text was rst written, Fermat's last theorem was still a conjecture; it has since been proved by Andrew Wiles.)

7. In this exercise, $Drv[v_0, v_1]$ is a formula that represents the set

$$\mathsf{Drv} = \{(a, b) : b \text{ is the Godel number of a derivation in } \mathcal{P} \text{ of}$$
$$\text{the formula whose Godel number is } a\}.$$

Among the following assertions, which are true for any closed formula F?

(a) $\mathbb{N} \models \exists v_1 Drv[\underline{\#F}, v_1] \Rightarrow F$; (c) $\mathbb{N} \models F \Rightarrow \exists v_1 Drv[\underline{\#F}, v_1]$;

(b) $\mathcal{P} \vdash \exists v_1 Drv[\underline{\#F}, v_1] \Rightarrow F$; (d) $\mathcal{P} \vdash F \Rightarrow \exists v_1 Drv[\underline{\#F}, v_1]$.

8. Show that there exists a formula $F[v_0]$ of $\mathcal{L}_0$ such that $\mathbb{N} \models \neg\exists v_0 F[v_0]$ and $\neg\exists v_0 F[v_0]$ is not derivable in $\mathcal{P}$. Conclude from this that, for every formula $G[v_0, v_1, \ldots, v_n]$, there exists a formula $H[v_0, v_1, \ldots, v_n]$ such that

$$\forall v_0 \forall v_1 \ldots \forall v_n (G[v_0, v_1, \ldots, v_n] \Leftrightarrow H[v_0, v_1, \ldots, v_n])$$

is true in $\mathbb{N}$ but is not derivable in $\mathcal{P}$.

9. Show that if F is a closed formula and if

$$\mathcal{P} \vdash \exists v_0 Drv[\underline{\#F}, v_0] \Rightarrow F,$$

then

$$\mathcal{P} \vdash F.$$

(See Exercise 7; we could apply the second incompleteness theorem to the theory $\mathcal{P} \cup \{\neg F\}$.)

10. This exercise uses the notion of **elementary extension**, which will be introduced in Chapter 8 on model theory.

Let $\mathcal{M}$ be a non-standard model of $\mathcal{P}$, and let A be a subset of the underlying set, M, of $\mathcal{M}$. We say that a function f from M^p into M is **denable with parameters from** A if there exists a formula $F[v_0, v_1, \ldots, v_p]$ of $\mathcal{L}_0$ with parameters in A such that, for all $a_1, a_2, \ldots, a_p$ belonging to M, we have

$$\mathcal{M} \models \forall v_0 (F[a_0, a_1, \ldots, a_p] \Leftrightarrow v_0 \simeq f(a_0, a_1, \ldots, a_p)).$$

(a) Let $\mathcal{N}$ be a substructure of $\mathcal{M}$ whose underlying set, N, is closed under functions that are denable with parameters from N; in other words, it is such that for all $p \in \mathbb{N}$, for every function f from M^p into M that is denable with parameters from N, and for all $a_1, a_2, \ldots, a_p$ belonging to N,

$$f(a_1, \ldots, a_p) \in N.$$

Show that $\mathcal{N}$ is an elementary substructure of $\mathcal{M}$ (and is therefore a model of $\mathcal{P}$).

(b) Next, we say that a subset X of M^p is **denable with parameters from** A if there exists a formula $G[v_1, \ldots, v_p]$ of $\mathcal{L}_0$ with parameters in A such that, for all $a_1, a_2, \ldots, a_p$ belonging to M,

$$(a_0, a_1, \ldots, a_p) \in X \quad \text{if and only if} \quad \mathcal{M} \models G[a_1, \ldots, a_p].$$

Show that the collection of subsets of M that are denable with parameters from A forms a Boolean subalgebra of the algebra of all subsets of M.

Show that if f and g are functions from M into M that are denable with parameters from A, then the set $\{a \in M\colon f(a) = g(a)\}$ is denable with parameters from A.

(c) Let F and G be two maps from M into M. Dene the maps Sf, $f + g$, and $f \times g$ from M into M by

$$Sf(x) = f(x) + 1;$$
$$(f + g)(x) = f(x) + g(x);$$
$$(f \times g)(x) = f(x) \times g(x).$$

Show that the set of functions denable with parameters from A is closed under these operations.

(d) Let $\mathcal{B}$ be the Boolean algebra of subsets of M that are denable with parameters from A, let $\mathcal{U}$ be an ultralter on this algebra, and let $\mathcal{F}$ be the set of functions from M into M that are denable with parameters from M.

Show that the relation $\approx$ on $\mathcal{F}$ dened by

$$f \approx g \quad \text{if and only if} \quad \{a \in M : f(a) = g(a)\} \in \mathcal{U}$$

is an equivalence relation and that if $f \approx f'$ and $g \approx g'$, then

$$Sf \approx Sf', \qquad f + g \approx f' + g', \qquad f \times g \approx f' \times g'.$$

If $f \in \mathcal{F}$, we let $f/\mathcal{U}$ denote the equivalence class of f relative to $\approx$ and we let $\mathcal{F}/\mathcal{U}$ denote the set of equivalence classes relative to $\approx$. We then observe that it is possible to dene the operations S, $+$, and $\times$ on $\mathcal{F}/\mathcal{U}$. The zero element, 0, of $\mathcal{F}/\mathcal{U}$ will be, by denition, the equivalence class of the constant function equal to 0. This allows us to treat $\mathcal{F}/\mathcal{U}$ as an $\mathcal{L}_0$-structure.

(e) For every element $a \in M$, let $\bar{a}$ denote the element of $\mathcal{F}/\mathcal{U}$ that is the equivalence class relative to $\approx$ of the constant function equal to a.

Show that the map from $\mathcal{M}$ into $\mathcal{F}/\mathcal{U}$ that sends $a \in M$ into $\bar{a} \in \mathcal{F}/\mathcal{U}$ is a homomorphism of $\mathcal{L}_0$-structures.

(f) Show that for every $p \in \mathbb{N}$, for every formula $F[v_1, \ldots, v_p]$ of $\mathcal{L}_0$, and for all $f_1, f_2, \ldots, f_p$ in $\mathcal{F}$, we have

$$\mathcal{F}/\mathcal{U} \models F[f_1/\mathcal{U}, f_2/\mathcal{U}, \ldots, f_p/\mathcal{U}]$$

if and only if

$$\{a \in M : \mathcal{M} \models F[f_1(a), f_2(a), \ldots, f_p(a)]\} \in \mathcal{U}.$$

Conclude that the map from $\mathcal{M}$ into $\mathcal{F}/\mathcal{U}$ that sends $a \in M$ into $\bar{a} \in \mathcal{F}/\mathcal{U}$ is elementary (see Chapter 8).

(g) Suppose that $\mathbb{N}$ is an elementary substructure of $\mathcal{M}$. Show that if f is a function from M into M that is denable with parameters from $\mathcal{M}$ and if $a \in M$, then there exists $b \in M$ such that

$$\mathcal{M} \models \forall v_0(v_0 < a \Rightarrow f(v_0) < b).$$

(h) Let $\mathcal{M}$ be a proper elementary extension of $\mathbb{N}$. Show that there exists a proper elementary extension $\mathcal{N}$ of $\mathcal{M}$, with base set N, which satises

$$\text{for all } a \in N, \quad \text{there exists } b \in M \text{ such that } \mathcal{N} \models a < b.$$

11. Let $\mathcal{L}$ be a nite language and let $\mathcal{M}$ be an $\mathcal{L}$-structure whose underlying set is M. We say that $\mathcal{M}$ is **strongly undecidable** if every theory in the language $\mathcal{L}$ that has $\mathcal{M}$ as a model is undecidable.

(a) Show that $\mathbb{N}$ is strongly undecidable.

(b) Let $G_0[v_0]$, $G_1[v_0]$, $G_2[v_0, v_1]$, $G_3[v_0, v_1, v_2]$, $G_4[v_0, v_1, v_2]$ be ve xed formulas of $\mathcal{L}$ and consider the theory T_0 that has the following formulas as axioms:

(1) $\forall v_0(G_1[v_0] \Rightarrow G_0[v_0])$;

(2) $\forall v_0 \forall v_1(G_2[v_0, v_1] \Rightarrow (G_0[v_0] \wedge G_0[v_1]))$;

(3) $\forall v_0 \forall v_1 \forall v_2(G_3[v_0, v_1, v_2] \Rightarrow (G_0[v_0] \wedge G_0[v_1] \wedge G_0[v_2]))$;

(4) $\forall v_0 \forall v_1 \forall v_2(G_4[v_0, v_1, v_2] \Rightarrow (G_0[v_0] \wedge G_0[v_1] \wedge G_0[v_2]))$;

(5) $\exists! v_0 G_1[v_0]$;

(6) $\forall v_1(G_0[v_1] \Rightarrow \exists! v_0 G_2[v_0, v_1])$;

(7) $\forall v_1 \forall v_2((G_0[v_1] \wedge G_0[v_2]) \Rightarrow \exists! v_0 G_3[v_0, v_1, v_2])$;

(8) $\forall v_1 \forall v_2((G_0[v_1] \wedge G_0[v_2]) \Rightarrow \exists! v_0 G_4[v_0, v_1, v_2])$.

If $\mathcal{M}$ is a model of T_0, we dene the $\mathcal{L}_0$-structure $\mathcal{N}$ in the following way:

- the base set of $\mathcal{N}$ is the set $N = \{a \in M : \mathcal{M} \models G_0[a]\}$;
- the constant symbol $\underline{0}$ is interpreted by the unique element a of $\mathcal{M}$ that satises $G_1[a]$;
- the symbol $\underline{S}$ is interpreted by the function that associates with each $a \in N$ the unique element b such that $\mathcal{M} \models G_2[b, a]$;
- the symbol $\underline{+}$ is interpreted by the function that associates, with two elements a and b of N, the unique element c such that $\mathcal{M} \models G_3[c, a, b]$;
- the symbol $\underline{\times}$ is interpreted by the function that associates, with two elements a and b of N, the unique element c such that $\mathcal{M} \models G_4[c, a, b]$.

We will say that $\mathcal{N}$ **is denable in** $\mathcal{M}$ (one must take care not to confuse this with the notion of a denable subset).

Show that, for every formula $F[v_1, v_2, \ldots, v_p]$ of $\mathcal{L}_0$, there exists a formula $F^*[v_1, v_2, \ldots, v_p]$ of $\mathcal{L}$ such that if $\mathcal{M}$ is a model of T_0 and $\mathcal{N}$ is the $\mathcal{L}_0$-structure dened in $\mathcal{M}$ and if $a_1, a_2, \ldots, a_p$ are elements of N, then

$$\mathcal{N} \models F[a_0, a_1, \ldots, a_p] \quad \text{if and only if} \quad \mathcal{M} \models F^*[a_0, a_1, \ldots, a_p].$$

Show that F^* can be determined from F in an effective way [which means that there exists a primitive recursive function α such that if $n = \#F$, then $\alpha(n) = \#F^*$].

(c) Let T be a theory of $\mathcal{L}$ that includes T_0. Set

$$T^- = \{F : F \text{ is a closed formula of } \mathcal{L}_0 \text{ and } T \vdash F^*\}.$$

Show that if G is a closed formula of $\mathcal{L}_0$, the following three conditions are equivalent:

(1) $G \in T^-$;

(2) $T^- \vdash G$;

(3) $T \vdash G^*$.

(d) Show that if $\mathbb{N}$ is denable in $\mathcal{M}$, then $\mathcal{M}$ is strongly undecidable.

(e) Show that the structure $\mathbb{Z}$ in the language $\mathcal{L} = \{\underline{0}, \underline{+}, \underline{\times}\}$ of ring theory is strongly undecidable. (Use Lagrange's theorem that every positive integer is the sum of four squares.) Show that the following theories are undecidable: the theory of rings, the theory of commutative rings, the theory of integral domains.

(f) Let $\mathcal{L}$ be the language that contains only the binary predicate symbol R. Consider the $\mathcal{L}$-structure $\mathcal{M}$ whose underlying set is $M = \mathbb{N} \cup (\mathbb{N} \times \mathbb{N})$ and in which $R^{\mathcal{M}}$ is equal to

$$\{(a, (a, b)) : a \in \mathbb{N},\ b \in \mathbb{N}\} \cup \{((a, b), b) : a \in \mathbb{N},\ b \in \mathbb{N}\}$$
$$\cup \{((a, b), (a + b, a \cdot b)) : a \in \mathbb{N},\ b \in \mathbb{N}\}.$$

Show that $\mathbb{N}$ is interpretable in $\mathcal{M}$. Show that the set of universally valid formulas of the language $\mathcal{L}$ is not recursive.

(g) This time, $\mathcal{L}$ is the language that contains a binary predicate symbol D and a binary function symbol $\underline{+}$. Let $\mathcal{M}$ be the $\mathcal{L}$-structure whose underlying set is $\mathbb{N}$ and in which $\underline{+}$ is interpreted by addition and D by the relation `divides' (Dxy is true if and only if x divides y).

Show that the element 1 and the relation $x = y \cdot (y + 1)$ are denable in $\mathcal{M}$. Show that $\mathcal{M}$ is strongly undecidable.

12. Let f be a total recursive function from $\mathbb{N}$ into $\mathbb{N}$ and let $F[v_0, v_1]$ be a Σ formula that represents it and is such that

$$\mathcal{P} \vdash \forall v_1 \exists v_0 F[v_0, v_1].$$

The purpose of this exercise is to show that there exist total recursive functions that are not provably total.

(a) Let $F[v_0, v_1, \ldots, v_k]$ be a Σ formula. Show that the set

$$\{(n_0, n_1, \ldots, n_k) : \mathbb{N} \models F[n_0, n_1, \ldots, n_k]\}$$

is recursively enumerable.

(b) Let f be a total function from $\mathbb{N}$ into $\mathbb{N}$. Show that the following two conditions are equivalent:

(i) f is recursive;

(ii) there exists a Σ formula that represents f.

(c) Show that there exists a partial recursive function h of two variables such that, for every integer n,

- if a is the Godel number of Σ formula, say $F[v_0, v_1]$, and if there exists an integer m such that $\mathcal{P} \vdash F[\underline{m}, \underline{n}]$, then

$$\mathcal{P} \vdash F[\underline{h(a, n)}, \underline{n}];$$

- if a is the Godel number of the formula $F[v_0, v_1]$ and if there does not exist an integer m such that $\mathcal{P} \vdash F[\underline{m}, \underline{n}]$, then $h(a, n)$ is not dened;
- $h(a, n) = 0$ otherwise.

(d) We now dene a function g from $\mathbb{N}^3$ into $\mathbb{N}$ in the following way: for every integer n,

- if a is the Godel number of a Σ formula $F[v_0, v_1]$ and if b is the Godel number of a derivation in $\mathcal{P}$ of the formula

$$\forall v_1 \exists v_0 F[v_0, v_1],$$

then $g(a, b, n) = h(a, n)$;
- $g(a, b, n) = 0$ otherwise.

Show that g is a total recursive function.

(e) Show that there exist total recursive functions that are not provably total.

13. This exercise should be worked after reading Chapter 7 on set theory. In particular, one needs to know what the cardinal number $2^{\aleph_0}$ is.

(a) Show that if T is a consistent theory obtained by adjoining a nite number of formulas to $\mathcal{P}$, then T is not complete.

(b) For every integer n and every sequence

$$s = (s(0), s(1), \ldots, s(n-1)) \in \{0, 1\}^n,$$

construct a closed formula F_s such that, for all s,

(i) $F_{(s(0),s(1),\ldots,s(n-1),1)} = \neg F_{(s(0),s(1),\ldots,s(n-1),0)}$;

(ii) $\mathcal{P} \cup \{F_\emptyset, F_{(s(0))}, F_{(s(0),s(1))}, \ldots, F_{(s(0),s(1),\ldots,s(n-1))}\}$ is a consistent theory.

(c) Show that there exist $2^{\aleph_0}$ theories that include $\mathcal{P}$ and that are pairwise inequivalent.

14. Some notions from set theory are required for this exercise as well. Knowledge of a certain amount of model theory is also required (elementary extensions and the method of diagrams).

Let $\mathcal{M}$ be an elementary extension of $\mathbb{N}$ and let X be a subset of $\mathbb{N}$. Recall (Chapter 3, Denition 3.96 that X is denable in $\mathcal{M}$ if there exists a formula F of $\mathcal{L}_0$ with one free variable and with parameters from $\mathcal{M}$ such that, for all $n \in \mathbb{N}$,

$$n \in X \quad \text{if and only if} \quad \mathcal{M} \models F[n].$$

(a) Show that if $\mathcal{M}$ is countable, then the set of subsets of $\mathbb{N}$ that are denable in $\mathcal{M}$ is countable.

(b) Show that for every subset X of $\mathbb{N}$, there exists a countable elementary extension $\mathcal{M}$ of $\mathbb{N}$ in which X is denable.

(c) Show that there exist $2^{\aleph_0}$ countable elementary extensions of $\mathbb{N}$ that are pairwise non-isomorphic.

15. (a) What is there that is paradoxical in the statement of Epimenides (see the introduction to this chapter).

(b) In a Carpathian village, there lives a barber who shaves all the men who do not shave themselves and only these. What can you say about this barber?

7 Set theory

The purpose of set theory, a subject created by G. Cantor at the beginning of the twentieth century, is to allow one to construct the whole of mathematics based, exclusively, on the concept of membership.

In this chapter, we present the axioms of ZermeloñFraenkel (ZF) *for a rst-order theory in a language that involves only two binary relation symbols, equality and membership. Except for the axiom of extensionality, which plays a special role, the axioms of* ZF *assert the existence of sets. No one will be surprised by some of these, such as the axiom of pairs or the axiom of unions, but other axioms may seem less natural. One must realize that they are the fruits of a compromise; they must, on the one hand, allow for the construction of all the sets required by mathematical practice while, on the other hand, they must not be contradictory, as may happen (indeed, as did happen, historically) if the axioms are introduced haphazardly.*

As for the axiom of choice, it is utterly incomprehensible at rst glance. In fact, it appears to be so obvious that it seems superuous to mention it. Concerning this point, one must clearly understand that our development is axiomatic and that, since this axiom is not derivable from the other axioms, it has to be added. Besides, it has some consequences that are altogether surprising, even paradoxical, such as the BanachñTarski theorem, which permits one to decompose a sphere S into two subsets that are each homeomorphic to S. In any case, whatever value one places on this axiom, it must be said that mathematicians commonly use it.

At the beginning of this chapter, we will undertake the task of translating familiar mathematical concepts into the language of set theory. We will quickly see how to dene relations and mappings; we will show how to dene objects that will (pretend to) play the role of integers and the reader will be able to convince himself that it is possible to construct $\mathbb{R}$, $\mathbb{C}$, *and any other structure he may need.*

Set theory provides a certain number of tools that are used in mathematics. First of all, there is Zorn's lemma. In Chapter 5, we saw how the integers can be used to enumerate sets that are apparently more complicated, such as $\mathbb{N} \times \mathbb{N}$, *or the set of nite sequences of integers, or even the set of recursive functions. The ordinals, which are a kind of generalization of the integers, can be used, provided we admit the axiom of choice, to enumerate any set. Cardinality is also an important concept; it permits us to count the number of elements in a set. Two sets have the same number of elements (we say 'have the same cardinality') if there exists a bijection from one onto the other. This entails distinguishing several 'orders of innity': for example, there is no bijection between* ℝ and ℕ. *This concept also produces some surprises: a set can have the same number of elements as one of its proper subsets.*

Finally, leaving classical mathematics behind, we will also undertake a brief study of models of the theory of sets. Here, the essential tool is the hierarchy of the V_α, *which will justify introducing a new axiom, the axiom of foundation. In particular, this axiom provides us with a negative answer to a natural question (does there exist a set that is an element of itself?) that the axioms of* ZF *alone are unable to answer. This will lead to some relative consistency results; for example, if the theory* ZF *is consistent, then so is the theory consisting of* ZF *plus the axiom of foundation.*

7.1 The theories Z and ZF

7.1.1 The axioms

We will present set theory as a rst-order theory. In addition to the usual symbol for equality, $\simeq$, the language L of this theory involves a symbol, $\in$, for a binary predicate called **membership**. As a matter of fact, we will introduce axioms for several set theories, of varying strengths. In the whole of this chapter, unless mentioned otherwise, $\mathcal{U}$ will denote a model of the theory ZF (see below) (prior to Denition 7.30, where the last axiom of ZF will be introduced, $\mathcal{U}$ will denote a model of those axioms that will have been introduced up to that point). The base set of $\mathcal{U}$ will be denoted by U and will be called the **universe**. When we say that a formula is true, it should be understood that we always mean true 'in $\mathcal{U}$'.

We must cope with a complication that we have already encountered several times: in mathematical texts, the words 'set', 'belongs', 'contains', etc., are in constant use with their intuitive meanings. However, the purpose of this chapter is to formalize these notions. So we see that we will need two levels of language and reasoning: on the one hand, for the formal language L in which we produce derivations that could, at least theoretically, be formalized in the sense of Chapter 4; on the other hand, for the metalanguage in which we will speak about L, about interpretations in $\mathcal{U}$ of the symbols of L, about theories that are expressed in L, and about models of these theories. For example, the formula $\exists v_0 \forall v_1 \neg v_1 \in v_0$ is part of the formal language; however, when we refer to the length of this formula or of the fact that it is derivable in ZF, we are using the metalanguage. In fact, these two languages apply to two different universes: the rst, to $\mathcal{U}$, the second, to the meta-universe, the universe that is familiar to all mathematicians and that involves, among others, the notions of integer, of nite sequence, and (even) of set. It is essential to avoid all confusion.

With this in mind, a certain number of words and symbols will be strictly reserved for use in the formal language. Thus, rst of all, the symbol $\in$ will always denote the membership relation between points of U (we will nonetheless permit the identication of $\in$ with what we would, in Chapter 3, have written $\bar{\in}^{\mathcal{U}}$). The word set will always mean a point in U (consequently, we forbid ourselves to speak of the set U). When we say that x is an element of y, this will always mean that x and y are sets (i.e. points in U) and that

$$\mathcal{U} \models x \in y.$$

But things do not stop here. We also wish to use set theory to formalize the whole of mathematics. We will be led to formally dene the concepts of relation, mapping (or function), and even natural number. As soon as these denitions are given, the corresponding words will be reserved for use by the formal language.

It will happen that we must use objects from the meta-universe: for example, the integers will be used to perform an induction on the length of formulas of L; in such circumstances, the adjective `intuitive' or the prex `meta' will be employed (intuitive integers, meta-relation, etc.).

With the exception of those in the last section, all theorems stated in this chapter are theorems of ZF or, when specied, of ZFC (ZF plus the axiom of choice). Accordingly this chapter is an axiomatic development of set theory. However, we will, naturally, adopt the usual attitude of mathematicians: our concern will be to convince the reader of the correctness of the theorems rather than to provide formal derivations for them. To avoid adding a complication related to language, we will, as we have already mentioned, dispense with the distinction between the symbol $\in$ and its interpretation in $\mathcal{U}$ or in other models that we will have occasion to deal with.

We will write $x \notin y$ as an abbreviation for the formula $\neg x \in y$. We will also freely use the following abbreviations:

$$\forall x \in y\ F \quad \text{for } \forall x(x \in y \Rightarrow F)$$
$$\exists x \in y\ F \quad \text{for } \exists x(x \in y \wedge F)$$

(here, x and y are symbols for variables and F is a formula).

The following is a list of axioms that we will shortly introduce and discuss: the axiom of extensionality, the axiom of pairs, the axiom of unions, the power set axiom (the axiom of subsets), the axioms of comprehension, and the axioms of replacement. The axiom of choice (AC) and the axiom of innity (Inf) will be introduced slightly later.

The axioms of extensionality, pairs, unions, subsets, comprehension, and innity constitute what is commonly known as **Zermelo set theory**, denoted by Z; the axioms of extensionality, pairs, unions, subsets, replacement, and innity constitute a stronger theory called **ZermeloñFraenkel set theory** and denoted by ZF. Each of the axioms of Z or of ZF, with the exception of extensionality, permits the construction of some set from some other sets; the axiom of extensionality guarantees the uniqueness of the result.

The theories obtained from Z and ZF by deleting the axiom of innity will be denoted by Z^- and ZF^-, respectively. Finally, ZFC denotes the theory consisting of ZF plus the axiom of choice.

- The **axiom of extensionality** asserts that two sets which have the same elements are equal:

$$\forall v_0 \forall v_1 (\forall v_2 (v_2 \in v_0 \Leftrightarrow v_2 \in v_1) \Rightarrow v_0 \simeq v_1).$$

Let a and b be two sets. We say that a is a subset of b (or that a is included in b) if every element of a is an element of b. In other words, if, in $\mathcal{U}$, a and b satisfy

$$\forall v_0 (v_0 \in a \Rightarrow v_0 \in b).$$

This formula will be abbreviated by $a \subseteq b$ while $a \subsetneq b$ is the formula

$$a \subseteq b \wedge a \neq b.$$

When we wish to prove that two sets are equal, we typically show that $a \subseteq b$ and $b \subseteq a$ and invoke the axiom of extensionality.

- The **axiom of pairs**:

$$\forall v_0 \forall v_1 \exists v_2 \forall v_3 (v_3 \in v_2 \Leftrightarrow (v_3 \simeq v_0 \vee v_3 \simeq v_1)).$$

Given two sets a and b, there exists a set whose only elements are a and b. According to the axiom of extensionality, there is only one such set; we denote it by $\{a, b\}$ and call it **the pair** a, b.

It can happen that a and b are the same. In this case, we obtain a set that has only one element; we denote it by $\{a\}$ instead of $\{a, a\}$ and call it **singleton** a.

We should remark that

$$\{a, b\} = \{a', b'\} \quad \text{if and only if}$$
$$(a = a' \text{ and } b = b') \text{ or } (a = b' \text{ and } b = a'),$$

and that

$$\{a\} = \{a'\} \quad \text{if and only if} \quad a = a'.$$

- The **axiom of unions**:

$$\forall v_0 \exists v_1 \forall v_2 (v_2 \in v_1 \Leftrightarrow \exists v_3 (v_3 \in v_0 \wedge v_2 \in v_3)).$$

Given a set a, this axiom asserts the existence of a set whose elements are the elements of elements of a; in other words, this set is the union of all the sets that are elements of a. As always, the axiom of extensionality guarantees the uniqueness of such a set; we will denote it by $\bigcup_{x \in a} x$ or, more simply, by $\bigcup a$.

Before proceeding further, we should discuss a few consequences of these rst three axioms. Let a and b be two sets. Thanks to the axiom of pairs, we can form the set $c = \{a, b\}$, then, with the axiom of unions, $\bigcup c$. This set is called the **union of** a **and** b and is denoted by $a \cup b$. It satises

$$\forall v_0(v_0 \in a \cup b \Leftrightarrow (v_0 \in a \vee v_0 \in b)).$$

Next consider three sets a, b, and c. We can form the sets $\{a, b\}$ and $\{c\}$, then the union of these last two sets $\{a, b\} \cup \{c\}$, which we will denote by $\{a, b, c\}$. We may then observe that

$$\forall v_0(v_0 \in \{a, b, c\} \Leftrightarrow (v_0 \simeq a \vee v_0 \simeq b \vee v_0 \simeq c))$$

is true. This process can be iterated: if n is a positive integer (in the intuitive sense) and if $a_1, a_2, \ldots, a_n$ are sets, then there exists a set, denoted by $\{a_1, a_2, \ldots, a_n\}$, that satises

$$\forall v_0(v_0 \in \{a_1, a_2, \ldots, a_n\} \Leftrightarrow (v_0 \simeq a_1 \vee v_0 \simeq a_2 \vee \cdots \vee v_0 \simeq a_n)).$$

We can also form $\bigcup\{a_1, a_2, \ldots, a_n\}$, which we will denote by $a_1 \cup a_2 \cup \cdots \cup a_n$ and which satises

$$\forall v_0(v_0 \in a_1 \cup a_2 \cup \cdots \cup a_n \Leftrightarrow (v_0 \in a_1 \vee v_0 \in a_2 \vee \cdots \vee v_0 \in a_n)).$$

- The **axiom of subsets** (the **power set axiom**) asserts that, for every set a, there exists a set b, unique by the axiom of extensionality, whose elements are precisely the subsets of a:

$$\forall v_0 \exists v_1 \forall v_2(v_2 \in v_1 \Leftrightarrow \forall v_3(v_3 \in v_2 \Rightarrow v_3 \in v_0)).$$

  We denote this set by $\wp(a)$.

- The **axiom scheme of comprehension** (the **comprehension scheme**). In this instance, we are dealing not with a single axiom but with an innity of axioms. Specically, this scheme comprises all formulas that can be written in the form

$$\forall v_1 \forall v_2 \ldots \forall v_{n+1} \exists v_{n+2} \forall v_0$$
$$(v_0 \in v_{n+2} \Leftrightarrow (v_0 \in v_{n+1} \wedge F[v_0, v_1, \ldots, v_n])),$$

  where n is an integer and $F[v_0, v_1, \ldots, v_n]$ is a formula of L.

What this scheme means, then, is that for any set a and for any formula $H[v_0]$ with one free variable and with parameters in $\mathcal{U}$, there exists a set, unique by the axiom of extensionality, whose elements are precisely those elements of a that satisfy H. We denote this set by $\{x \in a : H[x]\}$.

It is reasonable to ask why we burden ourselves with the set a. It would seem easier and more natural to consider the following scheme: for every formula $F[v_0]$

with one free variable and with parameters from $\mathcal{U}$, there exists a set whose elements are the sets that satisfy F;

$$\forall v_1 \forall v_2 \ldots \forall v_n \exists v_{n+1} \forall v_0 (v_0 \in v_{n+1} \Leftrightarrow F[v_0, v_1, \ldots, v_n]),$$

where n is an integer and $F[v_0, v_1, \ldots, v_n]$ is a formula of L.

There is in fact a good reason for not allowing this scheme: the resulting theory will be inconsistent. Indeed, if F is the formula $v_0 \notin v_0$, we would have

$$\exists v_1 \forall v_0 (v_0 \in v_1 \Leftrightarrow v_0 \notin v_0)$$

as an axiom. So there would exist a set a such that, for every set b,

$$b \in a \Leftrightarrow b \notin b.$$

In particular, for $b = a$, we would have

$$a \in a \Leftrightarrow a \notin a,$$

which is clearly contradictory.

The reader will undoubtedly have recognized the diagonal argument that logicians love. The unmistakable contradiction that results is known as **Russell's paradox**. This argument can be employed to prove that `the set of all sets' does not exist. Specically, we have

Theorem 7.1 *If $\mathcal{U}$ is a model of Z^-, then*

$$\mathcal{U} \models \neg \exists v_0 \forall v_1 (v_1 \in v_0).$$

Proof Suppose the contrary, let a be a set that satises

$$\forall v_0 (v_0 \in a),$$

and invoke the comprehension scheme with $n = 0$, $v_1 = a$ and with $F = v_0 \notin v_0$. As above, this produces a set b such that

$$\forall v_0 (v_0 \in b \Leftrightarrow v_0 \notin v_0)$$

and this leads, once more, to a contradiction when v_0 assumes the value b. ■

We have said that we would reserve the word `set' for points of $\mathcal{U}$. Nevertheless, it is sometimes convenient to speak of the collection (i.e. subset in the intuitive sense) of points in $\mathcal{U}$ that satisfy this or that rst-order property. We will use the word `class' for this purpose: if $F[v_0]$ is any formula of L with one free variable and with parameters in $\mathcal{U}$, we may refer to the class of sets a that satisfy $F[a]$. Thus, classes are nothing more than `intuitive subsets' of the structure $\mathcal{U}$ that are denable with parameters from $\mathcal{U}$ (see Denition 3.96). As a matter of fact, we could avoid introducing the notion of class but at the cost of considerably encumbering the exposition. To avoid misunderstandings, we will use upper case script letters ($\mathcal{U}$, $\mathcal{V}$, etc.) to denote classes. If $F[v_0]$ is a formula and $\mathcal{A}$ is the class of sets a that satisfy $F[a]$, we will say that a set b belongs to $\mathcal{A}$, or that $\mathcal{A}$ contains b, to mean that b satises F. We will accept the abuse of language that consists in identifying a set a with the class of sets b that belong to a. Except for the axiom of extensionality, all the axioms that we have stated thus far (and this will also be the case for the axioms of replacement) are declarations that certain classes are sets.

We will now introduce some consequences of the comprehension axioms. To begin with, when F is the formula $\neg v_0 \simeq v_0$, we obtain

$$\forall v_1 \exists v_2 \forall v_0 (v_0 \in v_2 \Leftrightarrow (v_0 \in v_1 \wedge \neg v_0 \simeq v_0)).$$

Now, whatever the set a is, there is no set v_0 that satises

$$(v_0 \in a \wedge \neg v_0 \simeq v_0).$$

As a result, there is a set that contains no elements (because the universe is not empty). By extensionality, there is a unique such set, which consequently does not depend on a. This set is called the **empty set** and is denoted by $\emptyset$.

Next, consider two sets a and b. The comprehension axiom with $F[v_0, v_1] = v_0 \in v_1$, with $v_2 = a$, and with $v_1 = b$, allows us to derive the existence of a set c such that

$$\forall v_0 (v_0 \in c \Leftrightarrow (v_0 \in a \wedge v_0 \in b)).$$

Uniqueness is, as usual, assured by the axiom of extensionality. This set c, whose elements are precisely those sets that are elements of both a and b, is called the **intersection of** a **and** b and is denoted by $a \cap b$.

Let a be a non-empty set. Then there exists a unique set b whose elements are those sets that belong to each element of a:

$$\forall v_0 (v_0 \in b \Leftrightarrow \forall v_3 (v_3 \in a \Rightarrow v_0 \in v_3)).$$

To prove the existence of such a set (uniqueness follows from extensionality), we choose an element c of a (a is non-empty) and observe that the formula

$$\forall v_3 (v_3 \in a \Rightarrow v_0 \in v_3)$$

is equivalent to the formula

$$v_0 \in c \wedge \forall v_3 (v_3 \in a \Rightarrow v_0 \in v_3).$$

We may then invoke the comprehension axiom

with $F = \forall v_3 (v_3 \in a \Rightarrow v_0 \in v_3)$, with $v_2 = c$, and with $v_1 = a$.

We will denote this set by $\bigcap_{x \in a} x$ or, more simply, by $\bigcap a$. With this notation, we should note that

$$\bigcap \{a\} = a, \qquad \bigcap \{a, b\} = a \cap b,$$

and, for any intuitive integer n,

$$\bigcap\{a_1, a_2, \ldots, a_n\} = a_1 \cap a_2 \cap \cdots \cap a_n.$$

Observe that the restriction 'a is non-empty' is essential; indeed, the formula

$$\forall v_2(v_2 \in \emptyset \Rightarrow v_1 \in v_2)$$

is satised by every set (because $v_2 \in \emptyset$ is always false) but we have seen that the set of all sets does not exist.

If a and b are sets, we will let $a - b$ denote the set of elements of a that are not elements of b:

$$a - b = \{x \in a : x \notin b\}.$$

If b is a subset of a, then $a - b$ is called the **complement of** b **in** a.

We will also dene the **symmetric difference** of two sets a and b:

$$a \mathbin{\triangle} b = (a - b) \cup (b - a).$$

Remark The fact that the connectives $\wedge$ and $\vee$ satisfy certain associative, commutative, and distributive properties implies that the corresponding properties hold for $\cap$ and $\cup$. For example,

$$a \cap b = b \cap a; \qquad (a \cap b) \cap c = a \cap (b \cap c);$$
$$a \cup b = b \cup a; \qquad (a \cup b) \cup c = a \cup (b \cup c);$$
$$a \cap (b \cup c) = (a \cap b) \cup (a \cap c); \qquad a \cup (b \cap c) = (a \cup b) \cap (a \cup c).$$

(See Exercise 2 from Chapter 2.)

- The **axiom scheme of replacement**. This comprises all formulas of the following form:

$$\begin{aligned}&\forall v_0 \forall v_1 \ldots \forall v_n\\ &\quad(\forall w_0 \forall w_1 \forall w_2((F[w_0, w_1, v_1, \ldots, v_n] \wedge F[w_0, w_2, v_1, \ldots, v_n])\\ &\quad\Rightarrow w_1 \simeq w_2) \Rightarrow \exists v_{n+1} \forall v_{n+2}(v_{n+2} \in v_{n+1} \Leftrightarrow \exists w_0(w_0 \in v_0\\ &\quad\wedge F[w_0, v_{n+2}, v_1, \ldots, v_n]))),\end{aligned}$$

  where n is an integer and $F[w_0, w_1, v_1, v_2, \ldots, v_n]$ is a formula of L.

These formulas deserve some explanation. To begin with, here is a denition.

Denition 7.2 *A formula $F[w_0, w_1, a_1, a_2, \ldots, a_n]$ of L with two free variables and with parameters from $\mathcal{U}$ is called* ***functional in*** w_0 ***in*** $\mathcal{U}$ *if the following formula is satised:*

$$\forall w_0 \forall w_1 \forall w_2((F[w_0, w_1, a_1, a_2, \ldots, a_n] \wedge F[w_0, w_2, a_1, a_2, \ldots, a_n])$$
$$\Rightarrow w_1 \simeq w_2).$$

Most of the time, we will not bother to specify `in $\mathcal{U}$'. If the formula $F[w_0, w_1]$ is functional in w_0, it allows us to dene an (intuitive) partial function from U into U, which we will call ϕ_F, in the following way: if b is a set and if there does not exist a set c such that $F[b, c]$, then ϕ_F is not dened at b; if such a c does exist, then it is unique and $\phi_F(b)$ is, by denition, equal to this set.

So the axiom schema of replacement asserts that if the formula $F[w_0, w_1]$ is functional in w_0 (while the variables $v_1, v_2, \ldots, v_n$ are replaced by parameters from $\mathcal{U}$) and if a is a set, then the class consisting of the values assumed by ϕ_F on the elements of a is in fact a set. We denote this set by $\{x : \exists v_0 \in a F[v_0, x]\}$.

It is not difcult to see that the replacement scheme implies the comprehension scheme. Let $F[v_0, a_1, a_2, \ldots, a_n]$ be a formula of L with parameters from $\mathcal{U}$ and b be a set. Using the axiom scheme of replacement, we will prove that there exists a set c whose elements are precisely those elements of b that satisfy $F[v_0, a_1, a_2, \ldots, a_n]$. Let $H[w_0, w_1, a_1, a_2, \ldots, a_n]$ denote the formula

$$w_0 \simeq w_1 \wedge F[w_0, a_1, a_2, \ldots, a_n].$$

This formula is obviously functional in w_0 and the set of values assumed by ϕ_H on the elements of b is the desired set.

The remaining axioms will be presented further on.

7.1.2 Ordered pairs, relations, and maps

Denition 7.3 *Let a and b be two sets. The set*

$$\{\{a\}, \{a, b\}\}$$

is called the ***ordered pair*** *a, b and is denoted by (a, b).*

The fact that (a, b) is a set can be justied by invoking the axiom of pairs three times. The rationale for this somewhat complicated denition and for the name given to (a, b) is in the following proposition (incidentally, any other denition that leads to this same property would serve just as well):

Proposition 7.4 *Suppose a, b, a', and b' are sets and that $(a, b) = (a', b')$; then $a = a'$ and $b = b'$.*

Proof By hypothesis, we have

$$\{\{a\}, \{a, b\}\} = \{\{a'\}, \{a', b'\}\}.$$

We distinguish two cases.

(1) $a = b$: then $\{\{a'\}, \{a', b'\}\} = (a, b) = \{\{a\}\}$ and, consequently, $\{a', b'\} \in \{\{a\}\}$.

We may then conclude $a = a' = b'$ using the remark that follows the statement of the axiom of pairs.

(2) $a \neq b$: then $\{a'\} \neq \{a, b\}$ (otherwise, as in (1), we would have $a = a' = b$); thus $\{a'\} = \{a\}$ and $a = a'$ (again, invoking that same remark). Also, $\{a, b\} = \{a', b'\}$ and $b = b'$. ■

This proposition justies the name `ordered pair'. If (a, b) is an ordered pair, then a is, by denition, its **rst projection** (we also say **rst coordinate**) and b is its **second projection**.

Denition 7.5 *Let a and b be sets. The set*

$$\{(x, \emptyset) : x \in a\} \cup \{(y, \{\emptyset\}) : y \in b\}$$

*is called the **disjoint union of** a **and** b and will be denoted by* $a \uplus b$.

So the disjoint union of a and b is the union of the two sets

$$\{(x, \emptyset) : x \in a\} \quad \text{and} \quad \{(y, \{\emptyset\}) : y \in b\}.$$

We should view these sets, intuitively, as copies of a and b respectively. The main feature of these copies is that they are necessarily disjoint (i.e. their intersection is empty); this is not always the case for a and b.

Notation We will write **0** in place of $\emptyset$ and **1** in place of $\{\emptyset\}$. This notation will be justied in due course.

Proposition 7.6 *Let a and b be two sets. Then there exists a set c such that*

$$\forall v_0 (v_0 \in c \Leftrightarrow \exists v_1 \exists v_2 (v_1 \in a \wedge v_2 \in b \wedge v_0 \simeq (v_1, v_2))).$$

In other words, given two sets a and b, there exists a set c whose elements are the ordered pairs whose rst coordinate is an element of a and whose second coordinate is an element of b. This set is called the **Cartesian product** of a and b and is denoted by $a \times b$.

Proof It is sufcient to observe that

$$c = \{u \in \wp(\wp(a \cup b)) : \exists v_0 \exists v_1 (v_0 \in a \wedge v_1 \in b \wedge u \simeq (v_0, v_1))\}.$$

So the proof uses the axiom of unions, the axiom of subsets, and an instance of the axiom of comprehension. ■

If a, b, and c are sets, the **triple** (a, b, c) is, by denition, the set $(a, (b, c))$. More generally, if n is an (intuitive) positive integer, we can dene the concept of n**-tuple** by induction: if $a_1, a_2, \ldots, a_n$ are sets, the n-tuple $(a_1, a_2, \ldots, a_n)$ is the set $(a_1, (a_2, a_3, \ldots, a_n))$. a_1 is the rst coordinate of this n-tuple, a_2 is its second coordinate, etc. Just as above, we can see that if $b_1, b_2, \ldots, b_n$ are sets, there exists a set whose elements are the n-tuples whose rst coordinate is an element of b_1, whose second coordinate is an element of b_2, etc., and this set is denoted by $b_1 \times b_2 \times \cdots \times b_n$. If, for i from 1 to n inclusive, all the b_i are equal to b, we will write b^n instead of $b \times b \times \cdots \times b$ and will refer to this as the **Cartesian power** rather than the Cartesian product.

Denition 7.7 (1) *Let a be a set and let n be an (intuitive) positive integer; an* n-***ary relation*** *on a is a subset of* a^n.

(2) *If n is an integer and R is an n-ary relation on a and b is a subset of a, the* ***restriction*** *of R to b is the set* $R \cap b^n$; *it is also denoted by* $R \restriction b$.

(3) *A* ***mapping*** *(or* ***map****) is a set all of whose elements are ordered pairs and which satises the following formula:*

$$Map(v_0) = \forall v_1 \forall v_2 \forall v_3(((v_1, v_2) \in v_0 \wedge (v_1, v_3) \in v_0) \Rightarrow v_2 \simeq v_3).$$

If f is a mapping, the set consisting of those sets that satisfy the formula $F[v_0] = \exists v_1 (v_0, v_1) \in f$ *is called the* ***domain*** *of f and is denoted by* $\mathrm{dom}(f)$. *The* ***image*** *(or* ***range****) of f is the set consisting of those sets that satisfy the formula* $G[v_0] = \exists v_1 (v_1, v_0) \in f$. *It is denoted by* $\mathrm{Im}\,(f)$. *A* ***mapping from*** *a* ***into*** *b is a mapping whose domain is a and whose image is included in b. If, in particular, its image is equal to b, the mapping is called* ***surjective from*** *a* ***onto*** *b. It is called* ***injective*** *if, for every d belonging to the image of f, there is a unique element c such that* $(c, d) \in f$.

We do need a slight justication for the denition of the domain and image of f; we should prove that they are sets. For example, the domain of f is the set

$$\left\{x \in \bigcup\bigcup f : \exists v_1 (x, v_1) \in f\right\}.$$

Let us insist on this point: a mapping is therefore a set of ordered pairs. By contrast, we will continue to use the word `function' with its intuitive meaning and will speak, for example, of functions from the universe $\mathcal{U}$ into $\mathcal{U}$. We can continue to translate all the usual notions concerning mappings into the language of set theory. We will be content with the following denitions:

- If f is a mapping and c belongs to the domain of f, then the unique set d such that $(c, d) \in f$ is called the **image of** c **under** f and is denoted by $f(c)$.
- The empty set is a mapping whose domain and image are both equal to the empty set. Thus, for every set b, $\emptyset$ is a mapping from $\emptyset$ into b. We will call this the **empty mapping**.
- If f is a mapping from a into b and g is a mapping from b into c, then the **composition mapping** $g \circ f$ is the set

  $$\{u \in a \times c : \exists v_0 \exists v_1 \exists v_2 (u \simeq (v_0, v_2) \wedge (v_0, v_1) \in f \wedge (v_1, v_2) \in g\}.$$

  It is a mapping from a into c.
- A **bijection** from a onto b is a surjective mapping from a onto b that is also injective. If f is a bijection from a onto b, the **inverse mapping** f^{-1},

dened by

$$f^{-1} = \{(v_0, v_1) \in b \times a : (v_1, v_0) \in f\},$$

is a bijection from b onto a.

- If f is a mapping from a into b and if c is a subset of a, we will write

$$\bar{f}(c) = \{x \in b : \exists y \in c f(y) \simeq x\};$$

$\bar{f}(c)$ is called the **direct image** of c under f, though we will usually shorten this and speak of the image of c under f (not to be confused with the image of f, which is the image of a under f). If there is no danger of confusion, we will simply write $f(c)$ instead of $\bar{f}(c)$.

In these same circumstances, if d is a subset of b, we dene the **inverse image** of d under f to be the set

$$\bar{f}^{-1}(d) = \{x \in a : f(x) \in d\}.$$

In this way, with any mapping from a into b, we can associate a mapping from $\wp(a)$ into $\wp(b)$ and a mapping from $\wp(b)$ into $\wp(a)$.

- Let a and b be two sets. The **exponentiation of** a **by** b, denoted by a^b (this is to be read as `a to the power b'), is the set of all mappings from b into a. This denition needs to be justied; we must show that this is a set. We will once more use the axiom of comprehension: a^b is the set of elements of $\wp(b \times a)$ that satisfy the following formula:

$$\forall v_1(v_1 \in b \Rightarrow \exists! v_2(v_1, v_2) \in v_0).$$

- Let I be a set. A mapping whose domain is I is also called a **family of sets indexed by** I.

This notion, which is in common use, is introduced only because the vocabulary and notation that it permits are more practical. If a is a family indexed by I and if $i \in I$, we generally write a_i rather than $a(i)$; the family a itself can be written as $(a_i : i \in I)$ or, preferably, $(a_i)_{i \in I}$.

- Let $a = (a_i)_{i \in I}$ be a family of sets; the **union** of this family, denoted by $\bigcup_{i \in I} a_i$, is the union of the elements of the image of a. In other words, for every set b, $b \in \bigcup_{i \in I} a_i$ if and only if there exists $i \in I$ such that $b \in a_i$.

If I is non-empty (as we noted in our discussion of the comprehension scheme, this restriction is essential), we can similarly dene the **intersection** of the family $(a_i)_{i \in I}$, which will be denoted by $\bigcap_{i \in I} a_i$. For every set b, $b \in \bigcap_{i \in I} a_i$ if and only if, for every $i \in I$, $b \in a_i$.

- Again let $a = (a_i)_{i \in I}$ be a family of sets. The **product** of this family, denoted by $\prod_{i \in I} a_i$, is the set of mappings f from I into $\bigcup_{i \in I} a_i$ which are such that, for all $i \in I$, $f(i) \in a_i$.

We are now in a position to state the **axiom of choice**: the product of a family of non-empty sets is non-empty. In other words,

(AC) Let $(a_i)_{i \in I}$ be a family of sets and assume that, for every $i \in I$, a_i is not empty; then $\prod_{i \in I} a_i$ is not empty.

We will not discuss the question of whether this axiom is justied or not. What is certain (though it will not be proved in this text) is that it cannot be derived in the theory ZF (nor, indeed, can its negation), assuming that ZF is consistent (see the section on inaccessible cardinals). Besides, this axiom is *required* to prove certain important theorems of mathematics [examples: for the fact that every vector space has a basis, for the HahnñBanach theorem, for Krull's theorem (see Chapter 2)].

7.2 Ordinal numbers and integers

7.2.1 Well-ordered sets

In this section, we introduce the notion of ordinal; this is a particularly important tool in set theory and in mathematics, generally. It can be viewed as a generalization of the notion of integer. We begin with some denitions.

Denition 7.8 *Suppose that X is a set and that R is a binary relation on X. We say that R is an* **order relation** *(or* **ordering***) on X (or that X is* **ordered by** *R) if*

- *R is transitive: if $(x, y) \in R$ and $(y, z) \in R$, then $(x, z) \in R$;*
- *R is irreexive: for every x, $(x, x) \notin R$.*

We say that R is a **total order relation** *(or* **total ordering***) (or that X is* **totally ordered by** *R) if, in addition,*

- *for all x and y belonging to X, if x and y are distinct, then either $(x, y) \in R$ or $(y, x) \in R$.*

A ***(totally)* ordered set** *is a pair (X, R) where R is a (total) order relation on X.*

Thus, the order relations considered in this chapter are strict (in the rest of this text, the opposite is generally the case). Antisymmetry [for every x and y belonging to X, $(x, y) \notin R$ or $(y, x) \notin R$] is a consequence of being irreexive and transitive; to see this, note that if (x, y) and (y, x) both belong to R, then, by transitivity, (x, x) would belong to R, violating irreexiveness. We will follow standard practice: if R is an order relation on X and if x and y are points in X, we will say that x **is less than** y **in** R to mean that $(x, y) \in R$ and we will write $x <_R y$; we will also use the notations $x >_R y$, $x \leq_R y$, and $x \geq_R y$. We will omit mentioning R if the context is unambiguous. If (X, R) and (Y, S) are ordered sets, an **isomorphism** from (X, R) onto (Y, S) is a bijective mapping f from X onto Y that satises

$$\text{for all } x \text{ and } y \text{ in } X, \quad x <_R y \text{ if and only if } f(x) <_S f(y).$$

If R is an order relation on X and if Y is a subset of X, then a **minimum element** (or **least element**) **of** Y (for R) is an element of Y that is less than or equal to every element of Y; a **minimal element of** Y (for R) is an element of Y that is not greater than any element of Y (if R is a total ordering, there is no difference between these two notions). The denitions of **maximum** (or **greatest**) **element** and **maximal element** are analogous. There can be at most one minimum (or maximum) element but there can be several minimal (or maximal) elements. A **lower bound for** Y is an element of X that is less than or equal to every element of Y; a **greatest lower bound for** Y is a maximal element of the set of lower bounds for Y. When the ordering is total, there is at most one greatest lower bound. The denitions of **upper bound for** Y and **least upper bound for** Y are analogous.

Now suppose that X is totally ordered by R. An **initial segment** of X is a subset Y of X with the following property:

$$\text{if } y \in Y \text{ and } x <_R y, \quad \text{then } x \in Y.$$

For example, X itself is an initial segment of X. A **proper initial segment** of X is an initial segment of X that is neither empty nor equal to X. If $x \in X$ and if x is not the minimum element of X, the set

$$S_x = \{y \in X : y <_R x\}$$

is a proper initial segment of X.

Remarks (1) If Y is an initial segment of X and if x is an element of X, we have $S_x \subsetneq Y$ (if $x \in Y$) or $Y \subseteq S_x$ (if $x \notin Y$); indeed, if $x \in Y$, then it is obvious that $S_x \subsetneq Y$; if $x \notin Y$, then for every element $y \in Y$, it is never the case that $x \leq y$ (for this would imply $x \in Y$), hence $Y \subseteq S_x$. Moreover, if x and y are elements of X such that $S_x = S_y$, then $x = y$; indeed, $x \notin S_x$ and $y \notin S_y$, i.e. $x \notin S_y$ and $y \notin S_x$, hence $y \leq_R x$ and $x \leq_R y$ (since $\leq_R$ is a total ordering of X).

(2) If x and y are elements of X such that $S_x = S_y$, then $x = y$; indeed, $x \notin S_x$ and $y \notin S_y$, so $x \notin S_y$ and $y \notin S_x$, which implies that $y \leq_R x$ and $x \leq_R y$ (since R is a total ordering of X).

(3) The set of initial segments of X is totally ordered by the relation $\subsetneq$; indeed, if Y and Z are two initial segments of X and if there exists an element x of Z such that $x \notin Y$, then, according to (1),

$$Y \subseteq S_x \subsetneq Z$$

and hence $Y \subsetneq Z$; in the opposite case, $Z \subseteq Y$.

(4) The set of initial segments of X with this ordering has a least element (the empty set) and a greatest element (X itself).

Denition 7.9 *Let X be a set and let R be a binary relation on X. We say that R is a* ***well-order relation****, or that R is a* ***well-ordering of*** *X, or that X is* ***well-ordered***

***by** R, if both the following hold:*

(1) *R is a total ordering of X;*

(2) *every non-empty subset of X has a least element.*

[Notice that in the presence of (2), (1) can be replaced by the statement that R is an ordering of X, for if x and y are distinct elements of X, then either $x \leq_R y$ or $y \leq_R x$ according as the least element of $\{x, y\}$ is x or y.] As intuitive examples of well-orderings, there are all the nite totally ordered sets as well as the set of integers with its usual order relation. We will see that there are many others.

Let X be a set that is well-ordered by a relation R. First, observe that if Y is a subset of X, then $R \restriction Y$ is a well-ordering of Y. If X is not empty, it has a least element x_0 (which we will call its `rst element'). If $X$ is not equal to $\{x_0\}$, then the set $X - \{x_0\}$ of elements of $X$ different from $x_0$ also has a least element $x_1$, which we will call the `second element' of X; this process can be continued.

Another property of well-ordered sets is given in the following proposition.

Proposition 7.10 *Let X be a well-ordered set and let Y be an initial segment of X. Then either $Y = X$ or there exists a (unique) element x of X such that $Y = S_x$.*

Proof Assume $Y \neq X$ and consider the least element of $X - Y$; call it x. We claim that $Y = S_x$. To see this, note that if $y \in Y$, then $y < x$ (otherwise $x \leq y$ and $x \in Y$) so $Y \subseteq S_x$; if $z \in S_x$, then $z < x$ and, because x is the least element of $X - Y$, $z \in Y$. ■

Denition 7.11 *Let X be a set. We say that X is **transitive** if every element of a set that belongs to X also belongs to X (in other words, if the following formula is satised):*

$$\forall v_0 \forall v_1 ((v_0 \in X \wedge v_1 \in v_0) \Rightarrow v_1 \in X).$$

So a set X is transitive if and only if every element of X is also a subset of X. This condition is also equivalent to the inclusion $\bigcup X \subseteq X$.

7.2.2 The ordinals

Denition 7.12 *Let α be a set. We say that α is an **ordinal** if the following properties are satised:*

(1) *α is transitive;*

(2) *the membership relation on α is a well-ordering*
[i.e. the set $\{(x, y) \in \alpha \times \alpha : x \in y\}$ is a well-ordering of α].

It is obvious that these properties can be expressed by formulas of L. We will write $On[v_0]$ to denote the formula of L expressing that v_0 is an ordinal. If α and β are ordinals, we will indifferently write $\alpha \in \beta$ or $\alpha < \beta$ (this latter notation, which we will justify shortly, leads naturally to the notations $\alpha \leq \beta$, $\alpha > \beta$, and $\alpha \geq \beta$).

Remark 7.13 If α is an ordinal, then $\alpha \notin \alpha$; the reason is that membership is a (strict) order relation on the elements of α: if we assume that α belongs to α, we conclude from this that α does not belong to α.

Remark 7.14 If α is an ordinal and if $\beta \in \alpha$, then β is an ordinal. Since α is transitive, $\beta \subseteq \alpha$, so the membership relation on β, which is equal to the restriction to β of the membership relation on α, is a well-ordering. It remains to show that β is transitive; so suppose that $\gamma \in \beta$ and that $\delta \in \gamma$; then β, γ, and δ are elements of α (since α is a transitive set) and, because $\in$ is a transitive relation on α, $\delta \in \beta$.

Remark 7.15 If α is an ordinal and if $\beta \in \alpha$, then $\beta = S_\beta$. This follows from the axiom of extensionality since the statements $x \in \beta$ and $x \in S_\beta$ are equivalent.

Remark 7.16 Let α and β be two ordinals. Then $\alpha \subseteq \beta$ if and only if $\alpha \leq \beta$. Indeed, if α is included in β, then, since α is transitive, α is an initial segment of β. Thus, either $\alpha = \beta$, or there exists $\gamma \in \beta$ such that $\alpha = S_\gamma$ (Proposition 7.10) and (Remark 7.15) $\alpha = \gamma$, hence $\alpha < \beta$. Conversely, if $\alpha \leq \beta$ (i.e. either $\alpha \in \beta$ or $\alpha = \beta$), and if $\gamma \in \alpha$, then by the transitivity of β, $\gamma \in \beta$; this shows that $\alpha \subseteq \beta$.

Proposition 7.17 *Let X be a transitive set of ordinals such that, for all distinct elements x and y of X, $x \in y$ or $y \in x$; then X is an ordinal.*

[We will see subsequently (Corollary 7.22) that the condition `$x \in y$ or $y \in x$' is superuous since it is always satised.]

Proof It sufces to verify the second condition from Denition 7.12. If $\alpha \in X$, then $\alpha \notin \alpha$ by Remark 7.13. If α, β, and γ are elements of X and if $\alpha \in \beta$ and $\beta \in \gamma$, then the transitivity of γ implies that $\alpha \in \gamma$. So the membership relation does dene an order relation on X and this ordering is total by hypothesis.

Let us prove that it is a well-ordering. Suppose Y is a non-empty subset of X; we will show that Y has a least element (for $\in$). Let $\alpha \in Y$:

ó If $\alpha \cap Y = \emptyset$, then α is the least element of Y: for if $\beta \in Y$, it is false that $\beta \in \alpha$ (since $\alpha \cap Y = \emptyset$), hence $\alpha \in \beta$ or $\alpha = \beta$.

ó If $\alpha \cap Y \neq \emptyset$, then, because α is an ordinal, $\alpha \cap Y$ has a least element β. If $\gamma \in \beta$, then $\gamma \in \alpha$ (since α is an ordinal) and $\gamma \notin \alpha \cap Y$ (because β is a least element of this set). Consequently, $\gamma \notin Y$ and thus no element of Y can be less than β. Since the ordering is total, this implies that β is the least element of Y. ∎

Corollary 7.18 *If α is an ordinal and if β is an initial segment of α, then β is an ordinal. If, in addition, β is distinct from α, then $\beta \in \alpha$.*

Proof The rst part of the corollary follows from the above proposition; β is transitive because it is an initial segment of α, and β is totally ordered by $\in$ since α is.

If we assume, in addition, that β is not equal to α, then by Proposition 7.10, there exists $\gamma \in \alpha$ such that $\beta = S_\gamma$, and by Remark 7.15, $\gamma = S_\gamma$, so $\beta \in \alpha$. ∎

It is clear that the empty set is an ordinal. It is even the least of all the ordinals. If α is a non-empty ordinal, we claim that $\emptyset \in \alpha$. For α contains a least element that we will call β. Every element of β also belongs to α (since α is transitive) and is strictly less than β; but this contradicts the minimality of β. So $\beta = \emptyset$.

The set $\{\emptyset\}$ is also an ordinal; this is easy to verify from the denition. The following sets are also ordinals:

$$\{\emptyset, \{\emptyset\}\} \quad \text{and} \quad \{\emptyset, \{\emptyset\}, \{\emptyset, \{\emptyset\}\}\}$$

More generally, we have the following corollary.

Corollary 7.19 *If α is an ordinal, then $\beta = \alpha \cup \{\alpha\}$ is also an ordinal.*

Proof We will once more use the preceding proposition. First, we will show that β is a transitive set. Assume $\gamma \in \beta$ and $\delta \in \gamma$; then either $\gamma \in \alpha$ and $\delta \in \alpha$ because α is transitive; or else $\gamma = \alpha$ and $\delta \in \alpha$ is then obvious. Also, if x and y are two distinct elements of β, then either they both belong to α (in which case we do have $x \in y$ or $y \in x$ since α is an ordinal) or else one of the two is equal to α and the other belongs to α (in which case we obtain the same conclusion). ■

The ordinal $\alpha \cup \{\alpha\}$ will be denoted by α^+ and will be called the **successor** of α. An ordinal is called a **limit ordinal** if it is not equal to $\emptyset$ and is not the successor of some other ordinal.

We have now arrived at a theorem which is slightly more difcult and whose corollaries are extremely important. Our argument will involve, without explicit mention, denitions and proofs by induction. We will analyse these more precisely and systematically in the section that follows this one.

Theorem 7.20 *Let X and Y be two sets that are well-ordered by the relations R and S, respectively. Then at least one of the following two situations holds:*

(a) *there exists one and only one initial segment Y_1 of Y and one and only one isomorphism f from (X, R) onto $(Y_1, S \upharpoonright Y_1)$;*

(b) *there exists one and only one initial segment X_1 of X and one and only one isomorphism g from (Y, S) onto $(X_1, R \upharpoonright X_1)$.*

Moreover, if both (a) and (b) hold, then $X_1 = X$ and $Y_1 = Y$ and the mappings f and g are inverses of one another.

Proof In this proof, we will consider initial segments of X and Y. We will always consider these as sets that are ordered by the restrictions of R or S. When we speak of isomorphisms between such sets, we always mean isomorphisms of ordered sets.

We will rst prove uniqueness. Suppose, for example, that Y_1 and Y_2 are two initial segments of Y and that f_1 and f_2 are isomorphisms from X onto Y_1 and Y_2, respectively. Consider the set

$$Z = \{x \in X : f_1(x) \neq f_2(x)\}.$$

We will prove that Z is empty. The argument is by contradiction. If Z is not empty, there is a least element x_0 (for R) in Z. Assume, for example, that $f_1(x_0) <_S f_2(x_0)$. As Y_2 is an initial segment of Y, $f_1(x_0) \in Y_2$ and there exists $x_1 \in X$ such that

$$f_2(x_1) = f_1(x_0) <_S f_2(x_0).$$

Since f_2 is an isomorphism, $x_1 <_R x_0$. But x_0 was chosen as the least element in Z, thus $f_1(x_1) = f_2(x_1)$; it follows from this that $f_1(x_1) = f_1(x_0)$, contradicting the fact that f_1 is injective.

The uniqueness for situation (b) is proved in a similar fashion. Suppose next that (a) and (b) both hold simultaneously. We can easily see that $g(Y_1)$ is an initial segment of X and hence that $g \circ f$ is an isomorphism from X onto one of its initial segments. But the identity mapping on X is also an isomorphism from X onto one of its initial segments (namely X itself), so by applying the uniqueness that has already been established, we see that $g \circ f$ is equal to the identity mapping on X. Since f is injective, it follows that f and g are inverses of each other.

What remains is to prove the existence. Consider the sets

$$A = \{(x, y) \in X \times Y : \text{there exists an isomorphism from } S_x \text{ onto } S_y\}$$

and

$$A^* = \{(y, x) \in Y \times X : \text{there exists an isomorphism from } S_x \text{ onto } S_y\}$$

Suppose that (x, y) and (x, z) both belong to A, so that there exist two isomorphisms from S_x onto S_y and S_z, respectively, which are initial segments of Y. We have just seen that this implies $S_y = S_z$, and hence $y = z$. In other words, A is a mapping whose domain, which we call A_1, is included in X and whose image, which we call A_2, is included in Y. In the same way, we prove that A^* is a mapping and, since $(x, y) \in A$ if and only if $(y, x) \in A^*$, it follows that the domain of A^* is A_2 and that its image is A_1; thus A and A^* are inverses of one another. So they are both bijections.

We will need the following observation: suppose that h is an isomorphism from a totally ordered set U onto a totally ordered set V; then it is very easy to verify that, for all $u \in U$, the image under h of the set $\{t \in U : t < u\}$ is equal to $\{v \in V : v < h(u)\}$. In other words, $\bar{h}(S_u) = S_{h(u)}$.

This allows us to prove that A_1 is an initial segment of X. For if $x \in A_1$ and $z <_R x$, then there exists $y \in Y$ and an isomorphism f from S_x onto S_y. The restriction of f to S_z is an isomorphism from S_z onto an initial segment of S_y, which is also an initial segment of Y distinct from Y, and hence equal to some $S_{y'}$, with $y' \in Y$ (Proposition 7.10). This shows that $(z, y') \in A$, and therefore $z \in A_1$. Similarly, A_2 is an initial segment of Y.

We also see that A, not content to be bijective, is an isomorphism. Indeed, suppose that (x, y) and (z, t) belong to A and that $z <_R x$; so there exists an isomorphism f from S_x onto S_y and $f \upharpoonright S_z$ is an isomorphism from S_z onto an initial segment of Y which, by the uniqueness that has already been proved, must equal S_t. Hence, $t <_S y$.

If $A_1 = X$, the conclusion of the theorem is true because (a) holds; if $A_2 = Y$, then it is (b) that holds. Finally, let us prove by contradiction that the situation in which both $A_1 \neq X$ and $A_2 \neq Y$ cannot arise. If it did, then by Proposition 7.10, there would exist $x \in X$ and $y \in Y$ such that $A_1 = S_x$ and $A_2 = S_y$. But then, A is an isomorphism from S_x onto S_y, which proves that $(x, y) \in A$ and that $x \in A_1 = S_x$, which is absurd. ■

We will now apply this theorem specically to the ordinals. We have already observed that an initial segment of an ordinal is an ordinal (see Corollary 7.18). In addition, we have the following:

Proposition 7.21 *Suppose that α and β are ordinals and that f is an isomorphism from α onto β. Then $\alpha = \beta$ and f is the identity on α.*

Proof Consider the set $X = \{x \in \alpha : f(x) \neq x\}$. If this set is not empty, it has a least element x_0. Let us examine $f(x_0)$.

If $y \in x_0$, then $y \in \alpha$ (α is transitive) and $f(y) \in f(x_0)$ (f is an isomorphism); so by the minimality of x_0, $y = f(y)$. It follows that $x_0 \subseteq f(x_0)$.

Conversely, if $y \in f(x_0)$, then $y \in \beta$ (β is transitive) and there exists $z \in \alpha$ such that $y = f(z)$ (f is surjective onto β). Since f is an isomorphism, $z \in x_0$ and, by the minimality of x_0, $z = f(z) = y$. Thus $f(x_0) \subseteq x_0$ and, by extensionality, $x_0 = f(x_0)$. So we have arrived at a contradiction; our conclusion is that X is empty. This proves the proposition. ■

Corollary 7.22 *Suppose that α and β are ordinals. Then exactly one of the following holds:*

(1) $\alpha \in \beta$;

(2) $\beta \in \alpha$;

(3) $\alpha = \beta$.

Proof It is impossible that two of these properties could hold simultaneously; this is an easy consequence of Remark 7.13 and of the fact that ordinals are transitive sets.

Next, apply Theorem 7.20. Assume, for example, that there exists an isomorphism f from α onto an initial segment S of β. If this initial segment is distinct from β, then, according to Corollary 7.18, S is itself an ordinal and belongs to β. From Proposition 7.21, we conclude that $S = \alpha$, so $\alpha \in \beta$. If the initial segment S is equal to β, then f is an isomorphism from α onto β and we may apply Proposition 7.21 directly to conclude that $\alpha = \beta$. The argument is analogous in the case where there exists an isomorphism from β onto an initial segment of α. ■

Remark 7.23 Proposition 7.17 can now be rephrased; it asserts that every transitive set of ordinals is an ordinal.

We will now consider the class of ordinals (i.e. the class of sets that satisfy the formula $On[v_0]$). We will shortly prove that this class is not a set. Nonetheless, the membership relation on this class has all the properties of a well-ordering.

- (transitive) If α, β, and γ are ordinals and $\alpha \in \beta$ and $\beta \in \gamma$, then $\alpha \in \gamma$ because γ is a transitive set.
- (irreexive) If α is an ordinal, then $\alpha \notin \alpha$ by Remark 7.13.
- (total) If α and β are ordinals, then either $\alpha \in \beta$ or $\beta \in \alpha$ or $\alpha = \beta$ by Corollary 7.22.
- Let $F[v_0]$ be a formula with parameters from U and assume that there exist ordinals that satisfy $F[v_0]$. We must prove that there is a least such ordinal; more precisely, we require an ordinal α such that $F[\alpha]$ is true and, for every ordinal β, $F[\beta]$ implies $\alpha \in \beta$ or $\alpha = \beta$. To prove this, let γ be an ordinal that satises $F[\gamma]$ (by hypothesis, such a γ exists). Then the set

  $$\{\beta \in \gamma^+ : F[\beta]\}$$

  is not empty since it contains γ. So it contains a least element since γ^+ is an ordinal. This least element is the desired ordinal α.

We will say that the membership relation on the class of ordinals is a meta-well-ordering. In particular,

Remark 7.24 Every non-empty class of ordinals contains a least element.

Proposition 7.25 *The class of ordinals is not a set.*

Proof Suppose the contrary and let X denote the set of all ordinals. If $\alpha \in X$ and $\beta \in \alpha$, then $\beta \in X$ by Remark 7.14. Together with Remark 7.23, this allows us to apply Proposition 7.17 to conclude that $X \in X$ since X is an ordinal. This contradicts Remark 7.13. ■

Proposition 7.26 *If A is a set of ordinals, then*

$$\beta = \bigcup_{\alpha \in A} \alpha$$

is an ordinal. Moreover, β is the least upper bound of A.

Proof The fact that any union of transitive sets is a transitive set is more or less obvious; thus β is a transitive set of ordinals. So it follows from Remark 7.23 that β is an ordinal.

If we assume that $\alpha \in A$, then, by the denition of A and of β, $\alpha \subseteq \beta$ and, by Remark 7.16, $\alpha \leq \beta$. This proves that β is greater than or equal to every element of A. We can even prove that β is the least such ordinal; for if γ is greater than or equal to every element of A, then, for every $\alpha \in A$, $\alpha \subseteq \gamma$; this shows that $\beta \subseteq \gamma$ and hence that $\beta \leq \gamma$.

So we have established that β is the least upper bound of A. We will write $\beta = \bigcup_{\alpha \in A} \alpha = \sup A$. ■

Proposition 7.27 *Suppose that a set X is well-ordered by a relation R; then there exists one and only one ordinal α that is isomorphic to (X, R). Moreover, there is a unique isomorphism from α onto (X, R).*

Proof Both uniqueness properties were already proved in Theorem 7.20.

Our argument now proceeds by contradiction. By applying Theorem 7.20, we see that every ordinal is isomorphic to an initial segment of X. Consider the set

$$T = \{x \in \wp(X) : x \text{ is an initial segment of } X \text{ and } x \text{ is isomorphic to an ordinal}\}$$

and the formula

$$F[v_0, v_1] = v_0 \in T \wedge On[v_1] \wedge \text{ there exists an isomorphism from } v_0 \text{ into } v_1.$$

Again by Theorem 7.20, this formula is functional in v_0 (Denition 7.2). So the replacement scheme guarantees the existence of the set

$$O = \{\alpha : \exists v_0 \in T\, F[v_0, \alpha]\}.$$

But, by hypothesis, there exists, for every ordinal α, an isomorphism from α onto an initial segment of X; as a result, $\exists v_0(v_0 \in T \wedge F[v_0, \alpha])$. In other words, O is the set of all ordinals; but this contradicts Proposition 7.25. ■

Remark 7.28 Suppose that α is an ordinal and that X is a subset of α; we have seen that X is well-ordered by $\in$. Consequently, there exists an ordinal β and an isomorphism f from β onto X. We claim that $\beta \leq \alpha$. This will follow from the next lemma.

Lemma 7.29 *Let f be a strictly increasing mapping from an ordinal β into an ordinal α; then for every ordinal $\gamma \in \beta$, $f(\gamma) \geq \gamma$.*

Proof We will argue by contradiction. Let γ be the least ordinal such that $f(\gamma) < \gamma$. For every $\delta \in \gamma$, $\delta \leq f(\delta)$ and, moreover, $f(\delta) < f(\gamma)$ since f is strictly increasing; thus $\delta \in f(\gamma)$. It follows that $\gamma \subseteq f(\gamma)$ and, from Remark 7.16, that $\gamma \leq f(\gamma)$. ■

We will now introduce the last axiom of ZF, the axiom of innity.

Denition 7.30 *Let α be an ordinal. We say that α is **nite** if neither it nor any of its elements is a limit ordinal. An ordinal is **innite** if it is not nite.*

Observe that if α is a nite ordinal and $\beta \in \alpha$, then β is also a nite ordinal. For example, $\emptyset$, $\{\emptyset\}$, $\{\emptyset, \{\emptyset\}\}$ are nite ordinals. More generally, if α is a nite ordinal, then so is α^+. In fact, the existence of an innite ordinal requires the introduction of a new axiom called, naturally, the **axiom of innity**.

(Inf) There exists an innite ordinal.

This is clearly equivalent to `there exists a limit ordinal'; this is the formula

$$\exists v_0(On[v_0] \wedge \neg v_0 \simeq \emptyset \wedge \forall v_1 \neg v_0 \simeq v_1 \cup \{v_1\}).$$

In keeping with what was said at the beginning of this chapter, we assume from now on that this axiom is satised in $\mathcal{U}$.

In the nal section of this chapter, we will see that this axiom cannot be derived from the ones introduced earlier.

Notation The least innite ordinal will be denoted by ω.

Such an ordinal exists since, as we have observed, the class of ordinals is a well-ordered class. Note that ω is also equal to the set of nite ordinals; for if α is a nite ordinal, then no ordinal less than α is innite, so $\alpha \in \omega$; conversely, if $\alpha \in \omega$, then, by the minimality of ω, α is nite.

7.2.3 Operations on ordinal numbers

We are now going to dene a few operations on ordered sets and ordinals.

Let $\mathfrak{a} = (a, R)$ and $\mathfrak{b} = (b, S)$ be two ordered sets. We are going to dene a new ordered set called the **direct sum of $\mathfrak{a}$ and $\mathfrak{b}$** and denoted by $\mathfrak{a} \oplus \mathfrak{b}$. The base set of $\mathfrak{a} \oplus \mathfrak{b}$ is $a \uplus b$, the disjoint union of a and b (see Denition 7.5). Set $c = a \uplus b$. We dene a binary relation T on c as follows:

$$\text{for all } (x_0, y_0) \quad \text{and} \quad (x_1, y_1) \text{ in } c, \quad ((x_0, y_0), (x_1, y_1)) \in T$$

if and only if

$$\begin{aligned} y_0 = \mathbf{0} \quad &\text{and} \quad y_1 = \mathbf{1}; \\ \text{or } y_0 = y_1 = \mathbf{0} \quad &\text{and} \quad x_0 <_R x_1; \\ \text{or } y_0 = y_1 = \mathbf{1} \quad &\text{and} \quad x_0 <_S x_1. \end{aligned}$$

Intuitively, the set $a \uplus b$ consists of a copy of a and a copy of b. In the relation we have just dened, the elements of the copy of a are ordered among themselves as they already are in $\mathfrak{a}$ and are less than the elements from the copy of b, which are in turn, among themselves, ordered as they already are in $\mathfrak{b}$.

We need to prove that the relation T is an order relation. First, consider transitivity. Let (x_0, y_0), (x_1, y_1), and (x_2, y_2) be three elements of c and assume that $((x_0, y_0), (x_1, y_1))$ and $((x_1, y_1), (x_2, y_2))$ both belong to T. Several cases are possible:

- $y_2 = \mathbf{0}$. The denition of T then implies that $y_0 = y_1 = \mathbf{0}$, $x_0 <_R x_1$, and $x_1 <_R x_2$. Since R is transitive, it follows that $x_0 <_R x_2$ and hence that $((x_0, y_0), (x_2, y_2)) \in T$.

- $y_2 = \mathbf{1}$ and $y_0 = \mathbf{0}$. Then $((x_0, y_0), (x_2, y_2)) \in T$ derives from the deni-tion of T.
- $y_0 = y_2 = \mathbf{1}$. Then $y_1 = \mathbf{1}$, $x_0 <_S x_1$, and $x_1 <_S x_2$. Since S is transitive, it follows that $x_0 <_S x_2$ and hence that $((x_0, y_0), (x_2, y_2)) \in T$.

We leave to the reader the verications that T is irreexive and that T is total if both R and S are total.

Suppose that $\mathfrak{a}$ and $\mathfrak{b}$ are both well-orderings. We will show that this is also the case for $\mathfrak{a} \oplus \mathfrak{b}$. Let d be a non-empty subset of c; we must show that it contains a least element. Exactly one of the following must happen: either d contains some elements of the form $(x, \mathbf{0})$ where $x \in a$ or it does not. In the rst case, if x_0 is the least such x, then $(x_0, \mathbf{0})$ is the least element of d; in the second case, all the elements of d are of the form $(y, \mathbf{1})$ where $y \in b$, and if y_0 is the least y such that $(y, \mathbf{1}) \in d$, then $(y_0, \mathbf{1})$ is the least element of c.

Thus, if α and β are ordinals, $\alpha \oplus \beta$ is a well-ordering and, by Proposition 7.27, it is isomorphic to a unique ordinal. This discussion justies the following denition.

Denition 7.31 *Let α and β be ordinals. The unique ordinal that is isomorphic to $\alpha \oplus \beta$ is called the **ordinal sum of** α **and** β and is denoted by $\alpha + \beta$.*

We next turn to the product. Again, let $\mathfrak{a} = (a, R)$ and $\mathfrak{b} = (b, S)$ be two ordered sets. We dene a relation T on $c = a \times b$ as follows:

$$\text{for all } (x_0, y_0) \text{ and } (x_1, y_1) \text{ in } c, \quad ((x_0, y_0), (x_1, y_1)) \in T$$

if and only if one of the following two conditions holds:

$$y_0 <_S y_1, \quad \text{or}$$
$$y_0 = y_1 \quad \text{and} \quad x_0 <_R x_1.$$

Once again, we need to prove that T is an order relation. First, consider transitivity. Let (x_0, y_0), (x_1, y_1), and (x_2, y_2) be three elements of c and assume that $((x_0, y_0), (x_1, y_1))$ and $((x_1, y_1), (x_2, y_2))$ both belong to T. It follows from this that $y_0 \leq_S y_1 \leq_S y_2$. So there are two possible cases: either $y_0 <_S y_2$, in which case $((x_0, y_0), (x_2, y_2)) \in T$ derives from the denition of T; or else $y_0 = y_1 = y_2$ and, in this case, $x_0 <_R x_1 <_R x_2$ and, once again, $((x_0, y_0), (x_2, y_2)) \in T$.

As above, we leave to the reader the verications that T is irreexive and that T is total if both R and S are total. The set $a \times b$, ordered by T, will be denoted by $\mathfrak{a} \otimes \mathfrak{b}$.

Also as above, let us prove that if R and S are both well-orderings, then so is T. Let d be a non-empty subset of $a \times b$; then the set

$$\{y \in b : \text{there exists } x \in a \text{ such that } (x, y) \in d\}$$

is non-empty. Let y_0 be its least element and let x_0 be the least element of the set

$$\{x \in b : (x, y_0) \in d\}.$$

It is easy to see that (x_0, y_0) is the least element of d. This justies the following denition.

Denition 7.32 *Let α and β be ordinals. The unique ordinal that is isomorphic to $\alpha \otimes \beta$ is called the* **ordinal product of** *α* **and** *β and is denoted by $\alpha \times \beta$.*

Theorem 7.33 *Let α, β, and γ be ordinals. Then*

(i) $\alpha + (\beta + \gamma) = (\alpha + \beta) + \gamma$;

(ii) $\alpha \times (\beta \times \gamma) = (\alpha \times \beta) \times \gamma$;

(iii) $\alpha \times (\beta + \gamma) = (\alpha \times \beta) + (\alpha \times \gamma)$;

(iv) $\alpha + \mathbf{0} = \alpha = \mathbf{0} + \alpha$;

(v) $\alpha \times \mathbf{0} = \mathbf{0} \times \alpha = \mathbf{0}$;

(vi) $\alpha \times \mathbf{1} = \mathbf{1} \times \alpha = \alpha$;

(vii) $\alpha^+ = \alpha + \mathbf{1}$;

(viii) *if α and β are nite, then so are $\alpha + \beta$ and $\alpha \times \beta$;*

(ix) *if $\alpha^+ = \beta^+$, then $\alpha = \beta$;*

(x) *if α and β are nite ordinals, then $\alpha < \beta$ if and only if there exists a non-zero ordinal γ such that $\alpha + \gamma = \beta$.*

Proof The proofs of (i), (ii), and (iii) are all based on the same principle. For example, for (i), we prove that if $\mathfrak{a} = (a, R)$, $\mathfrak{b} = (b, S)$, and $\mathfrak{c} = (c, T)$ are three ordered sets, then $\mathfrak{a} \oplus (\mathfrak{b} \oplus \mathfrak{c})$ is isomorphic to $(\mathfrak{a} \oplus \mathfrak{b}) \oplus \mathfrak{c}$. Here is the isomorphism f from $\mathfrak{a} \oplus (\mathfrak{b} \oplus \mathfrak{c})$ onto $(\mathfrak{a} \oplus \mathfrak{b}) \oplus \mathfrak{c}$: if $x \in a \uplus (b \uplus c)$, then

- either $x = (y, \mathbf{0})$ with $y \in a$; in this case, set $f(x) = ((y, \mathbf{0}), \mathbf{0})$;
- or $x = ((y, \mathbf{0}), \mathbf{1})$ with $y \in b$; in this case, set $f(x) = ((y, \mathbf{1}), \mathbf{0})$;
- or $x = ((y, \mathbf{1}), \mathbf{1})$ with $y \in c$; in this case, set $f(x) = (y, \mathbf{1})$.

For (ii), with $\mathfrak{a}$, $\mathfrak{b}$, and $\mathfrak{c}$ as above, we prove that $\mathfrak{a} \otimes (\mathfrak{b} \otimes \mathfrak{c})$ is isomorphic to $(\mathfrak{a} \otimes \mathfrak{b}) \otimes \mathfrak{c}$. The isomorphism f is dened as follows:

$$\text{for all } x \in a, \ y \in b, \ \text{and } z \in c, \qquad f((x, (y, z))) = ((x, y), z).$$

For (iii), we need to dene an isomorphism f from $\mathfrak{a} \otimes (\mathfrak{b} \oplus \mathfrak{c})$ onto $(\mathfrak{a} \otimes \mathfrak{b}) \oplus (\mathfrak{a} \otimes \mathfrak{c})$. It is as follows:

- if $x = (y, (z, \mathbf{0}))$ with $y \in a$ and $z \in b$, then $f(x) = ((y, z), \mathbf{0})$;
- if $x = (y, (z, \mathbf{1}))$ with $y \in a$ and $z \in c$, then $f(x) = ((y, z), \mathbf{1})$.

Properties (iv) and (v) are more or less obvious.

Property (vi) results from the isomorphism f from α onto $a \otimes \mathbf{1}$ dened, for $\beta \in \alpha$, by $f(\beta) = (\beta, \mathbf{0})$.

Property (vii) results from the isomorphism f from α^+ onto $\alpha \oplus \mathbf{1}$ dened as follows:

- if $\beta \in \alpha$, $f(\beta) = (\beta, \mathbf{0})$;
- $f(\alpha) = (\emptyset, \mathbf{1})$.

To justify (viii), suppose that α and β are nite ordinals and, for a proof by contradiction, suppose that $\alpha + \beta$ is innite. Then consider the set

$$A = \{x \in \omega : \alpha + x \text{ is innite }\}.$$

This set is not empty (it contains β) so it has a least element that we will call x_0. We see that $x_0 \neq \mathbf{0}$ since $\alpha + \mathbf{0} = \alpha$ and α is nite. Since x_0 is nite, it is not a limit ordinal so there exists an ordinal y_0 such that $y_0^+ = y_0 + \mathbf{1} = x_0$. Then $\alpha + x_0 = (\alpha + y_0) + \mathbf{1} = (\alpha + y_0)^+$ [using (i) and (vii)]; $\alpha + y_0$ is nite (because x_0 is the least element of A) and we have seen, in the comments that follow Denition 7.30, that the successor of a nite ordinal is nite; thus $\alpha + x_0$ is nite. But this contradicts the fact that $x_0 \in A$.

This proves that the sum of two nite ordinals is a nite ordinal. Let us now, again with a proof by contradiction, establish that the product of two nite ordinals is a nite ordinal. Suppose that the set

$$B = \{x \in \omega : \alpha \times x \text{ is innite }\}$$

is not empty and consider its least element, x_0. By (v), we know that $x_0 \neq \emptyset$, so there exists an ordinal y_0 such that $x_0 = y_0 + \mathbf{1}$. From (iii) and (vi), we see that $\alpha \times x_0 = (\alpha \times y_0) + \alpha$. Now $\alpha \times y_0$ is nite because x_0 is the least element of B; also, by hypothesis, α is nite so the sum $(\alpha \times y_0) + \alpha$ is nite by what we just proved. But this contradicts the fact that $x_0 \in B$.

For (ix), it is more or less clear that α is the greatest element of α^+ (and β is the greatest element of β^+). Thus, if $\alpha^+ = \beta^+$, then $\alpha = \beta$.

We are left with the proof of (x). It should be clear that if γ is a non-zero ordinal, $\alpha + \gamma \geq \alpha + \mathbf{1} = \alpha^+ > \alpha$. This proves the implication from right to left. In the other direction, we argue by contradiction: let β be the least nite ordinal such that $\beta > \alpha$ and for which there does not exist an ordinal γ satisfying $\alpha + \gamma = \beta$. Since $\beta > a$, β is not zero and, since it is nite, there exists an ordinal δ such that $\beta = \delta + \mathbf{1}$. We now distinguish two cases:

$\delta > \alpha$: By the minimality of β, there exists an ordinal γ' such that $\delta = \alpha + \gamma'$; from (i), we then have $\beta = \delta + \mathbf{1} = \alpha + (\gamma' + \mathbf{1})$, which is a contradiction.

$\delta \leq \alpha$: It is easy to see that in this case, $\delta = \alpha$, so $\beta = \alpha + \mathbf{1}$, which is again a contradiction. ■

Remark 7.34 The proofs of properties (i) through (vi) depended only on the fact that α and β are ordered sets.

Remark 7.35 It is easy to see that the ordinal sum and ordinal product are not commutative operations. For example, $\mathbf{1}+\omega = \omega < \omega+\mathbf{1}$ and $\mathbf{2}\times\omega = \omega < \omega\times\mathbf{2}$.

This example also illustrates that we must pay attention to directions in the distributivity property (iii): we have $\mathbf{2} \times \omega = (\mathbf{1} + \mathbf{1}) \times \omega \neq \omega + \omega$. It can also be seen that property (x) is true for all ordinals.

Remark 7.36 From now on, we will abandon the notation α^+ and replace it with $\alpha + \mathbf{1}$ as (vii) authorizes us to do.

7.2.4 The integers

We now know enough to construct the integers. Consider the set ω and dene

- the mapping S from ω into ω whose value, at $n \in \omega$, is $n + \mathbf{1}$;
- the mapping $+$ from $\omega \times \omega$ into ω whose value at (n, p) is $n + p$ (we will call this mapping `addition');
- the mapping $\times$ from $\omega \times \omega$ into ω whose value at (n, p) is $n \times p$ (we will call this mapping `multiplication').

The structure so dened has the following properties:

(a) The element $\mathbf{0}$ is an identity element for addition [part (iv) of Theorem 7.33] and $\mathbf{1}$ is an identity element for multiplication [part (vi) of Theorem 7.33].

(b) If $n \in \omega$ and n is not equal to $\mathbf{0}$, then there exists a unique element p of ω such that $n = S(p)$; this is a consequence of the denition of the nite ordinals and of part (ix) of Theorem 7.33.

(c) There does not exist an element $p \in \omega$ such that $\mathbf{0} = S(p)$ (this is a direct consequence of the denitions).

(d) Addition and multiplication are associative [parts (i) and (ii) of Theorem 7.33].

(e) There is a distributive relationship between addition and multiplication [part (iii) of Theorem 7.33].

(f) For all n and p in ω, $n < p$ if and only if there exists a non-zero nite ordinal q such that $n + q = p$ [part (x) of Theorem 7.33].

The structure $\langle \omega, \mathbf{0}, S, +, \times \rangle$ will be denoted by $\mathbb{N}$ and the elements of ω will be called integers. This name is justied by properties (a)ñ(f) and by the fact that every non-empty set of ordinals (of integers, in particular) has a least element. All the theorems of arithmetic can be proved using only these properties. For example, the reader, inspired by the proof found in Chapter 6, can conrm that addition and multiplication are commutative operations.

From this point on, the word `integer' will mean `element of ω'. We have already agreed to write $\mathbf{0}$ instead of $\emptyset$ and $\mathbf{1}$ instead of $\{\emptyset\}$, and this convention is justied by Theorem 7.33. We must continue to carefully distinguish the integers (element of ω) from what we have called the intuitive integers and which we have already frequently used. As a general rule, we need the intuitive integers when we wish to speak about the formal language, for example, about the length of a formula or about the number of free variables it contains.

We should also point out that, from the point of view of the meta-universe, we can associate an element of ω with each intuitive integer by iterating the construction

$\mathbf{0} = \emptyset$, $\mathbf{1} = \{\emptyset\}$, $\mathbf{2} = \{\mathbf{0}, \mathbf{1}\}$, and so on. But there can very well be other elements of ω than these (known as `non-standard' integers).

This process can be continued and we can dene, in $\mathcal{U}$, all the usual mathematical structures: the positive and negative integers $\mathbb{Z}$, the rationals, the real numbers, etc. This process can be observed, for example, in *Number Systems and the Foundations of Analysis*, by Elliott Mendelson, Academic Press, 1973. It is important to retain, from all this, that the structures obtained in this way (the base sets and the operations) are sets, i.e. elements of $\mathcal{U}$.

Nothing prevents us from dening, in $\mathcal{U}$, the notions of nite sequence, of rst-order formula, of structure and of satisfaction of a formula in a structure. We could then express Peano's axioms and observe that $\mathbb{N}$ is a model of these axioms. But we will only exceptionally transcend this stage: unless otherwise mentioned, the words `formula', `model', etc. will retain their intuitive meanings.

7.3 Inductive proofs and denitions

7.3.1 Induction

The ordinals appear as a generalization of the integers. One of the interesting features of this intuition is that proof by induction on the ordinals becomes possible. In fact, we have already used this kind of proof several times without mentioning it. Here is an explanation of the principle.

Let $F[v_0]$ be a formula of L with one free variable and with parameters from U; suppose that, as part of a proof, we wish to show that $F[\alpha]$ is true for every ordinal α. We x an ordinal β and assume (the induction hypothesis) that $F[\gamma]$ is true for every ordinal $\gamma < \beta$. From this hypothesis, and from others that we may have at our disposal, we conclude that $F[\beta]$ is true. The conclusion, from all of this, is that for every ordinal α, $F[\alpha]$ is true. We may express this by the following formula:

$$\forall\alpha(On[\alpha] \Rightarrow (\forall\beta(\beta \in \alpha \Rightarrow F[\beta]) \Rightarrow F[\alpha])) \Rightarrow \forall\alpha(On[\alpha] \Rightarrow F[\alpha]). \qquad (*)$$

The following argument by contradiction justies this principle. Suppose that

$$\forall\alpha(On[\alpha] \Rightarrow (\forall\beta(\beta \in \alpha \Rightarrow F[\beta]) \Rightarrow F[\alpha]))$$

is true. If there is an ordinal for which $F[\alpha]$ is false, then, by Remark 7.24, there is a least such ordinal which we will call α_0. Precisely because it is the least, we know that for every β, $(\beta \in \alpha_0 \Rightarrow F[\beta])$ is true, so by hypothesis, $F[\alpha_0]$ is also true; but this contradicts our choice of α_0.

When we wish to invoke $(*)$, we have to prove $F[\alpha]$ under the assumption that $F[\beta]$ is true for all $\beta \in \alpha$. It often happens that such a proof splits into three cases according as α is equal to $\mathbf{0}$, or is a successor, or is a limit. The fact that we need to consider the case where α is a limit constitutes a novelty compared with proofs by induction on the integers.

We next proceed to denitions by induction and, to explain what we will be doing, let us reconsider the example of denitions by induction of functions from the integers into the integers. To dene such a function f by induction, we have to dene $f(0)$ and provide a way to determine $f(n+1)$ from $f(n)$. If we try to generalize this type of denition for the ordinals, we see that there will be a problem for limit ordinals: how, for example, to dene $f(\omega)$? In Exercise 13 from Chapter 5, we noted that in denitions by induction, we could, in the denition of $f(n+1)$, permit the use of all the values $f(i)$ for $0 \leq i \leq n$ that were already determined. We will take our inspiration from this kind of induction. To dene a function ϕ on the ordinals, it will sufce to know how, for every ordinal α, to dene $\phi(\alpha)$ from the values $\phi(\beta)$ for $\beta \in \alpha$, i.e. from the mapping whose domain is α and which, for every $\beta \in \alpha$, maps β to $\phi(\beta)$.

The function ϕ that we seek to dene is not a mapping since its domain is the class of all ordinals, which is not a set (see the remarks that follow Denition 7.7; also see Remark 7.38 below). It will be dened by a formula $G[v_0, v_1]$ such that

$$\forall v_0(On[v_0] \Rightarrow \exists! v_1 G[v_0, v_1])$$

and, if α is an ordinal, $\phi(\alpha)$ will be the unique set x that satises $G[\alpha, x]$. By contrast, the restriction of this function to an ordinal α is a mapping. To see this, note that the axiom of replacement can be used to dene the set

$$A = \{x : \text{there exists an ordinal } \beta < \alpha \text{ such that } G[\beta, x]\}$$

and the restriction of ϕ to α, which we will denote by $\phi \upharpoonright \alpha$, is equal to

$$\{(\beta, x) \in \alpha \times A : G[\beta, x]\}.$$

We should now explain what we mean by '$\phi(\alpha)$ can be dened from the values $\phi(\beta)$ for $\beta \in \alpha$'. This means that we can set down conditions on the function ϕ that link $\phi(\alpha)$ to the values $\phi(\beta)$ for $\beta < \alpha$ (exactly as in the case for recursive functions) and that are sufciently restrictive that $\phi(\alpha)$ is completely determined if $\phi \upharpoonright \alpha$ is known. These conditions will be expressed by a formula $F[v_0, v_1, v_2]$ such that

$$\begin{aligned}&\forall v_0 \forall v_1((On[v_0] \wedge v_1 \text{ is a mapping whose domain is } v_0)\\ &\quad \Rightarrow \exists! v_2 F[v_0, v_1, v_2]). \qquad (**)\end{aligned}$$

We then want a function ϕ that will satisfy

for every ordinal α, $\phi(a)$ is the unique set x
such that $F[\alpha, \phi \upharpoonright \alpha, x]$. (***)

Theorem 7.37 *Let $F[v_0, v_1, v_2]$ be a formula (possibly with parameters from $\mathcal{U}$) such that*

$$\begin{aligned}&\forall v_0 \forall v_1((On[v_0] \wedge v_1 \textit{ is a mapping whose domain is } v_0)\\ &\quad \Rightarrow \exists! v_2 F[v_0, v_1, v_2]); \qquad (**)\end{aligned}$$

then there exists one and only one function ϕ whose domain is the class of ordinals and is such that

$$\text{for every ordinal } \alpha,\ \phi(a) \text{ is the unique set } x \text{ such that } F[\alpha, \phi \restriction \alpha, x]. \qquad (***)$$

Proof Consider the following condition on a mapping f:

$$\text{the domain of } f \text{ is an ordinal and, for every } \beta \in \mathsf{dom}(f),\ F[\beta, f \restriction \beta, f(\beta)]. \qquad (\text{Ü})$$

It is entirely obvious that if f satises (Ü) and if $\beta \in \mathsf{dom}(f)$, then $f \restriction \beta$ also satises (Ü). Also, for every ordinal α, there can be at most one mapping f with domain α that satises this condition; to see this, we argue by contradiction. Suppose that f and f' are two different mappings, each with domain α, that satisfy (Ü). Let β be the least ordinal such that $f(\beta) \neq f'(\beta)$. Then, by the minimality of β, $f \restriction \beta = f' \restriction \beta$ and we have both $F[\beta, f \restriction \beta, f(\beta)]$ and $F[\beta, f \restriction \beta, f'(\beta)]$; this contradicts (∗∗).

The uniqueness asserted in the theorem follows from this; if ϕ and ϕ' are functions whose domain is the class of ordinals and that both satisfy (∗∗∗) and if α is an ordinal, then $\phi \restriction (\alpha + \mathbf{1})$ and $\phi' \restriction (\alpha + \mathbf{1})$ each satisfy (Ü); so they are equal and $\phi(\alpha) = \phi'(\alpha)$.

Let $G[v_0, v_1]$ be the following formula:

$$On[v_0] \wedge \exists f (f \text{ is a map} \wedge \mathsf{dom}(f) = v_0 + \mathbf{1} \wedge f \text{ satises } (\text{Ü}) \wedge v_1 = f(v_0)).$$

From the uniqueness that we have just proved, it is clear that this formula is functional in v_0. Let ϕ_G denote the partial function that this formula denes.

Suppose for a moment that α is an ordinal and that ϕ_G is dened on α; i.e. for every $\beta < \alpha$, there exists a set x such that $G[\beta, x]$. By the axiom of replacement, the image of α under ϕ_G is a set, A:

$$A = \{y : \text{ there exists an ordinal } \beta \in \alpha \text{ such that } G[\beta, y]\}$$

and the set

$$g = \{(\beta, y) \in \alpha \times A : G[\beta, y]\}$$

is a mapping whose domain is α. Let us prove that this mapping satises (Ü); we have to show that

$$\text{for every ordinal } \beta, \quad \beta < \alpha \text{ implies } F[\beta, g \restriction \beta, g(\beta)].$$

We will once again argue by contradiction: consider the least ordinal $\beta < \alpha$ for which $F[\beta, g \restriction \beta, g(\beta)]$ is false. By the minimality of β, $g \restriction \beta$ satises (Ü). Since $g(\beta)$ satises $G[\beta, g(\beta)]$ (by denition of g), we know that there exists a mapping

f whose domain is $\beta + \mathbf{1}$, which satises (Ü), and is such that $f(\beta) = g(\beta)$. It follows, on the one hand, that $f \restriction \beta = g \restriction \beta$ [because both satisfy (Ü)] and, on the other hand, that $F[\beta, f \restriction \beta, f(\beta)]$ [because f satises (Ü)]. This proves that $F[\beta, g \restriction \beta, g(\beta)]$, which contradicts our original assumption.

We are now in a position to prove by induction that, for every ordinal α, there exists a set x satisfying $G[\alpha, x]$. Suppose that this is true for every $\beta < \alpha$. We have just seen that $g = \phi_G \restriction \alpha$ satises (Ü). Moreover, by (∗∗), there exists a set a such that $F[\alpha, g, a]$. Let

$$f = g \cup \{(\alpha, a)\}.$$

It is clear that f is a mapping whose domain is $\alpha + \mathbf{1}$, which still satises (Ü), and such that $G[\alpha, a]$.

So the formula G denes a function ϕ_G whose domain is the class of all ordinals and it follows from what has been said that ϕ_G satises (∗∗∗). ■

Remark 7.38 At some point, we will need to dene by induction a mapping whose domain is a xed ordinal α (for example, ω for the recursive functions). In this case, condition (∗∗) is replaced by the following weaker condition:

$$\forall v_0 \forall v_1 ((On[v_0] \wedge v_0 < \alpha \wedge v_1 \text{ is a mapping whose domain is } v_0) \\ \Rightarrow \exists! v_2 F[v_0, v_1, v_2]).$$

We can then show that there exists one and only one mapping f whose domain is α and is such that, for every ordinal $\beta < \alpha$, $f(\beta)$ is the unique set x such that $F[\beta, f \restriction \beta, x]$. We may choose to adapt the previous proof or we may apply the previous theorem using the formula

$$F'[v_0, v_1, v_2] = (v_0 < \alpha \Rightarrow F[v_0, v_1, v_2]) \wedge (v_0 \geq \alpha \Rightarrow v_2 \simeq \emptyset).$$

Remark 7.39 In many instances where the above theorem is to be applied, the situation is in fact more complicated. Let us keep the notations from the previous proof. While dening the function $\phi(\alpha)$ by induction, we also prove by induction that the mapping $\phi \restriction \alpha$ satises a certain formula, say $P[v_0]$. The subtle point is that $\phi(\alpha)$ can be dened only if $\phi \restriction \alpha$ satises P; in other words, we are no longer guaranteed that the condition (∗∗) is satised, but only that

$$\forall v_0 \forall v_1 ((On[v_0] \wedge v_0 < \alpha \wedge v_1 \text{ is a mapping whose domain is } v_0 \wedge P[v_1]) \\ \Rightarrow \exists! v_2 F[v_0, v_1, v_2]).$$

Once more, we may reduce this situation to that of the preceding theorem by a pirouette: replace the formula F by

$$F' = (P[v_1] \Rightarrow F[v_0, v_1, v_2]) \wedge (\neg P[v_1] \Rightarrow v_2 \simeq \emptyset).$$

7.3.2 The axiom of choice

First, let us recall the statement of this axiom.

(AC) If $(a_i)_{i \in I}$ is a family of non-empty sets, then $\prod_{i \in I} a_i$ is non-empty.

The axiom of choice is equivalent, in the presence of the axioms of ZF, to other statements that are easier to exploit; we present these now.

Denition 7.40 *An ordered set (X, R) is called* ***inductive*** *if for every subset Y of X, if Y is totally ordered by R, then Y has an upper bound in X.*

It follows from this denition that if (X, R) is inductive, then X is not empty; for the empty set, which is a subset of X that is totally ordered by R, must have an upper bound and this furnishes us with an element of X. In practice, when we wish to show that (X, R) is inductive, it is generally more transparent to show that X is non-empty and that every non-empty subset Y of X that is totally ordered by R has an upper bound in X.

Theorem 7.41 *The following three statements are equivalent:*

(i) *(AC);*

(ii) *if the ordered set (X, R) is inductive, then it has at least one maximal element;*

(iii) *for every set X, there exists a well-ordering of X.*

Properties (ii) and (iii) are therefore theorems of ZFC. Property (ii) is commonly known as **Zorn's lemma** and (iii) as **Zermelo's theorem**.

Proof (i) implies (ii): The argument is by contradiction, so we assume the axiom of choice, on the one hand, and on the other hand that thebe exists an inductive set (X, R) that does not have a maximal element. Consider the set

$$T = \{Y \in \wp(X) : Y \text{ is totally ordered by } R\}.$$

For every element Y of T, there exists an element a in X that is an upper bound for Y; since a is not maximal in X, there exists $b \in X$ such that $a <_R b$. Thus, for every $Y \in T$, the set

$$\{b \in X : \text{ for every } y \in Y,\ y <_R b\}$$

is not empty. So by the axiom of choice, there exists a mapping k from T into X such that

$$\text{for every } Y \in T \text{ and for every } y \in Y,\quad y <_R k(Y).$$

We can now dene by induction on the class of ordinals a function f with values in X such that if α and β are ordinals and $\alpha < \beta$, then $f(\alpha) < f(\beta)$. Since this is the rst occasion we have to use this induction principle, we will proceed with particular care. Let $F[v_0, v_1, v_3]$ be the formula

$$On[v_1] \wedge v_3 \text{ is a mapping whose domain is } v_1$$
$$\wedge ((\mathsf{Im}(v_3) \in T) \Rightarrow v_0 \simeq k(\mathsf{Im}(v_3)) \wedge ((\mathsf{Im}(v_3) \notin T) \Rightarrow v_0 \simeq \emptyset)).$$

This formula is functional and the conditions required by Theorem 7.37 are satised. So we may conclude that there exists a function h which is such that for every ordinal α, $F[h(\alpha), \alpha, h \upharpoonright \alpha]$. At the same time, we can prove by induction that for every ordinal α, the following condition $(\ddot{\mathrm{U}})_\alpha$ is satised:

$$h(\alpha) \in X \quad \text{and} \quad (\beta \in \alpha \quad \text{implies} \quad (\beta) <_R h(\alpha)). \qquad ((\ddot{\mathrm{U}})_\alpha)$$

Suppose that $(\ddot{\mathrm{U}})_\beta$ is true for every $\beta < \alpha$; then $h[\alpha] = \{h(\beta) \colon \beta \in \alpha\}$ is included in X and is totally ordered by R. In other words, $h[\alpha] \in T$ and hence, according to the inductive denition of h, $h(\alpha) \in X$ and $h(\beta) <_R h(\alpha)$ for every $\beta \in \alpha$.

The contradiction in then easy to obtain; let $H[v_0, v_1]$ denote the formula

$$On[v_0] \wedge v_1 \in T \wedge \text{ there exists an isomorphism from } v_0 \text{ onto } v_1.$$

We have seen (Proposition 7.27) that an ordered set cannot be isomorphic to two different ordinals. Thus the formula H is functional in v_0. By the axiom of replacement, the image of T under the function that this formula denes is a set. But we have just seen that every ordinal is isomorphic to an element of T, so this image would be equal to the set of all ordinals. This contradicts Proposition 7.25.

(ii) implies (iii): The proof that follows is characteristic of the way in which Zorn's lemma is generally used.

Let X be a set; we must show that there exists a well-ordering of X.

Consider the set

$$\mathfrak{a} = \{(A, R) \in \wp(X) \times \wp(X \times X) \colon R \text{ is a well-ordering of } A\}$$

and the following binary relation $\mathfrak{s}$ on $\mathfrak{a}$:

$$\{((A_0, R_0), (A_1, R_1)) \in \mathfrak{a} \times \mathfrak{a} \colon$$
$$A_0 \subseteq A_1, \ A_0 \text{ is an initial segment of } (A_1, R_1) \text{ and } R_0 = R_1 \upharpoonright A_0\}.$$

The verication that $\mathfrak{s}$ is an order relation on $\mathfrak{a}$ is immediate. Let us prove that this relation is inductive. First of all, $\mathfrak{a}$ is not empty since it contains $(\emptyset, \emptyset)$. Let $\mathfrak{b}$ be a non-empty subset of $\mathfrak{a}$ that is totally ordered by $\mathfrak{s}$. Set

$$C = \{x \in X : \text{there exists } (A, R) \in \mathfrak{b} \text{ such that } x \in A\}, \quad \text{and}$$
$$T = \{(x, y) \in X \times X : \text{there exists } (A, R) \in \mathfrak{b} \text{ such that } (x, y) \in R\}.$$

Let us prove that T is a well-ordering of C. We begin by verifying that $T \subseteq C \times C$. Suppose that $(x, y) \in T$; then there exists (A, R) belonging to $\mathfrak{b}$ such that $(x, y) \in R$. Since (A, R) must also belong to $\mathfrak{a}$, R is a relation on A so x and y belong to A. From the denition of C, it now follows that they also belong to C.

To prove the transitivity of T, suppose that (x, y) and (y, z) belong to T. We must show that (x, z) also belongs to T. There exist (A_0, R_0) and (A_1, R_1) in $\mathfrak{b}$ such that $x <_{R_0} y$ and $y <_{R_1} z$. But $\mathfrak{b}$ is totally ordered by $\mathfrak{s}$; suppose, for example, that $(A_0, R_0) <_{\mathfrak{s}} (A_1, R_1)$ (the opposite case is treated exactly the same way). This means that $A_0 \subseteq A_1$ and that $R_0 = R_1 \restriction A_0$, which implies that $x <_{R_1} y$. Since R_1 is an order relation, $x <_{R_1} z$.

The fact that T is irreexive is easy to see. If $x \in C$, then for every $(A, R) \in \mathfrak{b}$, (x, x) does not belong to R, and hence (x, x) does not belong to T.

It remains to show that C is well-ordered by T (we have already observed that this implies C is totally ordered by C). Let D be a non-empty subset of C; choose a point d in D. There exists $(A, R) \in \mathfrak{b}$ such that $d \in A$; thus $A \cap D$ is not empty. So it has a least element for R; we will call it a. We will show that a is the least element of D for T. So suppose x is another element of D and that $(B, S) \in \mathfrak{b}$ is such that $x \in B$. Again, we use the fact that $\mathfrak{b}$ is totally ordered by $\mathfrak{s}$. If $(B, S) <_{\mathfrak{s}} (A, R)$, then $B \subseteq A$, $x \in A \cap D$, and, by denition of a, $a <_R x$; thus $a <_T x$. If $(A, R) <_{\mathfrak{s}} (B, S)$, then there are two cases: if $x \in A$, then $x \in A \cap D$ and $a <_R x$; if $x \notin A$, then since A is an initial segment of B for S, $a <_S x$. In both cases, $a <_T x$.

All this establishes that the set $\mathfrak{a}$ ordered by $\mathfrak{s}$ is inductive; so, according to (ii), it has a maximal element. Let (D, W) be such a maximal element. Our proof will conclude by showing that D is equal to the whole of X.

The argument is by contradiction. Choose a point a in $X - D$. Set $D' = D \cup \{a\}$ and extend the relation W to a relation W' on D' by declaring that a is greater than all the elements of D. Then D' is well-ordered by W' (the proof is the same as that for Corollary 7.19) and has D as an initial segment. This shows that $(D', W') >_{\mathfrak{s}} (D, W)$, which contradicts the maximality of (D, W).

(iii) implies (i): Let $(a_i)_{i \in I}$ be a family of non-empty sets. Set

$$a = \bigcup_{i \in I} a_{i.}$$

By property (iii), there exists a well-ordering, say <, of a. Then

$$b = \{(i, x) \in I \times a : x \text{is the least element of } a_i \text{ for the relation } <\}$$

is an element of $\prod_{i \in I} a_i$.

7.4 Cardinality

7.4.1 Cardinal equivalence classes

Denition 7.42 *Let x and y be two sets. We say that x is* ***subpotent to*** *y if there exists an injection from x into y. We say that x and y are* ***equipotent*** *if there exists a bijection from x onto y.*

Consider the formula

$$F[v_0, v_1] = \exists f (f \text{ is an injection from } v_0 \text{ into } v_1).$$

We may not speak of the relation dened by F because, *a priori*, the class of sets (x, y) that satisfy $F[x, y]$ is not a set (we will see, after the fact, that indeed it is not).We may nonetheless observe that the meta-relation dened by this formula is reexive and transitive, i.e. that the formulas

$$\forall v_0 F[v_0, v_0] \quad \text{and} \quad \forall v_0 \forall v_1 \forall v_2((F[v_0, v_1] \wedge F[v_1, v_2]) \Rightarrow F[v_0, v_2])$$

are satised. In the same way, if we consider the formula

$$G[v_0, v_1] = \exists f(f \text{ is a bijection from } v_0 \text{ into } v_1),$$

then the meta-relation dened by G is reexive, symmetric, and transitive. So it has all the properties of an equivalence relation (but it is not a relation). In particular, if x is a set, we may consider the class of elements that are equipotent with x. This will be called the **cardinal class of** x and will be denoted by $\mathsf{card}(x)$. If x is a set and if λ is a cardinal class, we will say that x **has cardinality** λ to mean that x belongs to λ. We can verify that when x is not the empty set, the cardinal class of x is not a set.

We do have the important theorem of **CantorñBernstein**.

Theorem 7.43 *If x is subpotent to y and y is subpotent to x, then x and y are equipotent.*

Proof We are given two sets x and y, an injection f from x into y, and an injection g from y into x. Our task is to construct a bijection h from x onto y.

Since f and g are injections, every element of x has at most one preimage in y under g and each element of y has at most one preimage in x under f. Starting with an arbitrary element $t_0 \in x$, we build a `chain' $t_0, t_1, t_2, \ldots$, whose elements are alternately in x and in y, in such a way that each element in the chain is followed by its unique preimage (under g or f, as the case may be) if this exists. We see that there are three possibilities: either the chain never ends, or it ends with an element of x that does not have a preimage under g (this happens, for example, if t_0 itself does not belong to the image of g; t_1 remains undened), or it ends with an element of y that does not have a preimage under f. From these three possibilities, there results a partition of x into (at most) three subsets, which we will label x_∞, x_x, and x_y, respectively.

If we start with an element $u_0 \in y$, we can construct, in exactly the same way, a chain $u_0, u_1, u_2, \ldots$ such that, for every n, u_{n+1} is the unique preimage of u_n (under f or g, according to the parity of n) if it exists and is undened otherwise. We let y_∞ denote the set of elements of y that generate a chain that does not end, let y_x denote the set of those whose chain ends with an element of x that does not have a preimage under g, and let y_y denote the set of those whose chain ends with an element of y that does not have a predecessor under f.

To do all this formally, we must rst dene, for every set a and for every map ϕ from a into a, the mapping ϕ^n from a into a by induction on the integer $n \in \omega$.

ϕ^0 is the identity mapping on a and, for every $k \in \omega$, ϕ^{k+1} is the composition $\phi \circ \phi^k$. We then set

$$\begin{aligned}
x_x &= \{t \in x : \text{there exists } k \in \omega \text{ and } u \in x - \mathsf{Im}(g) \\
&\qquad \text{such that } t = (g \circ f)^k(u)\}; \\
x_y &= \{t \in x : \text{there exists } k \in \omega \text{ and } v \in y - \mathsf{Im}(f) \\
&\qquad \text{such that } t = g((f \circ g)^k(v))\} \\
x_\infty &= x - (x_x \cup x_y); \\
y_x &= \{u \in y : \text{there exists } k \in \omega \text{ and } t \in x - \mathsf{Im}(g) \\
&\qquad \text{such that } u = f((g \circ f)^k(t))\}; \\
y_y &= \{u \in y : \text{there exists } k \in \omega \text{ and } v \in y - \mathsf{Im}(f) \\
&\qquad \text{such that } u = (f \circ g)^k(v)\}; \\
y_\infty &= y - (y_x \cup y_y).
\end{aligned}$$

The following facts are then clear; the image under f of every element of x_x belongs to y_x; the image under f of every element of x_∞ belongs to y_∞; every element of y_x has a preimage under f that belongs to x_x; every element of y_∞ has a preimage under f that belongs to x_∞; x_y is included in the image of g and the preimage under g of every element of x_y belongs to y_y. It follows from these remarks that the restriction of f to the set $x_\infty \cup x_x$ is a bijection from $x_\infty \cup x_x$ onto $y_\infty \cup y_x$ (denote it by ϕ) and that the restriction of g to y_y is a bijection from y_y onto x_y (denote it by ψ). The map h from x into y dened by $h = \phi \cup \psi^{-1}$ is then a bijection from x onto y. ■

We should point out that any of the sets x_∞, x_x, and x_y may be empty (at the same time as, respectively, y_∞, y_x, and y_y), but that this in no way affects the preceding construction.

Observe that if x, x', y, and y' are sets, if x and x' are equipotent, and if y and y' are equipotent, and if x is subpotent to y, then x' is subpotent to y' (because the composition of two injections is an injection). This allows us to dene an order on cardinal classes: if λ and μ are cardinal classes, then we say that λ **is less than or equal to** μ, and will write $\lambda \leq \mu$, if there exist sets x and y in the classes λ and μ, respectively, such that x is subpotent to y. If λ, μ, and ν are cardinal classes, then

- $\lambda \leq \lambda$;
- $\lambda \leq \mu$ and $\mu \leq \lambda$ implies that $\lambda = \mu$ (this is the CantorñBernstein theorem);
- $\lambda \leq \mu$ and $\mu \leq \nu$ implies that $\lambda \leq \nu$.

We can prove that (Exercise 3) the fact that this ordering is total is equivalent to the axiom of choice.

7.4.2 Operations on cardinal equivalence classes

We now wish to dene a certain number of operations on cardinal classes. To do this, we will require the following proposition.

Proposition 7.44 (1) *If x is equipotent with y and if z is equipotent with t, then $x \uplus z$ is equipotent with $y \uplus t$, $x \times z$ is equipotent with $y \times t$, and x^z is equipotent with y^t.*

(2) *Let x, y, z, and t be sets and suppose that x is subpotent to y and that z is subpotent to t. Then $x \uplus z$ is subpotent to $y \uplus t$, $x \times z$ is subpotent to $y \times t$, and, if y is not empty, x^z is subpotent to y^t.*

Proof Let us begin by proving (2). Let f be an injection from x into y and let g be an injection from z into t. It is easy to verify that the map k, which we will dene in a moment, is an injection from $x \uplus z$ into $y \uplus t$. If $a \in x \uplus z$, then either there exists $b \in x$ such that $a = (b, 0)$ or there exists $c \in z$ such that $a = (c, 1)$. In the rst case, we set $k(a) = (f(b), 0)$; in the second case, we set $k(a) = (g(b), 1)$. This shows that $x \uplus z$ is subpotent to $y \uplus t$.

The proof that $x \times z$ is subpotent to $y \times t$ is analogous; it sufces to verify (this is easy) that the map h from $x \times z$ into $y \times t$ dened by

$$h((a, b)) = (f(a), g(b)) \quad \text{for all } a \in x \text{ and } \text{ for all } b \in t$$

is injective.

The case of exponentiation is slightly more delicate. We will begin by proving a special case; suppose that z is equipotent with t and that g is a bijection from z onto t. Recall that x^z is the set of all mappings from z into x and that y^t is the set of all mappings from t into y. If $h \in x^z$, set $i_h = f \circ h \circ g^{-1}$; i_h is therefore a map from t into y. We will show that the map from x^z into y^t whose value at h is i_h is injective. To do this, we suppose that h and k are two distinct elements of x^z and show that i_h is different from i_k. There exists an element $a \in z$ such that $h(a) \neq k(a)$. Set $b = g(a)$. Then $i_h(b) = f(h(a))$ and $i_k(b) = f(k(a))$. Since f is injective and since $h(a) \neq k(a)$, we have $i_h(b) \neq i_k(b)$.

Let us now turn to the general case and suppose that g is only an injection. Let u denote the image of z under g. Then g is a bijection from z onto u and we have just seen that this implies that x^z is subpotent to y^u. Let us next prove that y^u is subpotent to y^t; choose an element c in y (y is not empty). If h is a mapping from u into y, we can dene a mapping j_h from t into y as follows:

$$\text{if } a \in u, \quad \text{then } j_h(a) = h(a);$$
$$\text{if } a \in t - u, \quad \text{then } j_h(a) = c.$$

It is clear that if $j_h = j_k$, then $h = k$ so we do have an injective map from y^u into y^t.

Note that with our denitions, $0^0 = 1$ whereas $0^1 = 0$; so the hypothesis that y is not empty is essential.

(1) We could repeat the previous proofs, taking into consideration that if the maps f and g are bijective, then the various maps from $x \uplus z$ into $y \uplus t$, from $x \times z$ into $y \times t$, and from x^y into y^t are also bijective. Lazy readers could also deduce from (2) that $x \uplus z$ is subpotent to $y \uplus t$ and that $y \uplus t$ is subpotent to $x \uplus z$ and then invoke the CantorñBernstein theorem. [This same reasoning would also apply for the product and for exponentiation, though for exponentiation the (trivial) case in which y is empty would have to be treated separately.] ■

All of the above justies the following denition.

Denition 7.45 *Let λ and μ be two cardinal classes; then $\lambda + \mu$, $\lambda \times \mu$, and λ^μ are, respectively, the cardinal classes of $x \uplus y$, $x \times y$, and x^y, where x is a set of cardinality λ and y is a set of cardinality μ.*

The next brief remark will make our life much simpler.

Remark 7.46 Let x and y be two sets. We can dene an injection f from $x \cup y$ into $x \uplus y$; let $a \in x \cup y$. If $a \in x$, set $f(a) = (a, 0)$; otherwise, $a \in y$ and we set $f(a) = (a, 1)$. This shows that

$$\mathsf{card}(x \cup y) \leq \mathsf{card}(x) + \mathsf{card}(y).$$

If x and y are disjoint, then f is a bijection from $x \cup y$ onto $x \uplus y$ and

$$\mathsf{card}(x \cup y) = \mathsf{card}(x) + \mathsf{card}(y).$$

The reader can have fun proving that, in the general case,

$$\mathsf{card}(x \cup y) + \mathsf{card}(x \cap y) = \mathsf{card}(x) + \mathsf{card}(y).$$

Here is another consequence of Proposition 7.44.

Corollary 7.47 *Let λ, μ, ν, and κ be cardinal classes and assume that $\lambda \leq \mu$ and that $\nu \leq \kappa$. Then $\lambda + \nu \leq \mu + \kappa$, $\lambda \times \nu \leq \mu \times \kappa$ and, if μ is not equal to* **0**, *$\lambda^\mu \leq \mu^\kappa$.*

Caution: This becomes false if we replace $\leq$ by the strict inequality $<$ (see Remark 7.73 below and Exercise 14).

Proposition 7.48 *Let λ, μ, and ν be cardinal classes. Then*

(1) $\lambda + (\mu + \nu) = (\lambda + \mu) + \nu$;

(2) $\lambda + \mu = \mu + \lambda$;

(3) $\lambda \times (\mu \times \nu) = (\lambda \times \mu) \times \nu$;

(4) $\lambda \times \mu = \mu \times \lambda$;

(5) $\lambda \times (\mu + \nu) = (\lambda \times \mu) + (\lambda \times \nu)$;

(6) $\lambda^\nu \times \mu^\nu = (\lambda \times \mu)^\nu$;

(7) $\lambda^{\mu+\nu} = \lambda^\mu \times \lambda^\nu$;

(8) $(\lambda^\mu)^\nu = \lambda^{\mu\times\nu}$.

Proof Let x, y, and z be sets whose respective cardinalities are λ, μ, and ν and assume that these three sets are pairwise disjoint.

(1) We have seen that $x \cup (y \cup z) = (x \cup y) \cup z$. Now the cardinality of the rst of these sets is $\lambda + (\mu + \nu)$ and that of the second set is $(\lambda + \mu) + \nu$ (according to Remark 7.46 since the sets x, y, and z are pairwise disjoint).

(2) Similarly, $x \cup y$, whose cardinality is $\lambda+\mu$, is equal to $y \cup x$, whose cardinality is $\mu + \lambda$.

(3) If, for all $a \in x$, $b \in y$, and $c \in z$, we dene $f((a, (b, c))) = ((a, b), c)$, then f is a bijection from $x \times (y \times z)$ onto $(x \times y) \times z$.

(4) If, for all $a \in x$ and $b \in y$, we dene $f((a, b)) - (b, a)$, then f is a bijection from $x \times y$ onto $y \times x$.

(5) One rst checks that $x \times y$ and $x \times z$ are disjoint sets; then one veries that the set $x \times (y \cup z)$ [whose cardinality is $\lambda \times (\mu + \nu)$] is equal to the set $(x \times y) \cup (x \times z)$ [whose cardinality is $(\lambda \times \mu) + (\lambda \times \nu)$].

(6) Let h be a map from z into x and let k be a map from z into y. Dene the map $i_{h,k}$ from z into $x \times y$ by setting $i_{h,k}(c) = (h(c), k(c))$ for all $c \in z$. It is easy to check that the map that sends $(h, k) \in x^z \times y^z$ to $i_{h,k}$ is a bijection from $x^z \times y^z$ onto $(x \times y)^z$.

(7) Let h be a mapping from y into x and let k be a mapping from z into x. Since y and z are disjoint, $h \cup k$ is a map from $y \cup z$ into x and it is easy to see that the mapping that sends (h, k) to $h \cup k$ is a bijection from $x^y \times z^y$ onto $x^{y\cup z}$.

(8) Let h be a map from $y \times z$ into x. For every element c of z, we can dene a map h_c from y into x by setting $h_c(b) = h(b, c)$ for all $b \in y$; this allows us to dene a map i_h from z into x^y by setting $i_h(c) = h_c$ for all $c \in z$. Finally, one can verify that the map that sends h to i_h is a bijection from $x^{y\times z}$ onto $(x^y)^z$. ■

Before going further, we should observe that, for every set x, $\wp(x)$ is equipotent with 2^x. To see this, we introduce the notion of characteristic function of a subset of x: if y is a subset of x, the **characteristic function** χ_y of y is the map from x into 2 dened for all $a \in x$ by

$$\chi_y(a) = \begin{cases} 1 & \text{if } a \in y; \\ 0 & \text{if } a \notin y. \end{cases}$$

The map whose value at y is χ_y is then a bijection from $\wp(x)$ onto 2^x.

The next theorem, known as **Cantor's theorem**, is important for it proves that there does not exist a greatest cardinality.

Theorem 7.49 *For every cardinal class λ, we have $2^\lambda > \lambda$.*

Proof Let x be a set of cardinality λ. It is easy to dene an injection from x into $\wp(x)$; for every $a \in x$, set $f(a) = \{a\}$. This shows that $\lambda \leq 2^\lambda$. We will use a diagonal argument to show that there does not exist a surjection from x onto $\wp(x)$. Let h be a map from x into $\wp(x)$. Consider

$$y = \{a \in x : a \notin h(a)\}.$$

Then y is an element of $\wp(x)$ and we will show that h is not surjective by showing that y does not belong to the image of h. The argument will be by contradiction; assume that $y = h(b)$ where $b \in x$. If $b \in y$, then, by denition of y, $b \notin h(b)$; but this is not possible since $h(b) = y$. If, on the other hand, $b \notin y$, then, still by denition of y, $b \in h(b) = y$, which is also a contradiction. ■

7.4.3 The nite cardinals

Denition 7.50 *A set is called* ***nite*** *if it is equipotent with an integer. An* ***innite*** *set is one that is not nite.*

Recall that an integer n is equal to the set of integers that are less than n. We have previously introduced the notion of nite ordinal (in Denition 7.30). Before doing anything else, we should verify that these two denitions agree. It is obvious that a nite ordinal (i.e. an integer) is equipotent with itself; so it is nite in the sense of the denition just given. The converse (i.e. that an innite ordinal in the sense of Denition 7.30 is an innite set in the sense of the denition just given) will follow from Corollary 7.52 below.

Theorem 7.51 *Let n be an integer and let f be a mapping from n into n. Then*

(1) *if f is injective, then f is bijective;*

(2) *if f is surjective, then f is bijective.*

Proof (1) By induction on n. If $n = \mathbf{0}$, this is clear since the only mapping from n into n is surjective. We suppose next that the result is true for n and we prove it for $n + 1$. Let f be an injection from $n + 1$ into $n + 1$. Consider the bijection h from $n + 1$ into $n + 1$ dened by

$$\begin{aligned} h(p) &= p \quad \text{if } p \neq n \text{ and if } p \neq f(n); \\ h(n) &= f(n); \\ h(f(n)) &= n. \end{aligned}$$

[This denition makes sense even if $f(n)$ is equal to n.] Then $g = h \circ f$ is also an injection from $n + 1$ into $n + 1$; moreover, $g(n) = n$ and, consequently, $g \restriction n$ is an injection from n into n. By the induction hypothesis, $g \restriction n$ is a bijection from n onto n. Thus, every integer less than n is in the image of g, as is n itself [since $g(n) = n$]. Thus g is a bijection from $n + 1$ onto $n + 1$; so f, which is equal to $h^{-1} \circ g$, is a bijection.

(2) Let f be a surjective map from n onto n. Let h be the map dened for all $p \in n$ by setting $h(p)$ equal to the least integer k such that $f(k) = p$. Then $f \circ h$ is

the identity mapping on n; this implies that h is injective. So by (1), h is a bijection from n onto n. Therefore, $f = f \circ h \circ h^{-1} = h^{-1}$, so f is also a bijection. ■

Remark In the section on cardinals, we will see a generalization of the argument used above for (2).

Corollary 7.52 *Let α be an ordinal, let n be an integer, and suppose that $\alpha > n$; then there does not exist an injective map from α into n.*

Proof We argue by contradiction. Assume that f is an injective map from α into n. Then $f \restriction n$ is also injective, so by the theorem, f is surjective onto n. Now if β belongs to α but not to n (such an element exists since $\alpha > n$), $f(\beta)$ belongs to the image of n under f; this contradicts the fact that f is injective. ■

Corollary 7.53 *If x is a nite set, then every injective (or surjective) map from x into x is bijective.*

Proof Because x is nite, there exists a nite ordinal α and a bijection f from x onto α. Let h be an injective map from x into x. Then $k = f \circ h \circ f^{-1}$ is an injective map from α into α; so by Theorem 7.51, it is a bijection. So also is h, which is equal to $f^{-1} \circ k \circ f$. The case in which h is surjective is treated the same way. ■

Proposition 7.54 *If x is nite, there exists one and only one integer that is equipotent with x.*

Proof There exists at least one such integer by denition. If x is equipotent to two integers m and n, then m and n are equipotent with each other and Corollary 7.52 shows that both $m > n$ and $m < n$ are impossible. So m and n must be equal. ■

Proposition 7.55 *If x is nite and if y is subpotent to x, then y is nite.*

Proof Let g be an injection from y into x and let f be a bijection from x onto an integer n. Let z denote the image of y under $f \circ g$. Then $f \circ g$ is a bijection from y onto z, so it sufces to show that z is nite. The set z, with the order relation induced on it by the order on n (n is an ordinal), is well-ordered, so it is isomorphic to an ordinal α (Proposition 7.27) that is necessarily less than or equal to n (This can be deduced from Lemma 7.29); so it is nite. ■

Proposition 7.56 *If there exists a surjective map f from a nite set x onto a set y, then y is nite.*

Proof There exists an integer n and a bijection h from n onto x; so $f \circ h$ is a surjective map from n onto y. We can then dene an injective map k from y into n: for $a \in y$, $k(a)$ is the least integer $m < n$ such that $(f \circ h)(m) = a$. Now apply Proposition 7.55. ■

Proposition 7.57

(1) *The union of two nite sets is a nite set.*
(2) *The product of two nite sets is a nite set.*
(3) *The union and the product of a nite family of nite sets are nite sets.*
(4) *If a and b are nite sets, then a^b is a nite set.*
(5) *If A is a nite set of nite ordinals (i.e. of integers), then* $\sup A$ *is a nite ordinal (i.e. an integer).*

Proof Assume that the nite sets a and b are equipotent with the integers n and m, respectively.

(1) We have seen (Remark 7.46) that $\mathsf{card}(a \cup b) \leq \mathsf{card}(a \uplus b)$ and (Proposition 7.44) that $\mathsf{card}(a \uplus b) = \mathsf{card}(n) + \mathsf{card}(m) = \mathsf{card}(n \uplus m)$. Also, according to item (viii) from Theorem 7.33, $n \uplus m$ is equipotent with an integer. Now apply Proposition 7.55.

(2) The situation for the product is even simpler; $\mathsf{card}(a \times b) = \mathsf{card}(n \times m)$ (Proposition 7.44) and $n \times m$ is nite by Theorem 7.33.

(3) Let I be a nite set of cardinality p and, for every $i \in I$, let a_i be a nite set of cardinality n_i. We will prove by induction on p that $\bigcup_{i \in I} a_i$ and $\prod_{i \in I} a_i$ are nite sets. This is obvious if $p = 0$. If $p \neq 0$, let j be an element of I and set $J = I - \{j\}$. The cardinality of J is $p - 1$ (this is easy to check), so by the induction hypothesis $\bigcup_{i \in J} a_i$ and $\prod_{i \in J} a_i$ are nite sets. The conclusion now follows from (1) and (2) once we have observed that

$$\bigcup_{i \in I} a_i = \left(\bigcup_{i \in J} a_i\right) \cup a_j \quad \text{and} \quad \mathsf{card}\left(\prod_{i \in I} a_i\right) = \mathsf{card}\left(\prod_{i \in J} a_i\right) \times \mathsf{card}(a_j).$$

(4) When we refer to Denition 7.7, we see that $a^b = \prod_{x \in b} a$; so (4) follows immediately from (3).

(5) follows from (3) and Remark 7.28. ■

If λ is the cardinal class of a nite set, then λ contains a canonical representative: the unique integer that belongs to λ. We will allow the abuse of language that identies n and $\mathsf{card}(n)$. This can sometimes be inconvenient (for example, does $n \times m$ denote the product of sets or the product of cardinals?), but, in principle, it will be clear from the context.

Later, for an arbitrary cardinal class, we will need to determine a canonical representative.

7.4.4 Countable sets

Denition 7.58 *A set is called* ***denumerable*** *if it is equipotent with ω. A set will be called* ***countable*** *if it is either nite or denumerable.*

The cardinal class of ω (and hence of every denumerable set) will be denoted by $\aleph_0$ (read as `aleph-zero'; aleph is the rst letter of the Hebrew alphabet). So

$\aleph_0$ is strictly greater than every nite cardinal. Note that if α is an ordinal, then its cardinality is either nite or greater than or equal to $\aleph_0$. This property extends, of course, to every well-ordered set.

Proposition 7.59 *If x is denumerable and if y is subpotent to x, then y is countable.*

Proof We imitate the proof of Proposition 7.55. y is equipotent to a subset z of ω. Because it is naturally well-ordered, z is itself equipotent to an ordinal α which, by Remark 7.28, is less than or equal to ω. If α is an integer, then y is nite. If not, then α is equal to ω and y is denumerable. ■

Remark 7.60 The analogue of Corollary 7.52 is false for denumerable sets; the mapping f from ω into ω dened by $f(n) = n+\mathbf{1}$ is injective, but it is not bijective since its image does not contain $\mathbf{0}$. This is the case for any innite ordinal α; by dening $f(\beta) = \beta$ for all $\beta \geq \omega$, we can extend f to a map from α into itself that is injective but not surjective. The next theorem illustrates that situation is in fact much worse.

Theorem 7.61

(1) *The union of two countable sets is countable.*
(2) *The product of two countable sets is countable.*
(3) *If X is countable, then $S = \bigcup_{n\in\omega} X^n$ is countable.*

Note: (3) Asserts that the set of nite sequences of elements from a countable set is countable.

Proof (1) Let X and Y be two countable sets. So there exist injections f and g from X and Y, respectively, into ω. To show that $X \cup Y$ is countable, we will construct an injection h from $X \cup Y$ into ω. Let $z \in X \cup Y$; if $z \in X$, we set $h(z) = \mathbf{2}f(z)$; if not, then $z \in Y$ and we set $h(z) = \mathbf{2}g(z) + \mathbf{1}$.

(2) Once again, let X and Y be two countable sets and let f and g be injections from X and Y, respectively, into ω. Here is an injection h from $X \times Y$ into ω: if $(x, y) \in X \times Y$, then $h((x, y)) = \alpha_2(f(x), g(y))$, where α_2 is the bijection from Denition 5.5.

(3) Let X be a countable set and let f be an injection from X into ω. We dene a map h from S into $\bigcup_{n\in\omega} \omega^n$ as follows: if $a \in S$, there is an integer n such that a is a map from n into X; so we set $h(a) = f \circ a$ [intuitively, if $a = (x_1, x_2, \ldots, x_n)$, then $h(a) = (f(x_1), f(x_2), \ldots, f(x_n))$]. It is easy to verify that h is an injection from S into $\bigcup_{n\in\omega} \omega^n$. In addition, the map Ω from Denition 5.5 is an injection from $\bigcup_{n\in\omega} \omega^n$ into ω. So the composition $\Omega \circ h$ is an injection from S into ω. ■

Remark The proof that we have just given uses some elementary facts of arithmetic. With all that we know about $\mathbb{N}$, these facts are easily proved.

Proposition 7.56 generalizes, with the same proof.

Proposition 7.62 *If there exists a surjective map from a denumerable set x onto a set y, then y is countable.*

Corollary 7.63 *We have*

(1) $\aleph_0 + \aleph_0 = \aleph_0$;

(2) $\aleph_0 \times \aleph_0 = \aleph_0$;

(3) *for every integer n different from zero,* $\aleph_0^n = \aleph_0$.

Proof We have just seen that $\aleph_0 + \aleph_0$, $\aleph_0 \times \aleph_0$, and $\aleph_0^n$ are at most denumerable. So it sufces to note that these are not nite cardinals (which is rather obvious). ■

As an example of how the notion of cardinality can be applied, we will prove that there are real numbers that are not algebraic. There are, incidentally, purely algebraic proofs of this fact, but the proof that follows will show that `most' real numbers are not algebraic. (As a consequence, for example, the set of algebraic numbers has Lebesgue measure zero.) Recall that a number is **algebraic** if it is the root of a non-zero polynomial whose coefcients are from $\mathbb{Z}$. The strategy of the proof is simple: we rst show that the cardinality of the set $\mathbb{R}$ of real numbers is $2^{\aleph_0}$ and, second, that the set of algebraic numbers is denumerable.

Proposition 7.64 $\mathsf{card}(\mathbb{R}) = 2^{\aleph_0}$.

Proof Consider the following function α from 2^ω into $\mathbb{R}$ dened for $h \in 2^\omega$ by

$$\alpha(h) = \sum_{n=0}^{\infty} \frac{h(n)}{2^{n+1}}.$$

In other words, the real number $0.h(0)h(1)\ldots h(n)\ldots$ is the binary expansion of $\alpha(h)$. The image of α is the interval $[0, 1]$, but α is not injective because, for example, $0.1000\ldots$ and $0.0111\ldots$ represent the same real number. Nonetheless, this phenomenon is rather limited. Let us say that an element $h \in 2^\omega$ is **eventually zero** if there exists an integer n such that $h(p) = 0$ for all $p > n$. Set

$$S = \{h \in 2^\omega \colon h \text{ is not eventually zero}\}.$$

We will assume the reader knows the fact that $\alpha \restriction S$ is a bijection from S onto $(0, 1]$. Let us dene a new map β from 2^ω into $\mathbb{R}$ by setting

$$\beta(h) = \begin{cases} 1 + \alpha(h) & \text{if } h \in S; \\ \alpha(h) & \text{otherwise} \end{cases}$$

It is not difcult to see that β is injective; this shows that $\mathsf{card}(\mathbb{R}) \geq 2^{\aleph_0}$. Also, the map f dened by

$$f(x) = \frac{1}{\pi}\text{Arctan}(x) + \frac{1}{2}$$

is a bijection from $\mathbb{R}$ onto (0,1). So we have

$$\text{card}(\mathbb{R}) = \text{card}((0,1)) \leq \text{card}((0,1]) = \text{card}(S) \leq 2^{\aleph_0}.$$

It then follows from the CantorñBernstein theorem that card$(\mathbb{R}) = 2^{\aleph_0}$. ■

The fact that card$(\mathbb{R}) = 2^{\aleph_0}$ is often expressed by saying that `$\mathbb{R}$ has the **power of the continuum**'.

Proposition 7.65 *The set A of real numbers that are algebraic is denumerable.*

Proof The set $\mathbb{Z}$ is the union of two denumerable sets (the non-negative integers and the negative integers) so it is denumerable. The set S of nite sequences of elements of $\mathbb{Z}$ is also denumerable [item (3) from Theorem 7.61]. For every $s = (a_0, a_1, \ldots, a_n) \in S$, set

$$Z(s) = \begin{cases} \emptyset & \text{if for all } i \text{ from 0 to } n \text{ inclusive, } a_i = 0; \\ \{x \in \mathbb{R} : a_0 + a_1 x + \cdots + a_n x^n = 0\} & \text{otherwise.} \end{cases}$$

There exists a surjective map from S onto the set $\{Z(s) \colon s \in S\}$, so this set is denumerable by Proposition 7.62. Moreover, for every $s \in S$, the set $Z(s)$ is nite; also, A is the union of the family $(Z(s) \colon s \in S)$. Let f be a bijection from ω onto S. We dene a map g from $\omega \times \omega$ into A as follows:

- if $p \neq 0$ and $Z(f(n))$ has at least p elements, then $g(n, p)$ is the pth element of $Z(f(n))$ (under the ordering induced by that of $\mathbb{R}$);
- otherwise, $g(n, p) = 0$.

We then observe that g is surjective. Because $\omega \times \omega$ is denumerable, it follows from Proposition 7.62 that A is denumerable. ■

Exercise 23 also proves that $\mathbb{R}$ is not denumerable using a direct diagonal argument.

7.4.5 The cardinal numbers

From here on, certain theorems will require the axiom of choice. We will indicate this by placing (AC) at the beginning of the statement.

Denition 7.66 *An ordinal that is not equipotent to a strictly smaller ordinal will be called a **cardinal**. Such an ordinal is sometimes also called an **initial ordinal**.*

For example, the nite ordinals are cardinals (Corollary 7.52); ω is also a cardinal. By contrast, $\omega + \mathbf{1}$, $\omega + \omega$, and $\omega \times \omega$ are not cardinals (Corollary 7.63). An innite cardinal must be a limit ordinal; for if α is an innite ordinal, then the map f, dened as follows, is a bijection from $\alpha + \mathbf{1}$ onto α: ■

$$\begin{aligned} f(\beta) &= \beta + \mathbf{1} && \text{if } \beta \in \omega; \\ f(\beta) &= \beta && \text{if } \omega \leq \beta < \alpha; \\ f(\alpha) &= \mathbf{0}. \end{aligned}$$

Let α be an ordinal; it is clear that the class of ordinals β such that α is subpotent to β is not empty (it contains α) and that the least element of this class is a cardinal. More generally, let x be a set and suppose that there exists an ordinal α such that x is subpotent to κ. Then there exists one and only one cardinal κ such that x is equipotent to α. Uniquenessis obvious since two distinct cardinals cannot be equipotent. Now let α be the least ordinal such that x is subpotent to α. We see that this ordinal is necessarily a cardinal; we call this the **cardinality of** x. It follows from these denitions that the cardinality of an ordinal α is an ordinal that is less than or equal to α.

If α and β are cardinals, then $\alpha > \beta$ is equivalent to $\mathsf{card}(\alpha) > \mathsf{card}(\beta)$. This is no longer true if α and β are arbitrary ordinals (for example, take $\alpha = \omega + \mathbf{1}$ and $\beta = \omega$).

Using Proposition 7.27, we can rephrase Zermelo's theorem in the following way:

Theorem 7.67 (AC) *Every set is equipotent with some ordinal.*

For the remainder of this paragraph, we assume that the axiom of choice is satised. Under this assumption, we see that every set is equipotent with a cardinal, i.e. that every set has a cardinality. This implies, incidentally, that the relation $\leq$ on cardinal classes is total (see Exercise 4, in which the converse of this is also proved). We see that for the cardinal class of a set x, we can advantageously substitute the cardinality of the set, which is, in a way, a canonical representative of this class. We will allow the abuse of language that consists in not clearly distinguishing the cardinal class of a set X (which, we recall, is not a set) from the cardinality of the set X (which is an ordinal). This does present some inconveniences (the same ones that result from identifying the nite cardinals and the nite ordinals, as explained in the last paragraph of the earlier section devoted to the nite cardinals). Any ambiguities that might appear are generally resolved by the context: when we are computing a cardinality, then it is cardinal arithmetic that should be used; if it is a calculation with ordinals (this happens more rarely), ordinal arithmetic should be used. In any case, we will always be precise if there is the slightest chance of doubt.

Since we are dealing here with consequences of the axiom of choice, we offer a very useful proposition.

Proposition 7.68 (AC) *Suppose there exists a surjective mapping f from a set a onto a set b. Then* $\mathsf{card}(b) \leq \mathsf{card}(a)$.

Proof Invoke Zermelo's theorem to produce a well-ordering R of a. We can then dene an injection h from b into a by setting $h(x) =$ the least element (with respect to R) of the set $\{y \in a\colon f(y) = x\}$; this denition is legitimate because the fact that, f is surjective implies that, for every $x \in b$, this set is non-empty. ■

The class of cardinals does not have a greatest element.

Theorem 7.69 *For every cardinal α, there exists an ordinal β such that* $\mathsf{card}(\beta) > \alpha$.

Proof First observe that with the axiom of choice, this theorem is obvious; it sufces to consider the cardinality of 2^α, which we already know is strictly greater than α (Theorem 7.49).

Without the axiom of choice, the proof is a bit more difcult. We will prove that the class of ordinals subpotent to α is a set. Consider the set

$$R = \{(X, r) \in \wp(\alpha) \times \wp(\alpha \times \alpha) : X \subseteq \alpha \text{ and } r \text{ is a well-ordering of } X\}.$$

We have seen (Proposition 7.27) that, for every $(X, r) \in R$, there exists one and only one ordinal β such that $(\beta, \in)$ is isomorphic to (X, r). By the axiom of replacement, the image of R under the function whose value for $(X, r) \in R$ is the ordinal β such that $(\beta, \in)$ is isomorphic to (X, r) is a set. This image is precisely the class of ordinals that are subpotent to α; for if $(\beta, \in)$ is isomorphic to $(X, r) \in R$, there is certainly a bijection from β onto X, which is an injection from β into α; conversely, if f is an injection from β into α, then f is a bijection from β onto its image that we will call X, and

$$r = \{(x, y) \in X \times X : f^{-1}(x) \in f^{-1}(y)\}$$

is a well-ordering of X that is isomorphic (via f) to $(\beta, \in)$.

It is clear that the set $\{\beta : \beta \text{ is an ordinal that is subpotent to } \alpha\}$ is transitive; it is therefore an ordinal, γ (Proposition 7.17). Observe, however, that γ is not subpotent to α (otherwise it would be an element of itself, which is not possible). It is consequently the least ordinal whose cardinality is strictly greater than that of α, i.e. γ is a cardinal that is strictly greater than α. ∎

If α is a cardinal, the least cardinal greater than α will be called the **cardinal successor of** α and will be denoted by α^+. (We used a similar notation earlier but have abandoned it: the successor of an ordinal α is now denoted by $\alpha + \mathbf{1}$.)

Also, as shown in the proposition that follows, the least upper bound of a set of cardinals is a cardinal.

Proposition 7.70 *If A is a set of cardinals, then* $\sup A$ *is a cardinal.*

Proof Set $\alpha = \sup A$. If $A = 0$ or $A = \{0\}$, then $\sup A = 0$, which is a cardinal. In all other cases, let β be a cardinal that is strictly less than α. By the denition of least upper bound, β is not an upper bound of A, so there exists an ordinal γ in A such that $\gamma > \beta$. Since γ belongs to A, we have $\alpha \geq \gamma$ and $\text{card}(\alpha) \geq \text{card}(\gamma) = \gamma > \beta$. The cardinality of α is therefore strictly greater than every ordinal that is strictly less than α; this shows that α is a cardinal. ∎

In particular, this proves that the class of cardinals is not a set; otherwise its least upper bound would be a maximum element, contradicting Theorem 7.69.

Next, we will dene a strictly increasing function from the class of ordinals into the class of innite cardinals. The Hebrew letter aleph, $\aleph$, is generally used to denote this function.

- $\aleph_0 = \omega$ (the smallest innite cardinal).

- If α is a successor ordinal, say $\alpha = \beta + \mathbf{1}$, then $\aleph_\alpha = \aleph_\beta^+$.
- If α is a limit ordinal, then $\aleph_\alpha = \sup\{\aleph_\beta : \beta < \alpha\}$ (this is a cardinal according to Proposition 7.70).

The fact that $\aleph$ is strictly increasing (hence injective) is obvious from the deni-tion. In particular, this implies that, for every ordinal α, $\aleph_\alpha \geq \alpha$ (see Lemma 7.29). It is clear that, for every ordinal α, $\aleph_\alpha$ is an innite cardinal.

We can also prove that for every innite cardinal λ, there exists an ordinal α such that $\aleph_\alpha = \lambda$, i.e. that the function $\aleph$ is surjective. To do this, observe that because $\aleph_{\lambda+\mathbf{1}} > \lambda$, there exists a least ordinal α such that $\aleph_\alpha > \lambda$. Note that α cannot equal $\mathbf{0}$; nor can α be a limit ordinal, otherwise, by denition of $\aleph_\alpha$, there would exist an ordinal $\gamma < \alpha$ such that $\aleph_\gamma > \lambda$, contradicting the minimality of α. So there must exist an ordinal β such that $\alpha = \beta + \mathbf{1}$ and

$$\aleph_\beta \leq \lambda < \aleph_{\beta+\mathbf{1}} = \aleph_\alpha.$$

The cardinality of λ is therefore at most $\aleph_\beta$ and, since it is a cardinal, $\lambda = \aleph_\beta$. So we see that the function $\aleph$ is a strictly increasing bijection from the class of ordinals onto the class of innite cardinals.

Suppose temporarily that the axiom of choice is satised and let us re-examine the proofs of Theorem 7.69. Given a cardinal λ, each of the two proofs supplied a cardinal that is strictly greater: from the rst proof we have the cardinality of 2^λ (which is abusively denoted by 2^λ), and from the second proof we have the cardinal successor of λ, denoted by λ^+. It is clear that $\lambda^+ \leq 2^\lambda$. But are these two cardinals equal? This question remains undecided by the axioms of ZF alone (this phrase may seem cryptic but it will be claried later in this chapter). The following property is known as the **generalized continuum hypothesis** (abbreviated as GCH).

$$\text{For every ordinal } \alpha, \quad \aleph_{\alpha+\mathbf{1}} = 2^{\aleph_\alpha}. \tag{GCH}$$

The adjective `generalized' refers to the fact that the special case in which $\alpha = \mathbf{0}$ is of particular importance and is known as the **continuum hypothesis** (CH).

$$2^{\aleph_\mathbf{0}} = \aleph_\mathbf{1}. \tag{CH}$$

The continuum hypothesis is equivalent to the following assertion: every innite subset of $\mathbb{R}$ is either equipotent to $\mathbb{N}$ or equipotent to $\mathbb{R}$. Recall that the word `continuum' usually refers to $\mathbb{R}$ or to its cardinality.

We will now show that in the presence of the axiom of choice, the operations of addition and multiplication of innite cardinals are not very interesting.

Theorem 7.71 (*AC*) *For every innite cardinal* λ,

(1) $\lambda \uplus \lambda$ *is equipotent with* λ;

(2) $\lambda \times \lambda$ *is equipotent with* λ.

Proof There is an obvious bijection between $\lambda \uplus \lambda$ and $\lambda \times \mathbf{2}$. Also,

$$\mathsf{card}(\lambda \times \lambda) \geq \mathsf{card}(\lambda \times \mathbf{2})$$

(Corollary 7.47). So if we have $\mathsf{card}(\lambda \times \lambda) = \mathsf{card}(\lambda)$, we must also have $\mathsf{card}(\lambda \uplus \lambda) = \mathsf{card}(\lambda)$. Thus, it sufces to prove (2).

The proof is by induction; assume that, for every $\mu < \lambda$, $\mu \times \mu$ is equipotent with μ. Dene an order relation $\leq_R$ on $\lambda \times \lambda$ as follows: if β, γ, β_1, and γ_1 are elements of λ, then

$$(\beta, \gamma) \leq_R (\beta_1, \gamma_1)$$
$$\text{if and only if } \begin{cases} \sup(\beta, \gamma) < \sup(\beta_1, \gamma_1), \text{ or} \\ \sup(\beta, \gamma) = \sup(\beta_1, \gamma_1) \text{ and } \beta < \beta_1, \text{ or} \\ \sup(\beta, \gamma) = \sup(\beta_1, \gamma_1) \text{ and } \beta = \beta_1 \text{ and } \gamma \leq \gamma_1. \end{cases}$$

We leave it for the reader to verify that $\leq_R$ is indeed an order relation. Let us prove that it is a well-ordering. Let X be a non-empty subset of $\lambda \times \lambda$. Consider

$$X_1 = \{(\beta, \gamma) \in X : \text{ for all } (\beta_1, \gamma_1) \in X, \ \sup(\beta, \gamma) \leq \sup(\beta_1, \gamma_1)\},$$

i.e. the set of elements (β, γ) of X such that $\sup(\beta, \gamma)$ is a minimum. X_1 is not empty and we consider, in succession,

$$X_2 = \{(\beta, \gamma) \in X_1 : \text{ for every } (\beta_1, \gamma_1) \in X_1, \ \beta \leq \beta_1\}, \quad \text{and}$$
$$X_3 = \{(\beta, \gamma) \in X_2 : \text{ for every } (\beta_1, \gamma_1) \in X_2, \ \gamma \leq \gamma_1\}.$$

Then X_3 has only a single element which is the least element of X.

According to Proposition 7.27, there exists an ordinal α and an isomorphism f from $(\alpha, \in)$ onto $(\lambda \times \lambda, \leq_R)$. We will show that this ordinal α cannot be greater than λ. If we assume the contrary, then $\lambda \in \alpha$. Set $f(\lambda) = (\beta_0, \gamma_0)$; so β_0 and γ_0 are ordinals belonging to λ and the restriction $f \upharpoonright \lambda$ of f to λ is a bijection from λ onto the set $Y = \{(\beta, \gamma) : (\beta, \gamma) <_R (\beta_0, \gamma_0)\}$. Let $\delta_0 = \sup(\beta_0, \gamma_0)$. The cardinality of δ_0 is strictly less than that of λ (because λ is a cardinal) and therefore, by the induction hypothesis, so is the cardinality of $\delta_0 \times \delta_0$. Also, Y is included in $\delta_0 \times \delta_0$; consequently $\mathsf{card}(Y) < \mathsf{card}(\lambda)$. This is the desired contradiction since f is a bijection from λ onto Y.

This shows that f is a bijection from a subset of λ onto $\lambda \times \lambda$; so

$$\mathsf{card}(\lambda \times \lambda) \leq \mathsf{card}(\lambda).$$

The inequality in the other direction is obvious. ■

Corollary 7.72 (*AC*) (1) *If X and Y are non-empty sets and at least one of them is innite, then*

$$\mathsf{card}(X \times Y) = \mathsf{card}(X \cup Y) = \sup(\mathsf{card}(X), \mathsf{card}(Y)).$$

(2) *If* $(X_i : i \in I)$ *is a family of sets and if one of the* X_i *is innite, then*

$$\mathsf{card}\left(\bigcup_{i \in I} X_i\right) \leq \sup(\sup\{\mathsf{card}(X_i) : i \in I\}, \mathsf{card}(I)).$$

(3) *A denumerable union of denumerable sets is denumerable.*

Proof (1) Set

$$\lambda = \sup(\mathsf{card}(X), \mathsf{card}(Y)).$$

It is clear that the identity map from X into itself is an injection from X into $X \cup Y$; thus $\mathsf{card}(X) \leq \mathsf{card}(X \cup Y)$. For the analogous reason, $\mathsf{card}(Y) \leq \mathsf{card}\,(X \cup Y)$; so $\mathsf{card}(X \cup Y) \geq \lambda$. Also, letting y be an arbitrary point of Y (which is non-empty), the map that, with each $x \in X$, associates (x, y) is an injection from X into $X \times Y$. It follows that $\mathsf{card}(X) \leq \mathsf{card}(X \times Y)$ and, just as before, $\mathsf{card}(X \times Y) \geq \lambda$.

In the opposite direction, according to Corollary 7.47, we have

$$\mathsf{card}(X \times Y) \leq \lambda \times \lambda$$

and, from the preceding theorem, $\lambda \times \lambda = \lambda$. Thus, $\mathsf{card}(X \times Y) \leq \lambda$. There is also an injection f from $X \cup Y$ into $X \uplus Y$ obtained by dening $f(x) = (x, \mathbf{0})$ if $x \in X$ and $f(x) = (x, \mathbf{1})$ otherwise. This shows that

$$\mathsf{card}(X \cup Y) \leq \mathsf{card}(X \uplus Y) \leq \lambda + \lambda = \lambda.$$

(2) Set $X = \bigcup_{i \in I} X_i$ and $\lambda = \sup(\sup\{\mathsf{card}(X_i) : i \in I\}, \mathsf{card}(I))$. By hypothesis, λ is innite. For every $x \in X$, the set $I_x = \{i \in I : x \in X_i\}$ is not empty, so by AC, there exists a mapping f from X into I such that, for every $x \in X$, $x \in X_{f(i)}$. Also, for every $i \in I$, the set $\{g : g$ is an injective map from X_i into $\lambda\}$ is not empty [since $\mathsf{card}(X_i) \leq \lambda$], so, again using the axiom of choice, we can nd a family $(g_i : i \in I)$ such that, for every $i \in I$, g_i is an injective map from X_i into λ. It is then easy to verify that the map from X into $I \times \lambda$ which, with $x \in X$, associates $(f(x), g_{f(x)}(x))$ is injective. This shows that the cardinality of X is less than or equal to $\lambda \times \lambda$, which is itself equal to λ.

(3) This is a more or less obvious consequence of (2). We emphasize it, rst, because it is important, but also to insist on the fact that its proof uses the axiom of choice. ■

Remark 7.73 A consequence of all the above is that if λ and μ are two innite cardinals with $\lambda > \mu$, then $\lambda + \mu = \lambda + \lambda$ and $\lambda \times \mu = \lambda \times \lambda$ (see the comment that follows Corollary 7.47).

Here is a little fact that is very useful.

Proposition 7.74 *(AC) Let* A *be a subset of an innite set* B *and suppose that*

$$\mathsf{card}(A) < \mathsf{card}(B).$$

Then $\mathsf{card}(B) = \mathsf{card}(B - A)$.

Proof Indeed, $B = A \uplus (B - A)$. So by Corollary 7.72, because it equals $\sup(\mathsf{card}(A), \mathsf{card}(B-A))$, $\mathsf{card}(B)$ is equal either to $\mathsf{card}(A)$ or to $\mathsf{card}(B-A)$. But by hypothesis, the possibility that $\mathsf{card}(A) = \mathsf{card}(B)$ is excluded. ∎

So we have seen that it is easy to control the cardinality of the union of an innite family of sets. The situation is altogether different for the innite product of a family of sets. The following theorem, due to **Konig**, is suggestive.

Theorem 7.75 *(AC) Let $(X_i : i \in I)$ and $(Y_i : i \in I)$ be two families of sets and assume that, for every $i \in I$, $\mathsf{card}(X_i) < \mathsf{card}(Y_i)$. Then*

$$\mathsf{card}\left(\bigcup_{i\in I} X_i\right) < \mathsf{card}\left(\prod_{i\in I} Y_i\right).$$

(*Caution*: These inequalities are strict!)

Proof Set $X = \bigcup_{i\in I} X_i$ and $Y = \prod_{i\in I} Y_i$ and let f be a mapping from X into Y. We will prove that f is not surjective. For every $x \in X$, we can write $f(x)$ in the form $(f(x)_i : i \in I)$ where, for every $i \in I$, $f(x)_i \in Y_i$. This allows us to dene, for every $i \in I$, a mapping f_i from X_i into Y_i by setting $f_i(x) = f(x)_i$ for every $x \in X_i$. Since $\mathsf{card}(X_i) < \mathsf{card}(Y_i)$, the map f_i cannot be surjective, so the set

$$B_i = \{y \in Y_i : \text{ for every } x \in X_i,\ f_i(x) \neq y\}$$

is not empty. Using the axiom of choice, we produce an element

$$b = (b_i : i \in I) \in \prod_{i\in I} B_i.$$

Then b cannot belong to the image of f, for if we suppose that $b = f(x)$ for some $x \in X$, then there exists an $i \in I$ such that $x \in X_i$ and $f_i(x) = b_i$, which contradicts the fact that $b_i \in B_i$. ∎

In fact, Konig's theorem is equivalent to the axiom of choice: if $(Y_i : i \in I)$ is a given family of non-empty sets and if we let $(X_i : i \in I)$ be the family in which each $X_i = \emptyset$, the hypotheses of Theorem 7.75 are satised; consequently,

$$\mathsf{card}\left(\prod_{i\in I} Y_i\right) > 0,$$

which means that $\prod_{i\in I} Y_i \neq \emptyset$.

We will see applications of Konig's theorem in Exercise 16.

7.5 The axiom of foundation and the reection schemes

7.5.1 The axiom of foundation

There is still at least one natural and important question that we have not discussed: does there exist a set x that is an element of itself?

Widespread intuition suggests a negative answer, but we wish to guard against approaching the problem from this angle. As usual, we will adopt an axiomatic point of view. We will introduce, and then exploit, a new axiom, the axiom of foundation denoted by AF, one of whose consequences is that there indeed does not exist a set that is an element of itself. We will use this axiom to illustrate certain relative consistency results.

The **axiom of foundation**: $\forall v_0(\neg v_0 \simeq \emptyset \Rightarrow \exists v_1(v_1 \in v_0 \wedge v_0 \cap v_1 \simeq \emptyset))$.

First, a remark.

Remark 7.76 AF implies

$$\forall v_0(v_0 \notin v_0).$$

Proof Let x be a set. Then $\{x\}$ is not empty so, by AF, there exists a set $y \in \{x\}$ such that $y \cap \{x\}$ is empty. But y can only be equal to x; consequently, $x \cap \{x\}$ is empty, which clearly implies that $x \notin x$. ■

The next proposition will not be used in the remainder of the text. But perhaps it will help explain the signicance of the axiom of foundation. Its proof uses the axiom of choice.

Proposition 7.77 (AC) *The axiom of foundation is equivalent to the following property:*

$$\text{There does not exist a family } (a_i)_{i\in\omega} \text{ such that,}$$
$$\text{for every } i \in \omega,\ a_{i+1} \in a_i. \qquad (*)$$

Proof We rst show (without using the axiom of choice) that AF implies $(*)$. Let $(a_i : i \in \omega)$ be a family of sets indexed by ω and consider the set $A = \{a_i : i \in \omega\}$. According to AF, there exists an element of A, say a_n, such that $a_n \cap A = \emptyset$. Thus $a_{n+1} \notin a_n$.

Conversely, suppose that AF is false. So there exists a non-empty set x such that, for every $y \in x$, $y \cap x$ is not empty. Using the axiom of choice, we see that there exists a mapping f from x into itself such that, for every $y \in x$, $f(y) \in y \cap x$. Let a_0 be an element of x. We dene the sequence $(a_i : i \in \omega)$ by induction on i; for every $i \in \omega$, $a_{i+1} = f(a_i)$. It is then clear that, for every $i \in \omega$, $a_{i+1} \in a_i$. ■

We will present another property that is equivalent to AF, but it requires slightly more work. By induction on the ordinal α, we dene a set V_α by

$$V_\alpha = \bigcup_{\beta<\alpha} \wp(V_\beta).$$

Thus V_0, which is the union of the empty family, is equal to $\emptyset$. We may also compute

$$V_1 = \{\emptyset\}, \quad V_2 = \{\emptyset, \{\emptyset\}\}, \quad V_3 = \{\emptyset, \{\emptyset\}, \{\{\emptyset\}\}, \{\emptyset, \{\emptyset\}\}\}, \quad \text{etc.}$$

We see that if β and α are ordinals with $\beta < \alpha$, then $V_\beta \subseteq V_\alpha$. Also,

$$\text{if } \alpha \text{ is a limit ordinal, then } V_\alpha = \bigcup_{\beta<\alpha} V_\beta;$$
$$\text{if } \alpha \text{ is equal to } \beta + \mathbf{1}, \text{ then } V_a = \wp(V_\beta).$$

Let $\mathcal{V}$ denote the class of sets x such that, for some ordinal α, $x \in V_\alpha$. For every set x in $\mathcal{V}$, the **rank** of x is the least ordinal α such that $x \in V_a$; it is denoted by $rk(x)$. Note that the rank of x is always a successor ordinal because, if α is a limit ordinal, $V_\alpha = \bigcup_{\beta<\alpha} V_\beta$. The reader can also prove by induction that, for every ordinal α, α belongs to $V_{\alpha+\mathbf{1}}$ but does not belong to V_a; in other words, $rk(\alpha) = \alpha + \mathbf{1}$.

Remark 7.78 Recall that a set x is transitive if, for every $y \in x$ and for every $z \in y$, $z \in x$. For every ordinal, V_a is a transitive set; for if $x \in V_\alpha$ and $y \in x$, then there exists $\beta < \alpha$ such that $x \subseteq V_\beta$, so $y \in V_\beta$. Note that, in passing, we have proved that if x and y are in $\mathcal{V}$ and if $x \in y$, then $rk(x) < rk(y)$.

Remark 7.79 Every set x is included in a transitive set.

Proof By induction on the integer n, dene a set x_n as follows:

$$x_0 = x;$$
$$\text{for every integer } n, \quad x_{n+1} = x_n \cup \left(\bigcup_{t\in x_n} t\right).$$

Set $\mathsf{cl}(x) = \bigcup_{n\in\omega} x_n$. To begin with, it is clear that $x \subseteq \mathsf{cl}(x)$. Also, $\mathsf{cl}(x)$ is a transitive set; for if $y \in \mathsf{cl}(x)$ and $z \in y$, then there exists an integer n such that $y \in x_n$, so $z \in x_{n+1}$. ■

Moreover, $\mathsf{cl}(x)$ is the smallest transitive set that includes x; for if t is a transitive set and $x \subseteq t$, then by induction on $n \in \omega$, one can see that $x_n \subseteq t$ and thus that $\mathsf{cl}(x) \subseteq t$. The set $\mathsf{cl}(x)$ is called the **transitive closure** of x.

Theorem 7.80 *The axiom of foundation is equivalent to the following property:*

For every set x, there exists an ordinal α such that $x \in V_\alpha$.
(In other words, the class $\mathcal{V}$ is equal to the whole universe.) (∗∗)

Proof We will rst show that (∗∗) implies the axiom of foundation. Let x be a non-empty set; we must nd an element y in x such that $y \cap x = \emptyset$. By hypothesis, every set has a rank; choose an element y in x whose rank is minimal. If $t \in y$, then $rk(t) < rk(y)$ (see Remark 7.78) and, by the minimality of $rk(y)$, $t \notin x$.

We will now prove the converse. We begin with an observation: if every element of a set x belongs to $\mathcal{V}$, then x itself belongs to $\mathcal{V}$. Indeed, by the axiom of replacement, the image of x under the function rk is a set,

$$Y = \{\beta : \text{ there exists } y \in x \text{ such that } rk(y) = \beta\}.$$

Let $\alpha = \sup Y$. Then $x \subseteq V_\alpha$ and, consequently, $x \in V_{\alpha+1}$.

Let x be an arbitrary set. Set $y = \mathsf{cl}(x)$ and consider

$$z = \{t : t \in y \text{ and } t \text{ is not in } \mathcal{V}\}.$$

We will show that z is empty; this will imply that every element of x is in $\mathcal{V}$ and hence, by the observation just made, that x is in $\mathcal{V}$. If z is not empty, then by AF there exists an element t in z such that $t \cap z = \emptyset$. Let $u \in t$. First, $u \notin z$ since $t \cap z = \emptyset$; next, because y is transitive, $u \in y$ so, by the denition of z, u is in $\mathcal{V}$. In other words, every element of t is in $\mathcal{V}$, so t itself is in $\mathcal{V}$. This contradicts the fact that $t \in z$. ■

7.5.2 Some relative consistency results

Relative consistency theorems have the following form: given two theories T_1 and T_2 (very often, T_2 includes T_1), if T_1 is consistent, then so is T_2. In the examples that we will give, T_1 will be ZF. The principle underlying the proofs of these theorems is simple: starting with a model of T_1, we construct a model of T_2.

We begin with a parenthetical remark. One may prefer to prove relative consistency theorems that are expressed in the following form: if a contradiction cannot be derived from T_1, then a contradiction cannot be derived from T_2. Obviously, the completeness theorem implies that there is no difference between these two formulations. However, this second formulation has an advantage, a compelling one, when we wish to deal with the foundations of mathematics: specically, it lets us express the results using notions from nitistic mathematics (formulas are nite sequences of symbols, proofs are nite sequences of formulas, etc.). As a matter of fact, the proofs that we will be giving, though this is not obvious, will nonetheless provide algorithms for converting any formal derivation of a contradiction F (for example, $\mathbf{0} = \mathbf{1}$) in T_2 into a formal derivation of F in T_1. We will not insist further on this point.

We may observe that the set of integers, with the appropriate functions, is a model of Peano's axioms; we have thereby proved that

if ZF is consistent, then so is Peano arithmetic, PA.

This result establishes, in passing, that we cannot hope, due to Godel's second incompleteness theorem (Chapter 6), to have an absolute consistency theorem; for example, that ZF is consistent. To put this relationship between ZF and PA behind us, let us simply state that ZF is much stronger than PA. The consistency of PA can be expressed by a formula of set theory, and, moreover, this formula is derivable in ZF (because the structure $\mathbb{N}$ is a point in $\mathcal{U}$ and the fact that $\mathbb{N}$ is a model of PA is a theorem of ZF). The consistency of ZF can be expressed by a formula of arithmetic (since, obviously, ZF is a recursive theory) but, by contrast, this formula is not derivable in PA (otherwise, in view of the consistency theorem displayed above, we would conclude that PA can prove its own consistency, in violation of Godel's theorem).

If $\mathcal{A}$ is a class (or a set, considered as the class of its elements), we will consider the substructure $\langle \mathcal{A}, \in \rangle$ of $\mathcal{U}$ whose universe is $\mathcal{A}$. The models of thevarious theories whose consistency will be discussed will be structures of this kind.

Denition 7.81 *Let $D[v_0]$ be a formula and let $\mathcal{A}$ be the class of sets x that satisfy $D[x]$. Given a formula F, we dene the formula $F^{\mathcal{A}}$, called F* ***relativized to*** *$\mathcal{A}$, by (intuitive) induction on the complexity of F.*

If F is atomic, then $F^{\mathcal{A}} = F$.
If F is equal to $\neg G$, then $F^{\mathcal{A}} = \neg G^{\mathcal{A}}$.
If F is equal to $(G \alpha H)$, where α is a binary propositional connective, then

$$F^{\mathcal{A}} = (G^{\mathcal{A}} \alpha H^{\mathcal{A}}).$$

If F is equal to $\exists v G$, where v is a symbol for a variable, then

$$F^{\mathcal{A}} = \exists v (D[v] \wedge G^{\mathcal{A}}).$$

If F is equal to $\forall v G$, where v is a symbol for a variable, then

$$F^{\mathcal{A}} = \forall v [D[v] \Rightarrow G^{\mathcal{A}}].$$

It is then no trouble to prove, always by an (intuitive) induction on the height of F, that, for every formula $F[v_1, v_2, \ldots, v_n]$ and for all $x_1, x_2, \ldots, x_n$ in $\mathcal{A}$,

$$\mathcal{U} \models F^{\mathcal{A}}[x_1, x_2, \ldots, x_n] \quad \text{if and only if} \quad \langle \mathcal{A}, \in \rangle \models F[x_1, x_2, \ldots, x_n].$$

On several occasions, we will need to invoke some of the following remarks.

Remark 7.82 (i) If $\mathcal{A}$ is a transitive class, then $\langle \mathcal{A}, \in \rangle$ satises the axiom of extensionality. To see this, let x and y be distinct elements of $\mathcal{A}$. By the axiom of extensionality in $\mathcal{U}$, there exists a set z that belongs to one and not the other; let us say $z \in x$. Since $z \in x$ and $x \in \mathcal{A}$ and $\mathcal{A}$ is transitive, z belongs to $\mathcal{A}$; thus there exists an element of $\mathcal{A}$, namely z, that belongs to x but not to y. This proves the axiom of extensionality for $\langle \mathcal{A}, \in \rangle$.

(ii) If α is an ordinal and x and y belong to V_α, then $\{x, y\}$ belongs to $V_{\alpha+\mathbf{1}}$. If δ is a limit ordinal and x and y belong to V_δ, then there exists an ordinal $\alpha < \delta$ such that x and y belong to V_α, so $\{x, y\}$ belongs to V_δ. This proves that if δ is a limit ordinal, then $\langle V_\delta, \in \rangle$ satises the axiom of pairs. A similar argument shows that $\langle \mathcal{V}, \in \rangle$ satises the axiom of pairs.

(iii) If $x \in V_\alpha$, then $\wp(x) \subseteq V_{\alpha+\mathbf{1}}$ and $\wp(x) \in V_{\alpha+\mathbf{2}}$; this proves that if δ is a limit ordinal, then $\langle V_\delta, \in \rangle$ satises the power set axiom.

(iv) For every α, $\langle V_\alpha, \in \rangle$ satises the axiom of unions. To see this, suppose $x \in V_\alpha$; then the rank of x is a successor ordinal, hence of the form $\beta + \mathbf{1}$; if $y \in \bigcup x$, then there exists $z \in x$ such that $y \in z$. We have seen (Remark 7.78) that $rk(z) \leq \beta$ and $rk(y) < \beta$. This implies that $\bigcup x \subseteq V_\beta$ and $\bigcup x \in V_\alpha$.

(v) For every ordinal α and for every $a \in V_\alpha$, we have

$$\langle V_\alpha, \in\rangle \models On[a] \quad \text{if and only if} \quad \mathcal{A} \models On[a].$$

The proof presents no problems. We see that an element a of V_α is transitive if and only if it is transitive in $\langle V_\alpha, \in\rangle$, because the various properties expressing that the membership relation is a well-ordering are true in $\mathcal{U}$ if and only if they are true in $\langle V_\alpha, \in\rangle$. The proof that

$$\langle \mathcal{V}, \in\rangle \models On[a] \quad \text{if and only if} \quad \mathcal{A} \models On[a]$$

is similar.

There are many other properties that are inherited in this fashion. For example, the reader can just as easily show that if α is a limit ordinal and $a \in V_\alpha$, then the following three assertions are equivalent:

$$\mathcal{U} \models a \text{ is a cardinal;}$$
$$\langle V_\alpha, \in\rangle \models a \text{ is a cardinal;}$$
$$\langle \mathcal{V}, \in\rangle \models a \text{ is a cardinal.}$$

Theorem 7.83 *If ZF is consistent, then ZF + AF is consistent.*

Proof Let $\mathcal{U}$ be a model of ZF and let $\mathcal{V}$ be the class of sets x that satisfy the formula

$$F[x] = \text{ there exists an ordinal } \alpha \text{ such that } x \in V_\alpha.$$

We will show that $\langle \mathcal{V}, \in\rangle$ is a model of ZF + AF.

- The axioms of extensionality, pairs, unions, and subsets were proved in the previous remark.
- The axioms of replacement. Let x be an element of $\mathcal{V}$ and let $G[v_0, v_1]$ be a formula that, in $\langle \mathcal{V}, \in\rangle$, is functional in v_0; in other words,

  $$\langle \mathcal{V}, \in\rangle \models \forall v_0 \forall v_1 \forall v_2((G[v_0, v_1] \wedge G[v_0, v_2]) \Rightarrow v_1 \simeq v_2).$$

  Then $H = F[v_0] \wedge F[v_1] \wedge G^{\mathcal{V}}$ is functional in v_0 (in $\mathcal{U}$). Let b be the image of x under the function that this formula denes, i.e.

  $$b = \{z : (\exists y \in x) H[y, z]\}.$$

  Then b belongs to $\mathcal{V}$ since all its elements belong to $\mathcal{V}$; this is what we needed to prove since b is also the image of x under the function dened by the formula G in $\mathcal{V}$.
- The axiom of innity. We have seen that ω is an ordinal in $\mathcal{V}$ and it is easy to see that it is neither empty nor a successor. We can see that it is, in $\mathcal{V}$, the least innite ordinal.

- The axiom of foundation. Let x be a set that belongs to $\mathcal{V}$; then all the elements of x are also in $\mathcal{V}$. Let y be an element of x whose rank α is minimum; then, if u is an element of y, its rank is strictly less than α (Remark 7.78) so u does not belong to x; i.e. $x \cap y$ is empty. ∎

Remark 7.84 In the preceding proof, one might think that the axiom of foundation is *obviously* satised in $\mathcal{V}$. This is not quite true; the class $\mathcal{V}$ is dened by a formula which we have called $F[v_0]$. It needs to be veried that F has the same meaning in $\mathcal{U}$ and in $\mathcal{V}$; in other words, that

$$\mathcal{U} \models \forall v_0 (F[v_0] \Leftrightarrow F^{\mathcal{V}}[v_0]).$$

If, in $\mathcal{V}$, we dene the sets W_α by induction on α by setting

$$W_0 = \emptyset,$$
$$W_\alpha = \bigcup_{\beta<\alpha} \wp(W_\alpha),$$

then it is easy to verify that, for every ordinal α, $V_\alpha = W_\alpha$. This, together with Theorem 7.80 (and the fact that the ordinals are the same in $\mathcal{U}$ and in $\mathcal{V}$), provides another proof that $\langle \mathcal{V}, \in \rangle$ satises the axiom of foundation.

Remark 7.85 If we assume that $\mathcal{U}$ is a model of ZFC, then $\langle \mathcal{V}, \in \rangle$ is a model of ZFC + AF. Indeed, if $X = (x_i : i \in I)$ is a family of non-empty sets in $\mathcal{V}$, then I itself is in $\mathcal{V}$ (it is the domain of X considered as a mapping); thus I is included in $\bigcup\bigcup X$. Similarly, each x_i is in $\mathcal{V}$. This shows that all elements of $\prod_{i\in I} x_i$ are in $\mathcal{V}$.

The second relative consistency theorem which we will see shows that we cannot dispense with the axiom of innity.

Theorem 7.86 *If ZF is consistent, then ZF$^-$ + ¬Inf is also consistent.*

Proof We will prove that V_ω is a model of ZF$^-$ + ¬Inf. The axioms of extensionality, pairs, unions, and subsets have already been treated. Consider the axioms of replacement. Note that, for every integer n, V_n is a nite set (there is an obvious proof of this by induction on n). Thus every element of V_ω is a nite set. Conversely, if x is a nite set and all of its elements belong to V_ω, then x belongs to V_ω; this is because $X = \{rk(y) : y \in x\}$ = the image of x under the function rk is a set (by the axiom of replacement in $\mathcal{U}$) and is nite (Proposition 7.56); also, by hypothesis, it is a subset of ω. If we let $n = \sup X$, we see that $x \in V_{n+1}$.

Let x be an element of V_ω and $F[v_0, v_1]$ be a formula that, in $\langle V_\omega, \in \rangle$, is functional in v_0. Specically,

$$\langle V_\omega, \in \rangle \models \forall v_0 \forall v_1 \forall v_2 ((F[v_0, v_1] \wedge F[v_0, v_2]) \Rightarrow v_1 \simeq v_2).$$

Then $G = v_0 \in V_\omega \wedge v_1 \in V_\omega \wedge F^{V_\omega}$ is functional in v_0 (in $\mathcal{U}$). Let b be the image of x under the function that this formula denes in $\mathcal{U}$, i.e.

$$b = \{z : (\exists y \in x) G[y, z]\}.$$

Then b is a nite set (since there is a surjection from a subset of x onto b), all of whose elements belong to V_ω. Finally, as in the proof of Theorem 7.83, b is the image of x under the function dened by the formula F in V_ω and $b \in V_\omega$.

It is easy to show that V_ω does not satisfy the axiom of innity; if x is an element of V_ω, it is nite and there does not exist, either in $\mathcal{U}$ or, *a fortiori*, in V_ω, a mapping from x into x that is injective without being bijective [in contrast to what can happen in an arbitrary model of ZF (see Remark 7.60)]. ■

Remark It is very easy to show that the axiom of foundation is true in V_ω. The axiom of choice is also true there (even when it is not true in $\mathcal{U}$), but the proof of this is rather delicate.

7.5.3 Inaccessible cardinals

In this section, we will work in ZFC.

Denition 7.87 *Let λ be a cardinal:*

(i) *λ is a **strong limit** if, for every cardinal μ strictly less than λ, the cardinality of 2^μ is also strictly less than λ.*
(ii) *λ is **regular** if, for every subset X of λ of cardinality strictly less than λ,* $\sup(X) < \lambda$.
(iii) *λ is **inaccessible** if λ is a regular, strong limit cardinal strictly greater than $\aleph_0$.*

Let us see some examples. For every ordinal α, dene the cardinal $\beth_\alpha$ by induction:

$$\beth_0 = \aleph_0;$$
$$\beth_{\alpha+1} = 2^{\beth_\alpha};$$
$$\beth_\delta = \bigcup_{\alpha\in\delta} \beth_\alpha \quad \text{if } \delta \text{ is a limit ordinal.}$$

(The symbol $\beth$ is `beth', the second letter of the Hebrew alphabet.)

It is easily proved by induction on α that, for every ordinal α, $\aleph_\alpha \leq \beth_\alpha$, and if the generalized continuum hypothesis is assumed, then $\aleph_\alpha = \beth_\alpha$.

It is clear that $\beth_\omega$ is a strong limit cardinal. However, it is not regular; for if we set

$$X = \{\beth_n : n \in \omega\},$$

then X is countable, so its cardinality is strictly less than $\beth_\omega$, but

$$\sup(X) = \beth_\omega.$$

Also, for every ordinal α, $\aleph_{\alpha+1}$ is a regular cardinal (to prove this requires the axiom of choice; it can be shown that this fact is not a consequence of ZF). Let X be a subset of $\aleph_{\alpha+1}$ of cardinality at most $\aleph_\alpha$. The elements of X all have cardinality at most $\aleph_\alpha$; consequently, $\bigcup_{x \in X} x$, which is equal to sup(X), has cardinality at most $\aleph_\alpha$ [item (2) from Corollary 7.72].

$\aleph_{\alpha+1}$ is obviously not a strong limit cardinal since $2^{\aleph_\alpha} \geq \aleph_{\alpha+1}$. So we do not yet have any examples of inaccessible cardinals; this is not surprising, in view of the next theorem.

Theorem 7.88 *If ZFC is consistent, then ZFC + `there does not exist an inaccessible cardinal' is also consistent.*

Proof Let $\mathcal{U}$ be a model of ZFC. We wish to produce a model of ZFC in which there does not exist an inaccessible cardinal. We will assume that there does exist an inaccessible cardinal κ in $\mathcal{U}$ (otherwise there is nothing to prove). We will show that $\langle V_\kappa, \in\rangle$ is a model of ZFC; thus if κ is the least inaccessible cardinal in $\mathcal{U}$, then there is no inaccessible cardinal in V_κ.

The axioms of extensionality, pairs, unions, and subsets are proved in the usual fashion.

For the replacement scheme, we will rst show that if $x \in V_\kappa$, then $\text{card}(x) < \kappa$. If $x \in V_\kappa$, there exists an ordinal $\alpha < \kappa$ such that x is included in V_α; so it will sufce to show that if $\alpha < \kappa$, then $\text{card}(V_\alpha) < \kappa$. This is done by induction on α. It is obvious if $\alpha = \mathbf{0}$. If $\alpha = \beta + \mathbf{1}$, then $\text{card}(V_\alpha) = 2^{\text{card}(V_\beta)}$; by the induction hypothesis, $\text{card}(V_\beta) < \kappa$ so, because κ is a strong limit cardinal, $\text{card}(V_\alpha) < \kappa$. If α is a limit ordinal, then $V_\alpha = \bigcup_{\beta<\alpha} V_\beta$ and

$$\text{card}(V_\alpha) = \sup(\sup\{\text{card}(V_\beta) : \beta < \alpha\}, \alpha).$$

The fact that κ is regular together with the induction hypothesis permit us to conclude that $\text{card}(V_\alpha) < \kappa$.

Suppose now that x is a set of cardinality strictly less than κ and that all its elements belong to V_κ. Then x belongs to V_κ; to see this, consider

$$X = \{\alpha \in \kappa : \text{ there exists } y \in x \text{ such that } \alpha \text{ is the rank of } y\}.$$

The set X is the image of x under the rank function, so its cardinality is strictly less than κ (Proposition 7.68). By the regularity of κ, $\beta = \sup(X)$ is also strictly less than κ and x is included in V_β. This shows that $x \in V_{\beta+\mathbf{1}}$.

To show that the axioms of replacement are true in V_κ, it sufces to adapt the proof that we used for V_ω (replace V_ω by V_κ and `nite' by `of cardinality less than κ').

The axiom of innity is satised in $\langle V_\kappa, \in\rangle$ since ω is an innite ordinal in $\langle V_\kappa, \in\rangle$.

To prove that $\langle V_\kappa, \in\rangle$ satises the axiom of choice, we will show that the formula

> `for every set x, there exists an ordinal α
> and a bijection $j : \alpha \to x$'

is satised in V_κ. So let $x \in V_\kappa$ and let β ($< \kappa$) be the rank of x; thus, $x \in V_\beta$. Because $\mathcal{U} \models$ AC, there exists, in $\mathcal{U}$, an ordinal α and a bijection $j : \alpha \to x$. In particular, j is an injection from α into $V_{\beta,}$ which implies that $\mathrm{card}(\alpha) \leq \mathrm{card}(V_\beta)$. Since κ is inaccessible, $\mathrm{card}(V_\beta) < \kappa$, so it follows that $\alpha < \kappa$ and $\alpha \in V_\kappa$. Thus $j \subseteq \alpha \times x$. This means that $j \in \wp(\alpha \times x)$ and, since V_κ is a model of ZF, $j \in V_\kappa$. To conclude, it sufces to observe that the formula

$$`j : \alpha \to x \text{ is a bijection'},$$

which is true in $\mathcal{U}$, is then also true in V_κ.

It remains to verify that if κ is the least inaccessible cardinal, then

$$\langle V_\kappa, \in\rangle \models \text{ there does not exist an inaccessible cardinal.}$$

Recall, at this point, that the cardinals of V_κ are the same as those of $\mathcal{U}$ that are less than κ. Because κ is the least inaccessible cardinal, we already know that

$$\mathcal{U} \models \text{ there does not exist an inaccessible cardinal that belongs to } V_\kappa;$$

so it is sufcient to prove that, for every $x \in V_\kappa$,

$$\mathcal{U} \models x \text{ is an inaccessible cardinal}$$

if and only if

$$\langle V_\kappa, \in\rangle \models x \text{ is an inaccessible cardinal.}$$

It is not hard to check that, for every element $x \in V_\kappa$, if x is a cardinal, then

$$\mathcal{U} \models x \text{ is regular} \quad \text{if and only if} \quad \langle V_\kappa, \in\rangle \models x \text{ is regular,}$$

and

$$\mathcal{U} \models x \text{ is a strong limit} \quad \text{if and only if} \quad \langle V_\kappa, \in\rangle \models x \text{ is a strong limit.}$$ ■

There are many other relative consistency theorems. The most famous ones are

if ZF is consistent, then so is ZFC;
if ZF is consistent, then so is ZFC + GCH;
if ZF is consistent, then so is ZF +¬AC;
if ZF is consistent, then so is ZF + ¬CH.

The rst two are due to Godel; the third and fourth are due to P. Cohen. For details, see the textbook by T. Jech mentioned in the bibliography.

7.5.4 The reection scheme

In this section, we work in ZF + AF.

The **reection scheme** is the collection of formulas of the following form:

$$\forall v_0 \forall v_1 \ldots \forall v_n \exists \alpha (On[\alpha] \wedge v_0 \in V_\alpha \wedge v_1 \in V_\alpha \wedge \cdots \wedge v_n \in V_\alpha$$
$$\wedge (F[v_0, v_1, \ldots, v_n] \Leftrightarrow F^{V_\alpha}[v_0, v_1, \ldots, v_n])),$$

where n is an intuitive integer and $F[v_0, v_1, \ldots, v_n]$ is a formula of L.

Denition 7.89 *Let $F[v_0, v_1, \ldots, v_n]$ be a formula of L and let A be a set. We say that F* ***reects in*** *A if*

$$\mathcal{U} \models \forall v_0 \forall v_1 \ldots \forall v_n ((v_0 \in A \wedge v_1 \in A \wedge \cdots \wedge v_n \in A) \Rightarrow (F[v_0, v_1, \ldots, v_n] \Leftrightarrow F^A[v_0, v_1, \ldots, v_n])).$$

In other words, $F[v_0, v_1, \ldots, v_n]$ reects in A if and only if for all elements $a_0, a_1, \ldots, a_n$ of A,

$$\mathcal{U} \models F[a_0, a_1, \ldots, a_n] \quad \text{if and only if} \quad \langle A, \in \rangle \models F[a_0, a_1, \ldots, a_n].$$

Thus, the axiom scheme of reection asserts that, for every formula F, there exists an ordinal α such that F reects in V_α.

It follows trivially from the denition that if F_1 and F_2 both reect in A, then so do $\neg F_1$, $F_1 \wedge F_2$, $F_1 \vee F_2$, $F_1 \Rightarrow F_2$, and $F_1 \Leftrightarrow F_2$.

Theorem 7.90 *For every formula $F[v_0, v_1, \ldots, v_n]$ of L,*

$$\mathcal{U} \models \forall v_0 \forall v_1 \ldots \forall v_n \exists \alpha (On[\alpha] \wedge v_0 \in V_\alpha \wedge v_1 \in V_\alpha \wedge \cdots \wedge v_n \in V_\alpha \wedge (F[v_0, v_1, \ldots, v_n] \Leftrightarrow F^{V_\alpha}[v_0, v_1, \ldots, v_n])).$$

(In other words, the reection scheme is a consequence of ZF + AF.)

More precisely, what we have here is a `theorem scheme', i.e. a distinct theorem of ZF + AF for each formula F. As the theorem is expressed above (for every formula $F \ldots$), it is not a formula of L.

Proof We are going to prove that if F is a formula and if β is an ordinal, then there exists an ordinal $\alpha > \beta$ such that F reects in V_α. This sufces to prove the theorem, for if $a_0, a_1, \ldots, a_n$ are xed, then we need only choose β so that V_β contains all these points.

We will rst need the following lemma.

Lemma 7.91 *Let F be a formula and let* $(X_n : n \in \omega)$ be a sequence of sets that is increasing with respect to inclusion. We assume that, for every $n \in \omega$ and for every sub-formula G of F, G reects in X_n. Then F reects in $X = \bigcup_{n \in \omega} X_n$.

Proof The argument is by (intuitive) induction on the height of F. If F is atomic, this is obvious since F then reects in any set. Now suppose $F = F_1 \wedge F_2$. Since F_1 and F_2 are sub-formulas of F, they both reect in X_n for every $n \in \omega$ (by hypothesis). Now, by the induction hypothesis, F_1 and F_2 both reect in X. As remarked above, this implies that $F_1 \wedge F_2$ reects in X. The other propositional connectives are treated similarly.

It remains to deal with the quantiers. As an example, we will consider the case where F is of the form $\exists v_0 G[v_0, v_1, \ldots, v_k]$. As G is a sub-formula of F, G reects in each of the X_n, so by the induction hypothesis, G reects in X. Suppose, to begin, that $a_1, a_2, \ldots, a_k$ are elements of X and that

$$\langle X, \in \rangle \models \exists v_0 G[v_0, a_1, a_2, \ldots, a_k].$$

So there exists an element a_0 of X such that

$$\langle X, \in\rangle \models G[a_0, a_1, a_2, \ldots, a_k],$$

and, since G reects in X,

$$\mathcal{U} \models G[a_0, a_1, a_2, \ldots, a_k] \quad \text{and} \quad \mathcal{U} \models \exists v_0 G[v_0, a_1, a_2, \ldots, a_k].$$

Conversely, suppose that $a_1, a_2, \ldots, a_k$ are elements of X and that

$$\mathcal{U} \models \exists v_0 G[v_0, a_1, a_2, \ldots, a_k].$$

Then there exists an element $n \in \omega$ such that X_n contains $a_1, a_2, \ldots, a_k$ (here, we should note that k is an intuitive integer while n is an integer in the sense of $\mathcal{U}$). From the fact that F reects in X_n, we conclude that

$$\langle X_n, \in\rangle \models \exists v_0 G[v_0, a_1, a_2, \ldots, a_k],$$

so there exists a_0 in X_n such that

$$\langle X_n, \in\rangle \models G[a_0, a_1, a_2, \ldots, a_k].$$

Now we invoke that fact that G reects in X_n to deduce that

$$\mathcal{U} \models G[a_0, a_1, a_2, \ldots, a_k],$$

the fact that G reects in X to deduce that

$$\langle X, \in\rangle \models G[a_0, a_1, a_2, \ldots, a_k],$$

and, nally, that

$$\langle X, \in\rangle \models \exists v_0 G[v_0, a_1, a_2, \ldots, a_k]. \qquad ■$$

To nish the proof of the theorem, it is sufcient to prove the following property:

for every formula F and for every ordinal β,
there exists an ordinal α greater than β
such that F and all its sub-formulas reect in V_α. (*)

This will be done by an (intuitive) induction on the height of F. If F is atomic, this is obvious (just take $\alpha = \beta$) and, for $F = \neg G$, it is clear that a formula reects in a set if and only if its negation reects in this set. As an example, we will treat the case in which $F = F_1 \wedge F_2$, since the other binary propositional connectives are treated in exactly the same way. By induction on $n \in \omega$, we dene an increasing sequence of ordinals α_n as follows:

- $\alpha_0 = \beta$.
- If n is different from 0 and is even, then α_n is the least ordinal greater than α_{n-1} such that F_1 and all its sub-formulas reect in V_{α_n} (such an ordinal exists by the induction hypothesis).
- If n is odd, then α_n is the least ordinal greater than α_{n-1} such that F_2 and all its sub-formulas reect in V_{α_n} (same remark).

Observe that in order to dene the sequence $(\alpha_n : n \in \omega)$ by induction, we must be convinced in advance that if F is a xed formula of L, there exists a formula $G[v_0]$ of L such that, for every set x, $G[x]$ is equivalent to 'x is an ordinal and F and all its sub-formulas reect in V_x'.

Set $\alpha = \sup\{\alpha_n : n \in \omega\}$. We then note that V_α is the union of the family $\{V_{\alpha_n} : n \in \omega$ and n is even$\}$; so by the lemma, F_1 and all its sub-formulas reect in V_α. But V_α is also the union of the family $\{V_{\alpha_n} : n \in \omega$ and n is odd$\}$, so F_2 and all its sub-formulas also reect in V_α. Consequently, F and all its sub-formulas reect in V_α.

To conclude, we consider the case in which F is equal to $\exists v_0 G[v_0, v_1, \ldots, v_k]$. First, we prove that, for every ordinal γ, there exists an ordinal δ such that

$$\begin{aligned} \mathcal{U} \models \forall v_1 \forall v_2 \ldots \forall v_k ((v_1 \in V_\gamma \wedge v_2 \in V_\gamma \wedge \cdots \wedge v_k \in V_\gamma \\ \wedge \exists v_0 G[v_0, v_1, \ldots, v_k]) \Rightarrow \exists v_0 (v_0 \in V_\delta \wedge G[v_0, v_1, \ldots, v_k])). \end{aligned}$$

For this purpose, let $H[w, \alpha]$ be the following formula:

if there exist $v_0, v_1, v_2, \ldots, v_k$
such that $w = (v_1, v_2, \ldots, v_k)$ and $G[v_0, v_1, \ldots, v_k]$,
then α is the least ordinal such that there exists u satisfying
$u \in V_\alpha$ and $G[u, v_1, \ldots, v_k]$;
otherwise, $\alpha = \emptyset$.

We see that the formula H denes a function and, by the axiom of replacement, the image of the set of k-tuples of elements of V_γ is a set, Y. It sufces to choose δ so that V_δ includes Y.

Next, we dene a sequence $(\alpha_n : n \in \omega)$ by induction as follows:

- $\alpha_0 = \beta$.
- If n is different from 0 and is even, then α_n is the least ordinal greater than α_{n-1} such that G and all its sub-formulas reect in V_{α_n} (such an ordinal exists by the induction hypothesis).
- If n is odd, then α_n is the least ordinal such that

$$\begin{aligned} \forall v_1 \forall v_2 \ldots \forall v_k ((v_1 \in V_{\alpha_{n-1}} \wedge \cdots \wedge v_k \in V_{\alpha_{n-1}} \wedge \exists v_0 G[v_0, v_1, \ldots, v_k]) \\ \Rightarrow \exists v_0 (v_0 \in V_{\alpha_n} \wedge G[v_0, v_1, \ldots, v_k])). \end{aligned}$$

Let α be the least upper bound of the set $\{\alpha_n : n \in \omega\}$. Then, as before, V_α is the union of the family $(V_{a_n} : n \in \omega$ and n is even$)$ and, according to Lemma 7.91, G and all its sub-formulas reect in V_α. It remains to show that F itself reects in V_α.

Suppose, to begin with, that $a_1, a_2, \ldots, a_k$ belong to V_α and that

$$\langle V_\alpha, \in \rangle \models \exists v_0 G[v_0, a_1, a_2, \ldots, a_k].$$

So there exists an element a_0 in V_α such that

$$\langle V_\alpha, \in \rangle \models G[a_0, a_1, a_2, \ldots, a_k],$$

and, because G reects in V_α,

$$\mathcal{U} \models G[a_0, a_1, a_2, \ldots, a_k] \quad \text{and} \quad \mathcal{U} \models \exists v_0 G[v_0, a_1, a_2, \ldots, a_k].$$

Conversely, suppose that

$$\mathcal{U} \models \exists v_0 G[v_0, a_1, a_2, \ldots, a_k].$$

We know there exists an integer n, which can be assumed to be even, such that $a_1, a_2, \ldots, a_k$ belong to V_{α_n}; by denition of α_{n+1}, there exists an element a_0 of $V_{\alpha_{n+1}}$ such that

$$\mathcal{U} \models G[a_0, a_1, a_2, \ldots, a_k],$$

and, because G reects in V_a,

$$\langle V_\alpha, \in\rangle \models G[a_0, a_1, a_2, \ldots, a_k] \quad \text{and} \quad \langle V_\alpha, \in\rangle \models \exists v_0 G[v_0, a_1, a_2, \ldots, a_k].$$

We will conclude this chapter with an application of Theorem 7.90.

Proposition 7.92 *If the theory ZF is consistent, then it is not nitely axiomatizable.*

Proof We argue by contradiction. If ZF is nitely axiomatizable, then so is ZF + AF. Let F be a formula of L that is equivalent to ZF + AF. We will now work in a model $\mathcal{U}$ of ZF + AF (whose relative consistency was proved in Theorem 7.83). According to Theorem 7.90, there exists an ordinal α such that

$$\langle V_\alpha, \in\rangle \models F;$$

hence,

$$\mathcal{U} \models F^{V_\alpha}.$$

Let us return for a moment to the denition of the relativization of a formula (Denition 7.81). Two facts follow rather easily from this denition: rst, there exists a formula $G[v_0]$ of L such that, for every set A, F^A is equivalent to $G[A]$; second, if $A \subseteq B$, then $(F^A)^B = F^B$.

So there exists an ordinal β satisfying $G[V_\beta]$; let γ be the least such ordinal. Since $\langle V_\gamma, \in\rangle$ is a model of ZF + AF, it satises the reection scheme. Therefore,

$$\langle V_\gamma, \in\rangle \models \text{there exists an ordinal } \delta \text{ such that } F^{V_\delta}.$$

But as we have seen [item (v) from Remark 7.82], the ordinals of $\langle V_\gamma, \in\rangle$ are precisely the ordinals less than γ; we can also see (this follows easily from Remark 7.84) that, for every $x \in V_\gamma$,

$$\langle V_\gamma, \in\rangle \models x \in V_\delta \quad \text{if and only if} \quad \mathcal{U} \models x \in V_\delta.$$

This shows that

$$\mathcal{U} \models (F^{V_\delta})^{V_\gamma},$$

and, since $(F^{V_\delta})^{V_\gamma} = F^{V_\delta}$, $\langle V_\delta, \in\rangle$ is a model of F; this contradicts the minimality of γ. ∎

EXERCISES FOR CHAPTER 7

Unless otherwise specied, the context for all the exercises below is a universe $\mathcal{U}$ that is a model of ZF.

1. The notions of natural number, membership, function, and so on involved in this exercise are intuitive (not those of the universe $\mathcal{U}$). We will use them to construct a universe that satises some of the axioms of ZF.

Let W denote the set of nite subsets of $\mathbb{N}$.

(a) Let ϕ be a bijection from $\mathbb{N}$ onto W and let ε_ϕ be the binary relation on $\mathbb{N}$ dened as follows:

$$\text{for all integers } x \text{ and } y, \quad x\ \varepsilon_\phi\ y \text{ if and only if } x \in \phi(y).$$

Show that the universe $\mathcal{M}_\phi = \langle \mathbb{N}, \varepsilon_\phi \rangle$ satises all the axioms of ZF except the axiom of innity. Show that if, for all $x, y \in \mathbb{N}$, $x \in \phi(y)$ implies $x < y$, then $\mathcal{M}_\phi$ also satises the axiom of foundation.

(b) Show that the mapping ζ whose value for $A \in W$ is $\sum_{a \in A} 2^a$ [with the convention that $\zeta(\emptyset) = 0$] is a bijection from W onto $\mathbb{N}$. Let θ be the inverse mapping. Show that $\mathcal{M}_\theta$ is a model of ZF$^-$ and of AF.

(c) Find a bijection ϕ from $\mathbb{N}$ onto W such that $\mathcal{M}_\phi$ does not satisfy AF.

2. Show that the class On' dened in $\mathcal{U}$ by the formula

$$\forall y((y \subseteq x \wedge \neg y = x \wedge y \text{ is transitive}) \Rightarrow y \in x)$$

is the class of ordinals.

3. Let x be a set and $\Gamma(x)$ be the class of ordinals that are subpotent to x.

Show that $\Gamma(x)$ is an ordinal, that it is the least ordinal not subpotent to x, and that it is a cardinal. We call this **Hartog's cardinality of** x.

Characterize $\Gamma(x)$ assuming that $\mathcal{U}$ satises the axiom of choice.

4. This exercise is devoted to some statements that are equivalent to the axiom of choice.

A **choice function** on a set a is a mapping ϕ from the set of non-empty subsets of a into a such that, for every non-empty subset $x \subseteq a$, $\phi(x) \in x$.

Show that AC is equivalent (in the theory ZF) to each of the following statements:

(a) For every set a, there exists at least one choice function on a.

(b) If x and y are sets and if g is a surjective mapping from x onto y, then there exists a mapping h from y into x such that $g \circ h$ is the identity mapping from y into itself.

(c) For every set a whose elements are non-empty and pairwise disjoint, there exists a set b whose intersection with each of the elements of a is a singleton.

(d) For any sets a and b, either a is subpotent to b or b is subpotent to a.

[Exercise 3 may be used in proving the equivalence between (d) and AC].

Property (d) is known as the **trichotomy of cardinals** because it can also be expressed in the following way: given two cardinal classes λ and μ, one and only one of the three possibilities holds:

$$\lambda = \mu \quad \text{or} \quad \lambda < \mu \quad \text{or} \quad \mu < \lambda.$$

More simply, trichotomy is satised if and only if the ordering of the cardinal classes is a total ordering.

5. Show that in the theory ZF + AF, the axiom of choice is equivalent to each of the following three statements:

(a) If the set x has a well-ordering, then $\wp(x)$ has a well-ordering.

(b) For every ordinal α, $\wp(\alpha)$ has a well-ordering.

(c) Every totally ordered set has a well-ordering.

6. Without using the axiom of choice, show that for every non-empty set a, the following properties are equivalent:

(1) a includes a denumerable subset.

(2) a includes a denumerable subset b such that a and $a - b$ are equipotent.

(3) For every denumerable set b, a and $a \cup b$ are equipotent.

(4) For every nite set x, a and $a \cup x$ are equipotent.

(5) For every nite subset x of a, a and $a - x$ are equipotent.

(6) There exists a non-zero integer n such that, for every subset x of a that is subpotent to n, a and $a - x$ are equipotent.

(7) There exists a non-zero integer n such that, for every set x of cardinality n, a and $a \cup x$ are equipotent.

(8) for every t, a and $a \cup \{t\}$ are equipotent.

(9) There exists an element $t \in a$ such that a and $a - \{t\}$ are equipotent.

(10) There exists a subset of a that is non-empty, different from a, and equipotent to a.

(11) There exists a subset $b \subseteq a$ that is non-empty, different from a, and such that a is subpotent to b.

7. Determine the cardinality of each of the following sets:

$$x_1 = \{f \in \mathbb{N}^{\mathbb{N}} : (\forall n \in \mathbb{N})(\forall p \in \mathbb{N})(n < p \Rightarrow f(n) < f(p))\};$$

$$x_2 = \{f \in \mathbb{N}^{\mathbb{N}} : (\exists p \in \mathbb{N})(\forall n \in \mathbb{N})(f(n) \leq p)\};$$

$$x_3 = \{f \in \mathbb{Q}^{\mathbb{N}} : (\forall n \in \mathbb{N})(\forall p \in \mathbb{N})(n < p \Rightarrow f(n) < f(p))\};$$
$$x_4 = \{f \in \mathbb{Q}^{\mathbb{N}} : (\exists p \in \mathbb{Q})(\forall n \in \mathbb{N})(f(n) \leq p)\};$$
$$x_5 = x_3 \cap x_4;$$
$$x_6 = \{f \in \mathbb{Q}^{\mathbb{N}} : (\exists n \in \mathbb{N})(\forall p \in \mathbb{N})(n \leq p \Rightarrow f(n) = f(p))\};$$
$$x_7 = \{f \in \mathbb{R}^{\mathbb{N}} : (\forall r \in \mathbb{R})(\exists n \in \mathbb{N})(f(n) \geq r)\}.$$

8. Determine the cardinality of each of the following sets:

$E_0 =$ the set $\mathbb{Q}^{\mathbb{N}}$ of sequences of rational numbers;
$E_1 =$ the set $\mathbb{R}^{\mathbb{N}}$ of sequences of real numbers;
$E_2 =$ the set of sequences of rationals that converge to 0;
$E_3 =$ the set of convergent sequences of rationals;
$E_4 =$ the set of bounded sequences of rationals;
$E_5 =$ the set of unbounded sequences of rationals;
$E_6 =$ the set $\mathbb{R}^{\mathbb{Q}}$ of mappings from $\mathbb{Q}$ into $\mathbb{R}$;
$E_7 =$ the set of continuous mappings from $\mathbb{R}$ into $\mathbb{R}$;
$E_8 =$ the set of open intervals of $\mathbb{R}$;
$E_9 =$ the set of open subsets of $\mathbb{R}$ (with the usual topology).

9. Determine the cardinality of each of the following sets:

$$a_1 = \{f \in \omega^\omega : (\forall n \in \omega)(\forall p \in \omega)(f(n) \leq p)\};$$
$$a_2 = \{f \in \omega^\omega : (\forall n \in \omega)(\exists p \in \omega)(f(n) \leq p)\};$$
$$a_3 = \{f \in \omega^\omega : (\exists n \in \omega)(\forall p \in \omega)(f(n) \leq p)\};$$
$$a_4 = \{f \in \omega^\omega : (\exists n \in \omega)(\exists p \in \omega)(f(n) \leq p)\};$$
$$a_5 = \{f \in \omega^\omega : (\exists p \in \omega)(\forall n \in \omega)(f(n) \leq p)\};$$
$$a_6 = \{f \in \omega^\omega : (\forall p \in \omega)(\exists n \in \omega)(f(n) \leq p)\};$$
$$b_1 = \{f \in \omega^\omega : (\forall n \in \omega)(\forall p \in \omega)(f(n) \geq p)\};$$
$$b_2 = \{f \in \omega^\omega : (\forall n \in \omega)(\exists p \in \omega)(f(n) \geq p)\};$$
$$b_3 = \{f \in \omega^\omega : (\exists n \in \omega)(\forall p \in \omega)(f(n) \geq p)\};$$
$$b_4 = \{f \in \omega^\omega : (\exists n \in \omega)(\exists p \in \omega)(f(n) \geq p)\};$$
$$b_5 = \{f \in \omega^\omega : (\exists p \in \omega)(\forall n \in \omega)(f(n) \geq p)\};$$
$$b_6 = \{f \in \omega^\omega : (\forall p \in \omega)(\exists n \in \omega)(f(n) \geq p)\}.$$

10. Assume that the universe satises the axiom of choice. Let a and b be two innite sets whose cardinalities are λ and μ, respectively. Assume $\lambda > \mu$. Let g be an injective mapping from b into a. The reader should determine the cardinalities of each of the following sets:

$$\begin{aligned}
y_1 &= \{f \in b^a : \mathsf{card}(\bar{f}(a)) = 1\};\\
y_2 &= \{f \in b^a : (\forall x \in \wp(a))(\mathsf{card}(\bar{f}(x)) \leq 1)\};\\
y_3 &= \{f \in b^a : \mathsf{card}(\bar{f}^{-1}(b)) = \lambda\};\\
y_4 &= \{f \in b^a : \mathsf{card}(\bar{f}(a)) = 2\};\\
y_5 &= a - \bar{g}(b);\\
y_6 &= \{f \in b^a : (\forall y \in b)(f(g(y)) = y)\};\\
y_7 &= \{f \in b^a : \mathsf{card}(\bar{f}(a)) = \mu\}.
\end{aligned}$$

[Recall that if $f \in b^a$, then $\bar{f}$ and $\bar{f}^{-1}$ respectively denote the direct image mapping induced by f from $\wp(a)$ into $\wp(b)$ and the inverse image mapping from $\wp(b)$ into $\wp(a)$.]

11. Assume that the universe satisfies the axiom of choice. Let a be an infinite set and let λ be its cardinality. Set

$$\wp^*(a) = \{x \in \wp(a) : \mathsf{card}(x) = \mathsf{card}(a - x)\}.$$

(a) Show that for every integer n, if $n \neq 0$, we can find sets $a_1, a_2, \ldots, a_n$, each of cardinality λ, that constitute a partition of a (i.e. these sets are pairwise disjoint and $\bigcup_{1 \leq i \leq n} a_i = a$).

(b) Determine the cardinality of each of the elements of $\wp^*(a)$.

(c) Use the result from (a) for $n = 3$ to determine the cardinality of $\wp^*(a)$.

(d) Show that, for every set $a_1 \in \wp^*(a)$, there exists a bijection f from a onto a such that, for every $x \in a$, $f(x) = x$ if and only if $x \in a_1$.

(e) Determine the cardinality of the set of bijections from a onto a.

(f) Let b be an element of $\wp^*(a)$. Determine the cardinality of the set of bijections from a onto a whose restriction to b is the identity on b.

(g) What is the cardinality of the set of injections from a into $\wp(a)$?

12. Assume the universe satisfies the axiom of choice. We are given an infinite cardinal λ, an ordinal α, and a family of sets $(X_\beta)_{b \in \alpha}$ indexed by α that satisfies

$$\text{for every } \beta \in \alpha, \quad \mathsf{card}(X_\beta) < \lambda,$$

and

$$\text{for every } \beta \in \alpha \text{ and } \gamma \in \alpha, \quad \text{if } \beta < \gamma, \ X_\beta \subseteq X_\gamma.$$

Show that $\mathsf{card}(\bigcup_{\beta \in \alpha} X_\beta) \leq \lambda$.

13. Assume that the universe satises the axiom of choice. Show that for every family $(\lambda_\alpha)_{\alpha\in\kappa}$ of non-zero cardinals indexed by an innite cardinal κ, we have

$$\sum_{\alpha\in\kappa}\lambda_\alpha = \sup\Big(\kappa, \sup_{\alpha\in\kappa}(\lambda_\alpha)\Big).$$

14. Suppose that the universe satises the axiom of choice.

Let μ be an innite cardinal. By induction on the integers, we dene a sequence of cardinal $(\lambda_n)_{n\in\omega}$ by setting

- $\lambda_0 = \mu$;
- for every $n \in \omega$, $\lambda_{n+1} = 2^{\lambda_n}$.

Set $\lambda = \sum_{n\in\omega}\lambda_n$.

(a) Show that $\lambda^\mu = \mu^\lambda = \lambda^\lambda = 2^\lambda$.

(b) Show that, for every cardinal γ,

$$\text{if } \aleph_0 \le \gamma \le \lambda, \quad \text{then } \lambda^{\aleph_0} = \lambda^\gamma = \lambda^\lambda;$$
$$\text{if } \gamma \ge \lambda, \quad \text{then } \lambda^\gamma = 2^\gamma.$$

(c) Show that there exist cardinals α, β, γ, and δ such that

$$\alpha < \beta, \qquad \gamma < \delta, \quad \text{and} \quad \alpha^\gamma = \beta^\delta.$$

15. Let α and β be two ordinals. By denition, β is **conal with** α if and only if there exists a strictly increasing mapping f from β into α whose image does not have a strict upper bound. More precisely, this means that

- for all ordinals γ and δ belonging to β, if $\gamma < \delta$, then $f(\gamma) < f(\delta)$, and
- for every ordinal $\xi \in \alpha$, there exists $\gamma \in \beta$ such that $f(\gamma) \ge \xi$.

Remark This must not be confused with the notion of conal *in* : a subset Y of an ordered set $\langle X, \le\rangle$ is **conal in** X if, for every $x \in X$, there exists $y \in Y$ such that $x \le y$. Thus, for example, while ω is clearly not conal in $\aleph_\omega$, the mapping $n \mapsto \aleph_n$ witnesses that ω is conal with $\aleph_\omega$.

(a) Show that the (meta-)relation `is conal with' dened on the class On is reexive, transitive, and is not symmetric. With which ordinals is the ordinal **1** conal?

(b) Show that for every ordinal α, the class of ordinals β such that β is conal with α is a non-empty set. The least ordinal belonging to this set is called the **conality of** α and is denoted by $cof(\alpha)$. An ordinal that satises $cof(\alpha) = \alpha$ is called a **regular** ordinal.

Show that for every ordinal α, $cof(\alpha) \le \alpha$ and that $cof(\alpha)$ is a regular ordinal.

(c) Show that for all ordinals α and β, $\beta < cof(\alpha)$ if and only if every mapping from β into α is strictly bounded in α.

(d) Show that every regular ordinal is a cardinal. Show that for every cardinal λ, λ is a regular ordinal if and only if it is a regular cardinal in the sense of Denition 7.87.

(e) Assume that the universe satises the axiom of choice. Show that for every ordinal α, $\aleph_{\alpha+1}$ is regular. Show that if α is a limit ordinal, then $cof(\aleph_\alpha) = cof(\alpha)$.

(f) Determine the rst ordinal (respectively, the rst cardinal) strictly greater than ω with which ω is conal.

16. Assume that the universe satises the axiom of choice. This exercise presupposes the concepts and results from the previous exercise.

(a) Show that for every cardinal λ, $\lambda^{cof(\lambda)} > \lambda$ (use Konig's theorem).

(b) Show that ω is not conal with $\text{card}(2^\omega)$.

(c) Suppose that the universe satises the generalized continuum hypothesis (GCH), i.e. that for every ordinal α, $2^{\aleph_\alpha} = \aleph_{\alpha+1}$.

Let λ be an innite cardinal. Show that for every cardinal μ other than **0**, we have

$$\lambda^\mu = \begin{cases} \lambda & \text{if } \mu < cof(\lambda), \\ 2^\lambda & \text{if } cof(\lambda) \leq \mu \leq \lambda, \\ 2^\mu & \text{if } \lambda < \mu. \end{cases}$$

17. Let Φ be a denable strictly increasing function from the class On of ordinals into itself. We say that Φ is **continuous at a limit ordinal** α if $\Phi(\alpha) = \sup_{\beta \in \alpha} \Phi(\beta)$. Such a function Φ is called **continuous** if it is continuous at all limit ordinals.

An ordinal α such that $\Phi(\alpha) = \alpha$ is called a **xed point of** Φ.

(a) Show that every strictly increasing function Φ from On into On has the following property:

$$\text{for every ordinal } \alpha, \quad \Phi(\alpha) \geq \alpha.$$

(b) Show that if Φ is a strictly increasing function from On into On that is continuous at every limit ordinal whose conality is ω, then for every ordinal α, Φ has a xed point that is greater than α.

(c) Show that if Φ and Ψ are two strictly increasing functions from On into On that are continuous at every limit ordinal whose conality is ω, then for every ordinal α, Φ and Ψ have a common xed point greater than α.

(d) Suppose the universe satises the axiom of choice. Show that for every ordinal α, there exists an ordinal $\beta > \alpha$ such that $\text{card}(V_\beta) = \aleph_\beta = \beta$.

18. Assume the universe satises AC + GCH. (In fact, it can be proved that AC is true in every model of ZF + GCH.)

(a) Consider the function from On into On whose value at an ordinal α is $\aleph_0^{\aleph_\alpha}$. Is this function continuous at every limit ordinal? (For the denition of continuity, see Exercise 17.) Answer the same question for the function whose value at an ordinal α is $\aleph_\alpha^{\aleph_0}$.

(b) Let δ be an ordinal. Is the function whose value at an ordinal α is $\delta + \alpha$ (ordinal sum) continuous at every limit ordinal? Answer the same question for the functions whose value at an ordinal α are respectively $\alpha + \delta$, $\alpha \cdot \delta$, and $\delta \cdot \alpha$ (these ordinal operations are explained in Denitions 7.31 and 7.32).

19. Show that the axiom of foundation is equivalent (in the presence of the axioms of ZF) to the following axiom scheme:

for every formula F with one free variable in the language $\{\in, \simeq\}$,

$$\exists v_0 F[v_0] \Rightarrow \exists v_0 (F[v_0] \wedge \forall v_1 (v_1 \in v_0 \Rightarrow \neg F[v_1])).$$

20. In this exercise, we assume the axiom of choice. Let λ be an uncountable regular cardinal (see Denition 7.87). A subset X of λ is **closed conal** if

(1) it is closed: this means that for every subset X_0 of X that satises $\text{card}(X_0) < \lambda$, $\sup(X_0) \in X$ [note that because λ is regular, $\sup(X_0)$ is an ordinal that is strictly less than λ].

(2) it is conal in λ: this means that for every $\alpha \in \lambda$, there exists $\beta \in X$ such that $\alpha \leq \beta$.

(a) Show that the collection of closed conal subsets of λ forms a lter-base on λ (see Chapter 2).

(b) Show that if I is a non-empty set whose cardinality is strictly less than λ and if $(X_i : i \in I)$ is a family of closed conal subsets of λ, then $\bigcap_{i \in I} X_i$ is a closed conal subset of λ.

(c) A subset Y of λ is called **stationary** if it intersects every closed conal subset. Show that the following three properties are equivalent:

(1) there exists a pair of disjoint stationary sets;

(2) there exists at least one stationary set that does not include a closed conal set;

(3) the lter $\mathcal{F}$ generated by the collection of closed conal subsets is not an ultralter.

(d) Let $X = (X_\alpha : \alpha \in \lambda)$ be a sequence of subsets of λ. The set

$$\{\alpha \in \lambda : \alpha \in X_\alpha\}$$

is called the **diagonal intersection** of X and is denoted by $\Delta(X)$. Show that if X satises the following three conditions,

(1) for every $\alpha \in \lambda$, X_α is closed conal;

(2) for every $\alpha \in \lambda$ and $\beta \in \lambda$, if $\alpha \in \beta$, then $X_\beta \subseteq X_\alpha$;

(3) for every $\alpha \in \lambda$, if α is a limit ordinal, then $X_\alpha = \bigcap_{\beta \in \alpha} X_\beta$;

then $\Delta(X)$ is closed conal.

(e) Prove the following theorem (known as **Fodor's theorem**).

Theorem 7.93 *Let f be a mapping from λ into λ such that $\{\alpha \in \lambda : f(\alpha) < \alpha\}$ is stationary. Then there exists $\gamma \in \lambda$ such that $\bar{f}^{-1}(\gamma)$, the inverse image of γ under $\bar{f}$, is stationary.*

(f) Assume that $\lambda \geq \aleph_2$. Show that the set of ordinals in λ that have conality $\aleph_0$ (see Exercise 15) is stationary. Show that the set of ordinals in λ that have conality $\aleph_1$ is also stationary and is disjoint from the preceding set.

(g) Part (f) shows that for every regular cardinal λ strictly greater than $\aleph_1$, the conditions from part (c) are satised. The argument that we will offer in the next paragraph shows that these conditions are satised for $\aleph_1$ (indeed, the argument works for any successor cardinal).

For every denumerable ordinal α, let f_α be a surjective mapping from ω onto α and, for every $n \in \omega$, let h_n be the mapping from $\aleph_1$ into $\aleph_1$ dened by $h_n(\mathbf{0}) = \mathbf{0}$ and $h_n(\alpha) = f_\alpha(n)$ if $\alpha \neq \mathbf{0}$. Show that, for every $n \in \omega$, there exists $\beta_n \in \aleph_1$ and a stationary subset Y_n of $\aleph_1$ such that, for every $\gamma \in Y_n$, $f_\gamma(n) = \beta_n$. Show that there exists an integer n such that Y_n does not include a closed conal set.

21. (a) Show that if α is a limit ordinal strictly greater than ω, then $\langle V_\alpha, \in\rangle$ is a model of the theory Z.

(b) Conclude from (a) that if Z is a consistent theory, then the axioms of ZF cannot all be derivable from those of Z.

22. Suppose the universe satises the axiom of choice.

Consider the class $\mathcal{W}$ of sets x such that $\mathsf{cl}(x)$ (the transitive closure of x) is denumerable. Show that $\langle \mathcal{W}, \in\rangle$ is a model of all the axioms of ZF except the power set axiom.

23. Show directly, using a diagonal argument, that the half-open interval of reals, $\{x \in \mathbb{R} : 0 < x \leq 1\}$, is not denumerable.

8 Some model theory

Model theory studies the class of models of a given theory. We have already encountered two theorems that tend in this direction: the completeness theorem and the powerful compactness theorem, both of which assert that, under certain conditions, this class is not empty.

The central notion in this chapter and for the kind of model theory that we will develop here is the notion of elementary substructure. Intuitively, $\mathcal{M}$ is an elementary substructure of $\mathcal{N}$ if, obviously, $\mathcal{M}$ is a substructure of $\mathcal{N}$ and if, for every nite sequence s of elements of $\mathcal{M}$ and for every property $F[s]$ that is expressible by a rst-order formula, it is equivalent to verify that s satises F in $\mathcal{M}$ or that s satises F in $\mathcal{N}$. This notion will be our concern for the rst two sections; the important results will be the LowenheimñSkolem theorems and their corollaries which imply that a countable theory that has an innite model must have innite models of every innite cardinality.

We then continue with the interpolation theorem and the denability theorem. It is worth pausing to reect on the meaning of this latter theorem. When we wish to formalize a theory, we must rst prescribe its language; this amounts to deciding which notions should be taken as primitive and which others should be dened in terms of these (for example, in the case of arithmetic, 0, S, $+$, and $\times$ sufce; we can then dene the order relation, the notion of prime number, etc.). But how can we be sure that we have not introduced symbols that are unnecessary? The denability theorem provides a semantic criterion that answers this question.

The fourth section is devoted to reduced products and ultraproducts; these are algebraic operations that allow us to dene an L-structure from other L-structures. Ultraproducts are particularly important and permit a purely algebraic proof of the compactness theorem. In Section 8.5, we will prove some theorems of the following type: a theory T is equivalent to another theory of this or that form if and only if the class of its models is closed under this or that operation. These theorems are called preservation theorems. We will specically examine preservation under substructures, under unions of chains, and under reduced products. Finally, in the last section, we will study models of $\aleph_0$-categorical theories, i.e. theories whose countable models are all isomorphic.

The axiom of choice is necessary for nearly all the results in this chapter. So we assume, once and for all, that it is satised.

8.1 Elementary substructures and extensions

8.1.1 Elementary substructures

The following convention will be in effect for this entire chapter: calligraphic letters $\mathcal{M}, \mathcal{N}$, and so on will denote structures and the corresponding Latin letters M, N, and so on will denote the underlying sets of these structures. We systematically assume that the language includes the symbol for equality, $\simeq$, and that the structures respect equality. The denition which follows is very important and will be with us throughout the chapter.

Denition 8.1 *If L is a language, $\mathcal{M}$ is an L-structure, and $\mathcal{N}$ is a substructure of $\mathcal{M}$, we say that $\mathcal{N}$ is an* ***elementary substructure of*** *$\mathcal{M}$ (or, equivalently, that $\mathcal{M}$ is an* ***elementary extension of*** *$\mathcal{N}$) if, for every formula $F[v_1, v_2, \ldots, v_n]$ of L and for all elements $a_1, a_2, \ldots, a_n$ of N, we have*

$$\mathcal{M} \models F[a_1, a_2, \ldots, a_n] \quad \textit{if and only if} \quad \mathcal{N} \models F[a_1, a_2, \ldots, a_n].$$

We will write $\mathcal{N} \prec \mathcal{M}$ to assert that '$\mathcal{N}$ is an elementary substructure of $\mathcal{M}$'.

Recall that for $\mathcal{N}$ to be simply a substructure of $\mathcal{M}$, it is this same formal condition that must be satised, but only for atomic formulas (or, which amounts to the same thing, for formulas without quantiers). The rst question that comes to mind is to ask whether there are substructures that are not elementary substructures. Here are a few examples.

- In the language of groups, $\langle \mathbb{Z}, \mathbf{0}, + \rangle$ is a substructure of $\langle \mathbb{Q}, \mathbf{0}, + \rangle$ that is not an elementary substructure. For example, the formula

$$\forall v_0 \exists v_1 (v_1 + v_1 \simeq v_0)$$

  is satised in $\mathbb{Q}$ but not in $\mathbb{Z}$.
- Again in this same language, the structure $\langle 2\mathbb{Z}, \mathbf{0}, + \rangle$ of even integers is a substructure of $\langle \mathbb{Z}, \mathbf{0}, + \rangle$ and, moreover, these two structures are isomorphic and hence satisfy the same formulas that do not involve parameters. However, the formula $\exists v_0 (v_0 + v_0 \simeq 2)$ is satised in $\mathbb{Z}$ but not in $2\mathbb{Z}$; so this is not an elementary substructure of $\mathbb{Z}$. Here, in contrast with the previous example, we need a parameter from the smaller structure (namely, 2) to nd a formula that is true in one of the structures but not in the other.
- In the language of elds, $\mathbb{Q}$ is not an elementary substructure of $\mathbb{R}$; the formula $\exists v_0 (v_0 \times v_0 \simeq 2)$ is satised in $\mathbb{R}$ but not in $\mathbb{Q}$. Similarly, $\mathbb{R}$ is not an elementary substructure of $\mathbb{C}$; just consider the formula $\exists v_0 (v_0 \times v_0 \simeq -\mathbf{1})$ (here, all appearances to the contrary, we are not using a parameter).
- In the language for orderings, $[0, 1]$ is not an elementary substructure of $[0, 2]$; the formula $\forall v_0 (v_0 \leq 1)$ is satised in the rst but not in the second structure.

Before proceeding, we recall a denition from Section 3.5 of Chapter 3.

Denition 8.2 *Let $\mathcal{M}$ be an L-structure. The* **complete theory** *of $\mathcal{M}$, denoted by* $\mathsf{Th}(\mathcal{M})$*, is the theory*

$$\mathsf{Th}(\mathcal{M}) = \{F : F \text{ is a closed formula of } L \text{ and } \mathcal{M} \models F\}.$$

If $\mathcal{M}$ and $\mathcal{N}$ are two L-structures, we say that $\mathcal{M}$ and $\mathcal{N}$ are **elementarily equivalent** *if* $\mathsf{Th}(\mathcal{M}) = \mathsf{Th}(\mathcal{N})$*. When this is the case, we will write $\mathcal{M} \equiv \mathcal{N}$.*

It is clear that, concordant with the vocabulary, $\mathsf{Th}(\mathcal{M})$ is always a complete theory and that if two structures are isomorphic, then they are elementarily equivalent. It is an immediate consequence of the denitions that if $\mathcal{N} \prec \mathcal{M}$, then $\mathcal{M} \equiv \mathcal{N}$. The example of $\langle 2\mathbb{Z}, \mathbf{0}, +\rangle$ inside $\langle \mathbb{Z}, \mathbf{0}, +\rangle$ shows that it is possible for $\mathcal{N}$ to be a substructure of $\mathcal{M}$ and elementarily equivalent to $\mathcal{M}$ while not being an elementary substructure.

Remark 8.3 It also follows from the denitions that

- if $\mathcal{M}_1 \prec \mathcal{M}_2$ and $\mathcal{M}_2 \prec \mathcal{M}_3$, then $\mathcal{M}_1 \prec \mathcal{M}_3$;
- if $\mathcal{M}_1 \prec \mathcal{M}_3$, $\mathcal{M}_1 \subseteq \mathcal{M}_2$, and $\mathcal{M}_2 \prec \mathcal{M}_3$, then $\mathcal{M}_1 \prec \mathcal{M}_2$.

In general, it is rather difcult to show that a substructure is elementary. In the next example, we will illustrate a technique that is very useful for this type of problem.

Example 8.4 (*Dense ordering without endpoints*) Consider the following theory T in the language that contains only a single binary relation symbol, $<$:

(i) $\forall v_0(\neg v_0 < v_0)$;

(ii) $\forall v_0 \forall v_1((v_0 < v_1 \Leftrightarrow \neg v_1 < v_0) \vee v_0 \simeq v_1)$;

(iii) $\forall v_0 \forall v_1 \forall v_2((v_0 < v_1 \wedge v_1 < v_2) \Rightarrow v_0 < v_2)$;

(iv) $\forall v_0 \exists v_1 (v_0 < v_1)$;

(v) $\forall v_0 \exists v_1 (v_1 < v_0)$;

(vi) $\forall v_0 \forall v_1 \exists v_2 (v_0 < v_1 \Rightarrow (v_0 < v_2 \wedge v_2 < v_1))$.

The rst three axioms express that $<$ is a total ordering, the next two axioms that there is no maximum or minimum element, and the last that the ordering is dense, i.e. between any two distinct elements, there is always a third. It is clear that any model of T is necessarily innite. We will prove that if $\mathcal{M}$ and $\mathcal{N}$ are two models of T and if $\mathcal{M} \subseteq \mathcal{N}$, then $\mathcal{M} \prec \mathcal{N}$. We are dealing here with a very strong property of the theory T (see Exercise 8). We will begin by proving two lemmas.

Lemma 8.5 *Let $a_1, a_2, \ldots, a_n \in M$ and $b_1, b_2, \ldots, b_n \in N$ and assume that these two sequences satisfy the same atomic formulas in $\mathcal{M}$ and in $\mathcal{N}$, respectively; in other words, for every i and j in $\{1, 2, \ldots, n\}$,*

$$\mathcal{M} \models a_i \simeq a_j \quad \text{if and only if} \quad \mathcal{N} \models b_i \simeq b_j, \quad \text{and}$$
$$\mathcal{M} \models a_i < a_j \quad \text{if and only if} \quad \mathcal{N} \models b_i < b_j.$$

Then for every $a_0 \in M$, *there exists* $b_0 \in N$ *such that, for all* i *and* j *in* $\{0, 1, 2, \ldots, n\}$,

$$\mathcal{M} \models a_i \simeq a_j \quad \textit{if and only if} \quad \mathcal{N} \models b_i \simeq b_j, \quad \textit{and}$$
$$\mathcal{M} \models a_i < a_j \quad \textit{if and only if} \quad \mathcal{N} \models b_i < b_j;$$

and for every $b_0 \in N$, *there exists* $a_0 \in M$ *such that, for all* i *and* j *in* $\{0, 1, 2, \ldots, n\}$,

$$\mathcal{M} \models a_i \simeq a_j \quad \textit{if and only if} \quad \mathcal{N} \models b_i \simeq b_j, \quad \textit{and}$$
$$\mathcal{M} \models a_i < a_j \quad \textit{if and only if} \quad \mathcal{N} \models b_i < b_j.$$

Proof Assume that a_0 is given and that we must nd b_0. There are several cases to consider:

- a_0 is greater (in the sense of the ordering of $\mathcal{M}$) than all the a_i (for $1 \leq i \leq n$). Then choose b_0 in N to be any element that is greater (in the sense of the ordering of $\mathcal{N}$) than all the b_i $(1 \leq i \leq n)$. Such a point exists since there is no maximum element in $\mathcal{N}$.
- If a_0 is smaller than all the a_i (for $1 \leq i \leq n$), we take b_0 to be any element of N that is smaller than all the b_i $(1 \leq i \leq j)$.
- If a_0 is equal to one of the a_i, say a_k, then choose $b_0 = b_k$.
- In the remaining case, choose an index $p \in \{1, 2, \ldots, n\}$ such that a_p is the smallest (in the sense of the ordering of $\mathcal{M}$) of those a_i (for $1 \leq i \leq n$) that are greater than a_0 and choose an index $q \in \{1, 2, \ldots, n\}$ such that a_q is the greatest of the a_i (for $1 \leq i \leq n$) that are less than a_0. We have $a_q < a_0 < a_p$ and hence $b_q < b_p$. We then choose a b_0 in N which is strictly between b_q and b_p (this is possible because the ordering on $\mathcal{N}$ is dense).

Obviously, we do the same thing to nd a_0 when b_0 is given. ■

Lemma 8.6 *Let* $\mathcal{M}$ *and* $\mathcal{N}$ *be two models of* T, *let* $a_1, a_2, \ldots, a_n \in M$ *and* $b_1, b_2, \ldots, b_n \in N$, *and assume that these two sequences satisfy the same atomic formulas in* $\mathcal{M}$ *and in* $\mathcal{N}$, *respectively; then for every formula* $F[v_1, v_2, \ldots, v_n]$ *of* L,

$$\mathcal{M} \models F[a_1, a_2, \ldots, a_n] \quad \textit{if and only if} \quad \mathcal{N} \models F[b_1, b_2, \ldots, b_n]. \qquad (*)$$

With this lemma in hand, we may easily conclude that if $\mathcal{M}$ and $\mathcal{N}$ are two models of T and if $\mathcal{M}$ is a substructure of $\mathcal{N}$, then $\mathcal{M} \prec \mathcal{N}$; to see this, note that if $a_1, a_2, \ldots, a_n$ are elements of M, then the atomic formulas satised in $\mathcal{M}$ by this sequence are the same as those satised in $\mathcal{N}$ (because $\mathcal{M}$ is a substructure of $\mathcal{N}$). So the hypotheses of the lemma are satised; thus, for every formula $F[v_1, v_2, \ldots, v_n]$ of L,

$$\mathcal{M} \models F[a_1, a_2, \ldots, a_n] \quad \text{if and only if} \quad \mathcal{N} \models F[a_1, a_2, \ldots, a_n];$$

in other words, $\mathcal{M} \prec \mathcal{N}$.

So it remains to prove Lemma 8.6.

Proof We will assume that no universal quantiers occur in F (which we may assume, without loss of generality, subject to replacing F by an equivalent formula, as in Remark 3.57) and will prove the lemma by induction on F.

The hypothesis of the lemma says that the condition $(*)$ holds if F is an atomic formula. It is immediate that if $(*)$ holds for the formulas F_1 and F_2, then it also holds for the formulas $\neg F_1$, $F_1 \wedge F_2$, $F_1 \vee F_2$, $F_1 \Rightarrow F_2$, and $F_1 \Leftrightarrow F_2$. The remaining case is when

$$F[v_1, v_2, \ldots, v_n] = \exists v_0 G[v_1, v_2, \ldots, v_n].$$

Assume (in addition to the fact that the sequences $a_1, a_2, \ldots, a_n \in M$ and $b_1, b_2, \ldots, b_n \in N$ satisfy the same atomic formulas in $\mathcal{M}$ and in $\mathcal{N}$, respectively) that

$$\mathcal{M} \models F[a_1, a_2, \ldots, a_n].$$

So there exists an $a_0 \in M$ such that

$$\mathcal{M} \models G[a_0, a_1, a_2, \ldots, a_n].$$

We choose an element $b_0 \in N$ whose situation with respect to the b_i $(1 \leq i \leq n)$ is exactly the same as that of a_0 with respect to the a_i; in other words, such that the sequences $(a_0, a_1, a_2, \ldots, a_n)$ and $(b_0, b_1, b_2, \ldots, b_n)$ continue to satisfy the same atomic formulas in $\mathcal{M}$ and in $\mathcal{N}$, respectively (Lemma 8.5). Since the formula G has one fewer quantier than the formula F, we may apply the induction hypothesis to it and conclude that

$$\mathcal{N} \models G[b_0, b_1, b_2, \ldots, b_n],$$

and hence that

$$\mathcal{N} \models F[b_0, b_1, b_2, \ldots, b_n].$$

By interchanging the roles of $\mathcal{M}$ and $\mathcal{N}$, we can show in exactly the same way that if $\mathcal{N} \models F[b_0, b_1, b_2, \ldots, b_n]$, then $\mathcal{M} \models F[a_1, a_2, \ldots, a_n]$. ■

8.1.2 The TarskiñVaught test

The next result, known as the **TarskiñVaught test**, is, on occasion, a practical way to verify that a substructure is elementary.

Theorem 8.7 *Let $\mathcal{M}$ be a structure, let $\mathcal{N}$ be a substructure of $\mathcal{M}$, and assume that for every formula $F[v_0, v_1, \ldots, v_n]$ of L and for all elements $a_1, a_2, \ldots, a_n$ in N, if*

$$\mathcal{M} \models \exists v_0 F[v_0, a_1, a_2, \ldots, a_n],$$

then there exists a_0 in N such that

$$\mathcal{M} \models F[a_0, a_1, a_2, \ldots, a_n].$$

Then $\mathcal{N} \prec \mathcal{M}$.

The difference between this and Denition 8.1 is that satisfaction of formulas in only one of the two structures (the larger one) is involved.

Proof We will prove that for every formula $G[v_1, v_2, \ldots, v_n]$ and for all $a_1, a_2, \ldots, a_n$ in N,

$$\mathcal{N} \models G[a_1, a_2, \ldots, a_n] \quad \text{if and only if} \quad \mathcal{M} \models G[a_1, a_2, \ldots, a_n].$$

As we did earlier, we may assume (see Remark 3.57) that no universal quantier occurs in G and argue by induction on G. The cases which concern propositional connectives offer no difculty. So consider the case where $G[v_1, v_2, \ldots, v_n] = \exists v_0 F[v_0, v_1, v_2, \ldots, v_n]$ (thus F has one fewer quantier than G).

- If $\mathcal{N} \models \exists v_0 F[v_0, a_1, a_2, \ldots, a_n]$, then there exists $a_0 \in N$ such that $\mathcal{N} \models F[a_0, a_1, a_2, \ldots, a_n]$ and, by the induction hypothesis, we see that $\mathcal{M} \models F[a_0, a_1, a_2, \ldots, a_n]$; consequently,

$$\mathcal{M} \models \exists v_0 F[v_0, a_1, a_2, \ldots, a_n].$$

- Conversely, if $\mathcal{M} \models \exists v_0 F[v_0, a_1, a_2, \ldots, a_n]$, then by the hypothesis of the theorem, there exists $a_0 \in N$ such that $\mathcal{M} \models F[a_0, a_1, a_2, \ldots, a_n]$. By the induction hypothesis, we obtain $\mathcal{N} \models F[a_0, a_1, a_2, \ldots, a_n]$, and hence $\mathcal{N} \models \exists v_0 F[v_0, a_1, a_2, \ldots, a_n]$. ■

Remark 8.8 In fact, in applying the TarskiñVaught test, there is no need to rst verify that $\mathcal{N}$ is a substructure of $\mathcal{M}$; to see this, notice that if A is any subset of M that satises

for every formula $F[v_0, v_1, \ldots, v_n]$ of L
and for all elements $a_1, a_2, \ldots, a_n$ in N,
if $\mathcal{M} \models \exists v_0 F[v_0, a_1, a_2, \ldots, a_n]$,
then there exists a_0 in N such that $\mathcal{M} \models F[a_0, a_1, a_2, \ldots, a_n]$,

then A will be closed under the functions of the language; for if f is a k-ary function, it sufces to consider the formula $F = v_0 \simeq f v_1 v_2 \ldots v_k$. In other words, A is a substructure of $\mathcal{M}$ (A is non-empty since $\mathcal{M} \models \exists v_0\, v_0 \simeq v_0$).

We will give an example which applies the TarskiñVaught test. Before we begin, let us mention that the cardinality of a language L, denoted by $\mathsf{card}(L)$, is equal, by denition, to the cardinality of the set of formulas of L; so it is equal to $\sup(\aleph_0, \mathsf{card}(X))$, where X denotes the set of symbols for constants, functions, and relations of L. The cardinality of a structure is, naturally, the cardinality of its underlying set. The next theorem is known as the **downward LowenheimñSkolem theorem**.

Theorem 8.9 *Let $\mathcal{M}$ be an L-structure, let A be any subset of M, and suppose that* $\mathsf{card}(M) \geq \mathsf{card}(L)$. *Then there exists an elementary substructure $\mathcal{M}_0$ of $\mathcal{M}$ that includes A and whose cardinality is* $\sup(\mathsf{card}(A), \mathsf{card}(L))$.

Proof We may assume that $\mathsf{card}(A) \geq \mathsf{card}(L)$ without loss of generality by arbitrarily expanding A. Next, we note that if B is a subset of M and if $\mathsf{card}(B) \geq \mathsf{card}(L)$, then the substructure of $\mathcal{M}$ generated by B (i.e. the smallest subset N of M which includes B and is closed under the functions of the language) has the same cardinality as B (for every element of N is the interpretation of a term with parameters from B and the set of such terms is a set of nite sequences from $L \times B$, so its cardinality is less than or equal to that of B).

By induction on the integer n, we dene subsets $A_0 \subseteq A_1 \subseteq A_2 \subseteq \cdots \subseteq A_n \subseteq \cdots$ of M which all have cardinality equal to $\mathsf{card}(A)$.

- A_0 is the substructure generated by A.
- To dene A_{i+1} from A_i, we proceed as follows: for every formula $F[v_0, v_1, \ldots, v_n]$ of L and for every sequence $(a_1, a_2, \ldots, a_n)$ of elements of A_i, if $\mathcal{M} \models \exists v_0 F[v_0, a_1, a_2, \ldots, a_n]$, then we choose an element $c_{F,a_1,a_2,\ldots,a_n}$ of M such that

$$\mathcal{M} \models F[c_{F,a_1,a_2,\ldots,a_n}, a_1, a_2, \ldots, a_n];$$

then set

$$\begin{aligned} B_i = A_i \cup \{c_{F,a_1,a_2,\ldots,a_n} :\ & n \in \mathbb{N},\ F[v_0, v_1, \ldots, v_n] \text{ is a formula of } L, \\ & a_1, a_2, \ldots, a_n \text{ are in } M \text{ and } \mathcal{M} \models \exists v_0 F[v_0, a_1, a_2, \ldots, a_n]\}, \end{aligned}$$

and let A_{i+1} be the substructure of $\mathcal{M}$ generated by B_i. There are $\mathsf{card}(L)$ formulas F in L and $\mathsf{card}(A_i)$ sequences $(a_1, a_2, \ldots, a_n)$ in A_i of the appropriate length. So we need to add at most $\mathsf{card}(A_i)$ points to A_i to obtain B_i; this shows that

$$\mathsf{card}(A_{i+1}) = \mathsf{card}(B_i) = \mathsf{card}(A_i) = \mathsf{card}(A).$$

We set $\mathcal{M}_0 = \bigcup_{i \in \mathbb{N}} A_i$. It is clear that $\mathcal{M}_0$ is a substructure of $\mathcal{M}$ and that its cardinality is $\mathsf{card}(A)$ (see Corollary 7.72). We will now use the TarskiñVaught test to show that it is an elementary substructure.

Let $F[v_0, v_1, \ldots, v_n]$ be a formula of L, let $a_1, a_2, \ldots, a_n$ be elements of M_0, and suppose that

$$\mathcal{M} \models \exists v_0 F[v_0, a_1, a_2, \ldots, a_n].$$

We know that there exists an integer i such that A_i contains $a_1, a_2, \ldots, a_n$. By the construction of A_{i+1}, this set, and hence M_0, contains a point c such that $\mathcal{M} \models F[c, a_1, a_2, \ldots, a_n]$; this is exactly what is required by the hypothesis of the TarskiñVaught test. ■

8.2 Construction of elementary extensions

8.2.1 Elementary maps

Here is another important concept.

Denition 8.10 *Let $\mathcal{M}$ and $\mathcal{N}$ be two L-structures and let h be a mapping from M into N; the mapping h is called an* **elementary mapping** *if, for every formula $F[v_1, v_2, \ldots, v_n]$ of L and for all elements $a_1, a_2, \ldots, a_n$ of M, we have*

$$\mathcal{M} \models F[a_1, a_2, \ldots, a_n] \quad \textit{if and only if} \quad \mathcal{N} \models F[h(a_1), h(a_2), \ldots, h(a_n)].$$

It is immediate, when we consider the formula $v_0 \simeq v_1$, that an elementary mapping is injective. To insist on this fact, we will sometimes say **elementary embedding** rather than elementary mapping. If there exists an elementary mapping from $\mathcal{M}$ into $\mathcal{N}$, we will say that $\mathcal{M}$ can be **elementarily embedded in** $\mathcal{N}$. It is also clear that an elementary mapping is a monomorphism of L-structures. The converse is not true; to be convinced, recall the example of a substructure $\mathcal{M}$ of $\mathcal{N}$ that is not elementary (see the previous section). Nonetheless, we do have the next proposition.

Proposition 8.11 *Let h be a monomorphism from a structure $\mathcal{M}$ into a structure $\mathcal{N}$. Then h is an elementary mapping if and only if the image of h is an elementary substructure of $\mathcal{N}$.*

Proof Denote the image of h by $\mathcal{N}_1$; thus h is an isomorphism from $\mathcal{M}$ onto $\mathcal{N}_1$. So it follows from Theorem 3.72 that, for every formula $F[v_1, v_2, \ldots, v_n]$ of L and for all elements $a_1, a_2, \ldots, a_n$ of M, we have

$$\mathcal{M} \models F[a_1, a_2, \ldots, a_n] \quad \text{if and only if} \quad \mathcal{N}_1 \models F[h(a_1), h(a_2), \ldots, h(a_n)].$$

Suppose rst that $\mathcal{N}_1 \prec \mathcal{N}$. Then

$$\mathcal{N}_1 \models F[h(a_1), h(a_2), \ldots, h(a_n)] \quad \text{if and only if}$$
$$\mathcal{N} \models F[h(a_1), h(a_2), \ldots, h(a_n)],$$

which, together with the equivalence above, implies that h is elementary.

Conversely, suppose that h is elementary and let $b_1, b_2, \ldots, b_n$ be elements of N_1. There exist $a_1, a_2, \ldots, a_n$ in M such that $h(a_1) = b_1$, $h(a_2) = b_2$, $\ldots$, $h(a_n) = b_n$. For every formula $F[v_1, v_2, \ldots, v_n]$ of L, we have

$$\mathcal{M} \models F[a_1, a_2, \ldots, a_n] \quad \text{if and only if} \quad \mathcal{N} \models F[h(a_1), h(a_2), \ldots, h(a_n)],$$

and hence

$$\mathcal{N} \models F[b_1, b_2, \ldots, b_n] \quad \text{if and only if} \quad \mathcal{N}_1 \models F[b_1, b_2, \ldots, b_n]. \qquad \blacksquare$$

Corollary 8.12 *If there exists an elementary mapping from $\mathcal{M}$ into $\mathcal{N}$, then $\mathcal{M}$ and $\mathcal{N}$ are elementarily equivalent.*

Proof Indeed, if $\mathcal{N}_1$ denotes the image of this elementary mapping, then $\mathcal{N}_1 \equiv \mathcal{N}$ since $\mathcal{N}_1 \prec \mathcal{N}$ and $\mathcal{N}_1 \equiv \mathcal{M}$ by the isomorphism. ■

8.2.2 The method of diagrams

We will now describe the **method of diagrams** which allows us to construct extensions and elementary extensions. This method was, incidentally, already sketched in Section 3.5 of Chapter 3. Let $\mathcal{M}$ be an L-structure and consider the language L_M obtained by adjoining to L a constant symbol $\underline{a}$ for each element a of M. Then there is a natural enrichment of $\mathcal{M}$ to an L_M-structure which we will denote by $\mathcal{M}^*$; simply interpret $\underline{a}$ by a. Set

$$D(\mathcal{M}) = \{F[\underline{a}_1, \underline{a}_2, \ldots, \underline{a}_n] : F[v_1, v_2, \ldots, v_n] \text{ is a formula of } L, \\ a_1, a_2, \ldots, a_n \in M \text{ and } \mathcal{M} \models F[a_1, a_2, \ldots, a_n]\}.$$

We see that $D(\mathcal{M})$, called the **complete diagram of** $\mathcal{M}$, is the complete theory of $\mathcal{M}^*$. What is important is that any other model of $D(\mathcal{M})$, or, more precisely, the reduct to the language L of any other model of $D(\mathcal{M})$, is, up to isomorphism, an elementary extension of $\mathcal{M}$. Let us explain.

Let $\mathcal{N}^*$ be an L_M-structure that is a model of $D(\mathcal{M})$. Let $\mathcal{N}$ denote the reduct of $\mathcal{N}^*$ to the language L (thus $\mathcal{N}$ is obtained from $\mathcal{N}^*$ by ignoring the interpretations of the constant symbols $\underline{a}$ for $a \in M$). For each $a \in M$, let $g(a)$ denote the interpretation of $\underline{a}$ in $\mathcal{N}^*$. Thus, g is a mapping from M into N and, for every formula $F[v_1, v_2, \ldots, v_n]$ of L and for all elements $a_1, a_2, \ldots, a_n$ of M, we have

$$\mathcal{M} \models F[a_1, a_2, \ldots, a_n] \quad \text{if and only if} \\ \mathcal{N} \models F[g(a_1), g(a_2), \ldots, g(a_n)]. \qquad (*)$$

[The reason is that both these conditions are equivalent in turn to the fact that $F[\underline{a}_1, \underline{a}_2, \ldots, \underline{a}_n] \in D(\mathcal{M})$, the rst by denition of $D(\mathcal{M})$, the second because $\mathcal{N}^*$ is a model of $D(\mathcal{M})$, and because the symbols $\underline{a}_i$ are interpreted there by $g(a_i)$ and because $D(\mathcal{M})$ is a complete theory.]

In other words, g is an elementary mapping from $\mathcal{M}$ into $\mathcal{N}$. So we are not far from the goal we had set: $\mathcal{N}$ is not an elementary extension of $\mathcal{M}$ but is only an elementary extension of a structure that is isomorphic to $\mathcal{M}$, namely the image of g. To repair this imperfection, we will prove that we may assume that, for all $a \in M$, $g(a) = a$; to do this, we will proceed with a purely formal construction that is altogether uninteresting. This will be done in the next lemma, which will then be used several times in the remainder of this section. Before we start, recall (from Section 3.5 of Chapter 3) that the simple diagram of a structure $\mathcal{M}$, denoted by $\Delta(\mathcal{M})$, is the following theory in L_M:

$$\Delta(\mathcal{M}) = \{H[\underline{a}_1, \underline{a}_2, \ldots, \underline{a}_n] : H[v_1, v_2, \ldots, v_n] \text{ is a quantier-free formula of } L, a_1, a_2, \ldots, a_n \text{ are elements of } M, \text{ and } \mathcal{M} \models H[a_1, a_2, \ldots, a_n]\}.$$

Lemma 8.13 *If $\mathcal{M}$ is an L-structure, then every model of $\Delta(\mathcal{M})$ is isomorphic to an extension of $\mathcal{M}^*$ (in which each symbol $\underline{a}$, for $a \in M$, is therefore interpreted by a).*

Proof Let $\mathcal{N}$ be a model of $\Delta(\mathcal{M})$ and let g denote the map from M into N that, with each $a \in M$, associates the interpretation of $\underline{a}$ in $\mathcal{N}$. Let M_1 be a set that includes M and is such that $M_1 - M$ has the same cardinality as $N - g(M)$. The mapping g can then be extended to a bijection g_1 from M_1 onto N. We dene an L_M-structure $\mathcal{M}_1$, whose base set is M_1, by requiring that g_1 be an isomorphism from $\mathcal{M}_1$ onto $\mathcal{N}$; to do this, interpret each of the symbols $\underline{a}$, for $a \in M$, by the corresponding element a; then, if R is a p-ary relation symbol, the interpretation of R in $\mathcal{M}_1$ is

$$\{(a_1, a_2, \ldots, a_p) \in M_1^p : \mathcal{N} \models Rg_1(a_1)g_1(a_2)\ldots g_1(a_p)\}.$$

The interpretations of the constant symbols and the function symbols are dened analogously. It is quite clear that the structure $\mathcal{M}_1$ is an extension of $\mathcal{M}^*$. ■

If, as we assumed above, $\mathcal{N}$ is a model of $D(\mathcal{M})$ [and not merely of $\Delta(\mathcal{M})$], then $\mathcal{M}$ is an elementary substructure of the reduct of $\mathcal{M}_1$ to L; to see this, let $F[v_1, v_2, \ldots, v_n]$ be a formula of L and let $a_1, a_2, \ldots, a_n$ be points of M. We have

$\mathcal{M}_1 \models F[a_1, a_2, \ldots, a_n]$
if and only if
$\mathcal{N} \models F[g_1(a_1), g_1(a_2), \ldots, g_1(a_n)]$ (because g_1 is an isomorphism)
if and only if
$\mathcal{N} \models F[g(a_1), g(a_2), \ldots, g(a_n)]$ (because g_1 is an extension of g)
if and only if
$\mathcal{M} \models F[a_1, a_2, \ldots, a_n]$ (by condition $(*)$).

Here is a rst application of this technique.

Theorem 8.14 *Every innite structure $\mathcal{M}$ has a proper elementary extension (i.e. an elementary extension that is different from $\mathcal{M}$ itself).*

Proof Add a new constant symbol c to L_M and consider the following theory T' in the resulting language:

$$T' = D(\mathcal{M}) \cup \{\neg c \simeq \underline{a} : a \in M\}.$$

We begin by observing that, by virtue of the compactness theorem, this theory has a model since every nite subset of T' is included in a set of the form $D(\mathcal{M}) \cup \{\neg c \simeq \underline{a} : a \in A\}$, where A is a nite subset of M; to obtain a model of such a set, it sufces to enrich $\mathcal{M}^*$ by taking any point of M that does not belong to A as the interpretation of c (which is possible since A is nite and M is innite).

Let $\mathcal{N}^*$ be a model of T' and let $\mathcal{N}$ be its reduct to the language L. We have seen that we may assume that $\mathcal{N}$ is an elementary extension of $\mathcal{M}$. It is obvious that the interpretation of the symbol c in $\mathcal{N}^*$ cannot belong to M; this shows that $\mathcal{N} \neq \mathcal{M}$. ∎

To spare the notation, when we apply this method, we will dispense with the distinction between the structures $\mathcal{M}$ and $\mathcal{M}^*$. This is, of course, an abuse of language but it presents no danger.

The same idea, applied in a more daring manner, gives us the **upward LowenheimñSkolem theorem**:

Theorem 8.15 *Let $\mathcal{M}$ be an innite L-structure and let κ be a cardinal that satises $\kappa \geq \sup(\mathsf{card}(M), \mathsf{card}(L))$. Then there exists an elementary extension $\mathcal{N}$ of $\mathcal{M}$ whose cardinality is κ.*

Proof It will sufce, in fact, to construct an $\mathcal{N}$ of cardinality greater than or equal to κ such that $\mathcal{N} \succ \mathcal{M}$. Once this is done, we choose a subset A of N that includes M and that has cardinality κ, then construct $\mathcal{N}_0$, using the downward LowenheimñSkolem theorem (Theorem 8.9), such that $M \subseteq N_0$, $\mathcal{N}_0 \prec \mathcal{N}$ and $\mathsf{card}(N_0) = \kappa$. We have already noted (Remark 8.3) that this implies $\mathcal{M} \prec \mathcal{N}_0$.

For each $i \in \kappa$, we introduce a new constant symbol c_i and we consider the theory

$$T' = D(\mathcal{M}) \cup \{\neg c_i \simeq c_j : i \in \kappa,\ j \in \kappa \text{ and } i \neq j\}.$$

This theory is consistent since every nite subset of T' is included in a set of the form $D(\mathcal{M}) \cup \{\neg c_i \simeq c_j : i, j \in A \text{ and } i \neq j\}$, where A is a nite subset of κ; such a subset has a model, for it sufces to interpret the c_i, for $i \in A$, by pairwise distinct elements of M; this is possible since M is innite.

We complete the proof as before. Let $\mathcal{N}$ be a model of T' and assume, as we did above, that $\mathcal{N} \succ \mathcal{M}$. Then the interpretations in $\mathcal{N}$ of the c_i, for $i \in \kappa$, are distinct points, so the cardinality of $\mathcal{N}$ is at least equal to κ. ∎

The corollary which follows is an immediate consequence of the two Lowenheimñ Skolem theorems.

Corollary 8.16 *Let T be a theory in a language L and let κ be a cardinal that is greater than or equal to* $\text{card}(L)$. *If T has an innite model, then it has a model of cardinality κ.*

For the next theorem, known as **Vaught's theorem**, we will need the following denition.

Denition 8.17 *Let T be a theory and let κ be a cardinal. T is called* ***κ-categorical*** *if, rst of all, T has a model of cardinality κ and, second, if all the models of cardinality κ are isomorphic.*

Theorem 8.18 *Let T be a theory in a language L and assume that T does not have a nite model. Suppose that T is κ-categorical for a cardinal κ that is greater than or equal to* $\text{card}(L)$. *Then T is complete.*

Proof Assume the contrary and let F be a closed formula of L such that $T_1 = T \cup \{F\}$ and $T_2 = T \cup \{\neg F\}$ are both consistent. From the upward Lowenheimñ Skolem theorem, there exist models $\mathcal{M}_1$ of T_1 and $\mathcal{M}_2$ of T_2 of cardinality κ; $\mathcal{M}_1$ and $\mathcal{M}_2$ cannot be isomorphic, but this contradicts κ-categoricity. ■

Example 8.19 (*Dense orderings without endpoints*) Let us re-examine the theory T of dense orderings without endpoints which we presented earlier in Example 8.4. This theory clearly does not have any nite models. We will prove that it is $\aleph_0$-categorical; this will imply, together with Vaught's theorem, that this theory is complete.

Let $\mathcal{M}$ and $\mathcal{N}$ be two denumerable models of T. We will produce an isomorphism between these two structures by a \`**back-and-forth**' technique. We begin by invoking the denumerability of these two models to nd enumerations of M and N:

$$M = \{m_i : i \in \mathbb{N}\} \quad \text{and} \quad N = \{n_i : i \in \mathbb{N}\}.$$

By induction, we will dene two sequences

$$(a_p : p \in \mathbb{N}) \quad \text{and} \quad (b_p : p \in \mathbb{N})$$

such that, for every integer p,

$a_p \in M$,
$b_p \in N$, and
the sequences $(a_0, a_1, \ldots, a_{p-1})$ and $(b_0, b_1, \ldots, b_{p-1})$
satisfy the same atomic formulas in $\mathcal{M}$ and in $\mathcal{N}$, respectively.

To dene a_p and b_p, we distinguish two cases:

- If p is even, say $p = 2i$, we set $a_p = m_i$; we conclude from Lemma 8.5 that there exists a point in N which we will call b_p with the property that the sequences $(a_0, a_1, \ldots, a_p)$ and $(b_0, b_1, \ldots, b_p)$ continue to satisfy the same atomic formulas in $\mathcal{M}$ and in $\mathcal{N}$, respectively.

- If p is odd, say $p = 2i + 1$, we set $b_p = n_i$ and we choose a_p in M such that $(a_0, a_1, \ldots, a_p)$ and $(b_0, b_1, \ldots, b_p)$ satisfy the same atomic formulas in $\mathcal{M}$ and in $\mathcal{N}$, respectively.

Assume, for example, that $q \leq p$ and note that $a_p = a_q$ if and only if $b_p = b_q$ [because $(a_0, a_1, \ldots, a_p)$ and $(b_0, b_1, \ldots, b_p)$ satisfy the same atomic formulas]. So we are able to dene a mapping f from $\{a_k : k \in \mathbb{N}\}$ into $\{b_k : k \in \mathbb{N}\}$ by setting $f(a_k) = b_k$ for all $k \in \mathbb{N}$. By the way the a_p were chosen for even p, we see that $\{a_k : k \in \mathbb{N}\} = M$ and the choice of b_p for odd p guarantees that $\{b_k : k \in \mathbb{N}\} = N$; thus f is a bijection from M into N. It is an isomorphism from $\mathcal{M}$ into $\mathcal{N}$ because, for every p, the sequences $(a_0, a_1, \ldots, a_p)$ and $(b_0, b_1, \ldots, b_p)$ satisfy the same atomic formulas.

Example 8.20 (*Divisible torsion-free abelian groups*) In the language of groups, $\{\mathbf{0}, +\}$, we consider the following theory:

(i) $\forall v_0 \forall v_1 \forall v_2\, (v_0 + v_1) + v_2 \simeq v_0 + (v_1 + v_2)$;

(ii) $\forall v_0 \forall v_1\, v_0 + v_1 \simeq v_1 + v_0$;

(iii) $\forall v_0\, v_0 + \mathbf{0} \simeq v_0$;

(iv) $\forall v_0 \exists v_1\, v_0 + v_1 \simeq \mathbf{0}$;

(v) $\exists v_0\, (\neg v_0 \simeq \mathbf{0})$;

(vi) $\forall v_0\, (n \cdot v_0 \simeq \mathbf{0} \Rightarrow v_0 \simeq \mathbf{0})$ for every positive n;

(vii) $\forall v_0 \exists v_1\, v_0 \simeq n \cdot v_1$ for every positive n.

[In (vi) and (vii) above, $n \cdot v$ denotes the term $(\ldots((v + v) + v)\ldots) + v$ that contains n occurrences of the symbol v.]

We have here an innite theory [because of (vi) and (vii) which are, in fact, axiom schemes]. Axioms (i)ñ(iv) express the fact that we are dealing with an abelian group, axiom (v) asserts that it is not trivial, axiom (vi) that the group is **torsion-free**, and axiom (vii) that it is **divisible**. We will use Vaught's theorem to show that this theory is complete.

The group $\langle \mathbb{Q}, \mathbf{0}, + \rangle$ is a model of T, so T is consistent. But T has other models; for example, let V be a vector space over $\mathbb{Q}$ in which $\times$ denotes scalar multiplication. Then the group G that underlies V is a divisible torsion-free abelian group; for if $p \in \mathbb{N}$ and $a \in V$,

$$\begin{aligned} p \times a &= (\mathbf{1} + \mathbf{1} + \cdots + \mathbf{1}) \times a \\ &= \mathbf{1} \times a + \mathbf{1} \times a + \cdots + \mathbf{1} \times a \\ &= a + a + \cdots + a = p \cdot a \end{aligned}$$

and if $p \cdot a = \mathbf{0}$ with $p \neq 0$, then $\mathbf{0} = p^{-1} \times (p \cdot a) = p^{-1} \times (p \times a) = a$, which shows that G is torsion-free. As for divisibility, we certainly have $p \cdot (p^{-1} \times a) = a$.

There is, in fact, a bijective correspondence between vector spaces over $\mathbb{Q}$ and divisible torsion-free abelian groups. If V is a vector space over $\mathbb{Q}$, we have just seen that its underlying group, which we will denote by V^-, is a divisible torsion-free abelian group; conversely, if G is a divisible torsion-free abelian group, then there exists a unique vector space over $\mathbb{Q}$, which we denote by G^+, whose underlying group is G; in other words, $(G^+)^- = G$. If $a \in G$ and $r = p/q$ (with $p \in \mathbb{N}$, $q \in \mathbb{N}^*$), $r \times a$ must be that element x of G, which is unique, such that $q \cdot x = p \cdot a$, and if r is negative, $r \times a$ must equal $-((-r) \times a)$. Moreover, we can see that if G and G' are two divisible torsion-free abelian groups and h is a mapping from G into G', then h is an isomorphism of groups if and only if h is an isomorphism of vector spaces over $\mathbb{Q}$ from G^+ into G'^+.

In particular, this shows that T is not $\aleph_0$-categorical; the two-dimensional and the three-dimensional vector spaces over $\mathbb{Q}$ are both countable, but they give rise to models of T that are not isomorphic.

To show that T is $\aleph_1$-categorical (in fact, T is λ-categorical for every uncountable cardinal λ), it sufces to show that any two vector spaces over $\mathbb{Q}$ of cardinality $\aleph_1$ are necessarily isomorphic (the reason is that they must each have a basis of cardinality $\aleph_1$; this is a small exercise concerning cardinalities).

Yet another remark: there are complete theories that only have innite models and that are not categorical in any innite cardinality (see Exercises 4, 6, and 11).

To conclude this section, here is a procedure for constructing L-structures which, under the right hypotheses, yields elementary extensions. Let $\langle I, <\rangle$ be a totally ordered set and, for every $i \in I$, let $\mathcal{M}_i$ be an L-structure; assume in addition that if $i < j$, then $\mathcal{M}_i$ is a substructure of $\mathcal{M}_j$. Set $M = \bigcup_{i \in I} M_i$. We can easily, and in a unique fashion, build an L-structure $\mathcal{M}$, also denoted by $\bigcup_{i \in} \mathcal{M}_i$, whose base set is M and is such that each $\mathcal{M}_i$ is a substructure of $\mathcal{M}$. For example, if R is a p-ary relation symbol and if $a_1, a_2, \ldots, a_p$ are elements of M, we choose an index $i \in I$ such that all the points $a_1, a_2, \ldots, a_p$ belong to M_i (this is possible since the M_i are totally ordered by inclusion), we set

$$(a_1, a_2, \ldots, a_p) \in \bar{R}^{\mathcal{M}} \quad \text{if and only if} \quad (a_1, a_2, \ldots, a_p) \in \bar{R}^{\mathcal{M}_i}.$$

This decision does not depend on the choice of index i; if j is some other index for which $a_1, a_2, \ldots, a_p$ belong to M_j, then we have either $i \leq j$ or $j < i$. Thus, either $\mathcal{M}_i$ is a substructure of $\mathcal{M}_j$ or $\mathcal{M}_j$ is a substructure of $\mathcal{M}_i$. In both cases,

$$(a_1, a_2, \ldots, a_p) \in \bar{R}^{\mathcal{M}_i} \quad \text{if and only if} \quad (a_1, a_2, \ldots, a_p) \in \bar{R}^{\mathcal{M}_j}.$$

We do the same for the constant symbols and the function symbols.

We have arrived at a very useful theorem, due to Tarski, known as the **union of chains theorem**.

Theorem 8.21 *Let $\langle I, <\rangle$ be a totally ordered set and, for every $i \in I$, let $\mathcal{M}_i$ be an L-structure; assume in addition that if $i < j$, then $\mathcal{M}_i$ is an elementary substructure of $\mathcal{M}_j$. Set $M = \bigcup_{i \in I} M_i$. Then for every $j \in I$, $\mathcal{M}_j$ is an elementary substructure of $\mathcal{M}$.*

Proof By induction on the formula $F[v_1, v_2, \ldots, v_p]$, we will prove that, for all $i \in I$ and for all $a_1, a_2, \ldots, a_p \in M_i$,

$$\mathcal{M} \models F[a_1, a_2, \ldots, a_p] \quad \text{if and only if} \quad \mathcal{M}_i \models F[a_1, a_2, \ldots, a_p].$$

As mentioned several times previously, we may assume that F contains no occurrences of the universal quantier.

For atomic formulas, there is no problem since $\mathcal{M}_i$ is a substructure of $\mathcal{M}$. The cases which involve propositional connectives are obvious. So suppose that $F = \exists v_0 G[v_0, v_1, \ldots, v_p]$, that $i \in I$, and that $a_1, a_2, \ldots, a_p \in M_i$.

- If $\mathcal{M}_i \models \exists v_0 G[v_0, v_1, \ldots, v_p]$, then there exists a point a_0 in M_i such that $\mathcal{M}_i \models G[a_0, a_1, \ldots, a_p]$; it then follows by the induction hypothesis that $\mathcal{M} \models G[a_0, a_1, \ldots, a_p]$ and, consequently, $\mathcal{M} \models F[a_1, a_2, \ldots, a_p]$.
- If $\mathcal{M} \models \exists v_0 G[v_0, v_1, \ldots, v_p]$, then there exists a point a_0 in M such that $\mathcal{M} \models G[a_0, a_1, \ldots, a_p]$; so there exists $j \in I$, $j > i$, such that $a_0 \in M_j$ and, by the induction hypothesis, $\mathcal{M}_j \models G[a_0, a_1, \ldots, a_p]$. Thus $\mathcal{M}_j \models \exists v_0 G[v_0, a_1, \ldots, a_p]$. But since $\mathcal{M}_i \prec \mathcal{M}_j$, we also have $\mathcal{M}_i \models F[a_1, a_2, \ldots, a_p]$. ■

8.3 The interpolation and denability theorems

At the beginning of the previous section, we saw that if there exists an elementary mapping between two structures, then they are elementarily equivalent. The converse is false (see Exercise 4). But we do, nonetheless, have the following theorem.

Theorem 8.22 *Let $\mathcal{M}_1$ and $\mathcal{M}_2$ be two L-structures. Then $\mathcal{M}_1$ and $\mathcal{M}_2$ are elementarily equivalent if and only if there is some third L-structure into which they can both be elementarily embedded.*

Proof In one direction, this is clear: if $\mathcal{M}_1$ and $\mathcal{M}_2$ can both be elementarily embedded in $\mathcal{M}_3$, then we have $\mathcal{M}_1 \equiv \mathcal{M}_3 \equiv \mathcal{M}_2$.

Conversely, suppose that $\mathcal{M}_1$ and $\mathcal{M}_2$ are elementarily equivalent. We will use the method of diagrams. Consider the language L' obtained by adding to L

- a new constant symbol $\underline{a}$ for every element a of M_1;
- a new constant symbol $\bar{b}$ for every element b of M_2.

Caution: All of these symbols must be distinct and, in particular, if a belongs to both M_1 and M_2, one must take care that $\underline{a}$ is different from $\bar{a}$.

We introduce the diagrams:

$$D(\mathcal{M}_1) = \{F[\underline{a}_1, \underline{a}_2, \ldots, \underline{a}_n] : F[v_1, v_2, \ldots, v_n] \text{ is a formula of } L,$$
$$a_1, a_2, \ldots, a_n \in M_1 \text{ and } \mathcal{M}_1 \models F[a_1, a_2, \ldots, a_n]\}$$

and

$$D(\mathcal{M}_2) = \{F[\bar{b}_1, \bar{b}_2, \ldots, \bar{b}_n] : F[v_1, v_2, \ldots, v_n] \text{ is a formula of } L, \\ b_1, b_2, \ldots, b_n \in M_2 \text{ and } \mathcal{M}_2 \models F[b_1, b_2, \ldots, b_n]\}.$$

We have seen that $\mathcal{M}_1$ can be elementarily embedded in any model of $D(\mathcal{M}_1)$ and, similarly, $\mathcal{M}_2$ can be elementarily embedded in any model of $D(\mathcal{M}_2)$. So it will sufce to prove that $T' = D(\mathcal{M}_1) \cup D(\mathcal{M}_2)$ is a consistent theory. To do this, we will use the compactness theorem. Note that $D(\mathcal{M}_1)$ is closed under conjunction; thus, a nite subset of $D(\mathcal{M}_1)$ is equivalent to a formula $F[\underline{a}_1, \underline{a}_2, \ldots, \underline{a}_n]$ in $D(\mathcal{M}_1)$ and, if T' were contradictory, there would exist a formula such that

$$D(\mathcal{M}_2) \vdash \neg F[\underline{a}_1, \underline{a}_2, \ldots, \underline{a}_n].$$

But, since the $\underline{a}_i$ do not appear in $D(\mathcal{M}_2)$, we have (see Lemma 4.26)

$$D(\mathcal{M}_2) \vdash \forall v_1 \forall v_2 \ldots \forall v_n \neg F[v_1, v_2 \ldots, v_n].$$

Thus $\forall v_1 \forall v_2 \ldots \forall v_n \neg F[v_1, v_2 \ldots, v_n]$ is a closed formula of L which is true in $\mathcal{M}_2$; it should consequently be true in $\mathcal{M}_1$, but this contradicts the fact that $\mathcal{M}_1 \models F[a_1, a_2, \ldots, a_n]$. ■

The theorem which follows is called **Robinson's consistency lemma**; it provides us with a very pretty example of the construction of a model.

Theorem 8.23 *Let T be a complete theory in a language L and let L_1 and L_2 be two enrichments of L such that $L_1 \cap L_2 = L$. If T_1 and T_2 are two consistent theories (in the languages L_1 and L_2, respectively) that each include T, then $T_1 \cup T_2$ is consistent.*

Proof The proof makes use of the next three lemmas. Throughout, if $\mathcal{M}$ is an L_1-structure or an L_2-structure, then $\mathcal{M}^-$ will denote the reduct of $\mathcal{M}$ to L.

Lemma 8.24 *Let $\mathcal{M}$ be a model of T. Then there exists a model $\mathcal{B}$ of T_2 such that $\mathcal{M} \prec \mathcal{B}^-$.*

Proof We will use the method of diagrams. It sufces to show that $T_2 \cup D(\mathcal{M})$ is consistent. If we assume the opposite in order to obtain a contradiction, and invoke the compactness theorem together with the fact that $D(\mathcal{M})$ is closed under conjunction, we have a formula $F[\underline{a}_1, \underline{a}_2, \ldots, \underline{a}_n]$ of $D(\mathcal{M})$ such that $T_2 \vdash \neg F[\underline{a}_1, \underline{a}_2, \ldots, \underline{a}_n]$. The symbols $\underline{a}_i$ do not belong to the language L_2 in which T_2 is expressed. Therefore,

$$T_2 \vdash \forall v_1 \forall v_2 \ldots \forall v_n \neg F[v_1, v_2 \ldots, v_n].$$

The formula $\forall v_1 \forall v_2 \ldots \forall v_n \neg F[v_1, v_2 \ldots, v_n]$ is in L and T is included in T_2; thus the formula $\neg \forall v_1 \forall v_2 \ldots \forall v_n \neg F[v_1, v_2 \ldots, v_n]$ cannot be a consequence of T.

Since T is complete, $\forall v_1 \forall v_2 \ldots \forall v_n \neg F[v_1, v_2 \ldots, v_n]$ is a consequence of T. This yields a contradiction since $\mathcal{M}$ is a model of T and $\mathcal{M} \models F[a_1, a_2, \ldots, a_n]$. ∎

Lemma 8.25 *Let $\mathcal{A}_1$ be a model of T_1, let $\mathcal{M}$ be a model of T, and assume that $\mathcal{A}_1^- \prec \mathcal{M}$. Then there exists an L_1-structure $\mathcal{A}_2$ such that $\mathcal{A}_1 \prec \mathcal{A}_2$ and $\mathcal{M} \prec \mathcal{A}_2^-$ (and hence $\mathcal{A}_2$ is a model of T_1).*

Proof Again, we use the same method; it sufces to construct a model of $T' = D(\mathcal{M}) \cup D(\mathcal{A}_1)$, where, as we recall,

$$D(\mathcal{M}) = \{F[\underline{a}_1, \underline{a}_2, \ldots, \underline{a}_n] : F[v_1, v_2, \ldots, v_n] \text{ is a formula of } L,$$
$$a_1, a_2, \ldots, a_n \in M \text{ and } \mathcal{M} \models F[a_1, a_2, \ldots, a_n]\}$$

and

$$D(\mathcal{A}_1) = \{F[\underline{a}_1, \underline{a}_2, \ldots, \underline{a}_n] : F[v_1, v_2, \ldots, v_n] \text{ is a formula of } L_1,$$
$$a_1, a_2, \ldots, a_n \in A_1 \text{ and } \mathcal{A}_1 \models F[a_1, a_2, \ldots, a_n]\}.$$

We must insist on the fact that the language of T' is L_1 augmented by parameters from M. Unlike what was done for Theorem 8.22, we only introduce a new constant symbol $\underline{a}$ for each element a of A_1 (the base set of $\mathcal{A}_1$) which is used both for $D(\mathcal{M})$ and for $D(\mathcal{A})$; it is because of this that we will be able to consider a model of T' as an elementary extension of $\mathcal{A}_1$ and its L-reduct as an elementary extension of $\mathcal{M}$.

Again, the argument is by contradiction; we assume that T' is not consistent and deduce, from this, the existence of a formula of $D(\mathcal{M})$ that is in contradiction with $D(\mathcal{A}_1)$. This formula can be written in the form

$$F[\underline{a}_1, \underline{a}_2, \ldots, \underline{a}_n, \underline{a}_{n+1}, \ldots, \underline{a}_{n+p}],$$

where F is a formula of L, $a_1, a_2, \ldots, a_n \in A_1$ and $a_{n+1}, a_{n+2}, \ldots, a_{n+p} \in M - A_1$.

Since $D(\mathcal{A}_1) \vdash \neg F[\underline{a}_1, \underline{a}_2, \ldots, \underline{a}_n, \underline{a}_{n+1}, \ldots, \underline{a}_{n+p}]$ and since the $\underline{a}_i$, for $n{+}1 \leq i \leq n + p$, do not appear in $D(\mathcal{A}_1)$, it follows that

$$D(\mathcal{A}_1) \vdash \forall v_1 \forall v_2 \ldots \forall v_p \neg F[\underline{a}_1, \underline{a}_2, \ldots, \underline{a}_n, v_1, v_2, \ldots, v_p]$$

and hence that

$$\mathcal{A}_1 \models \forall v_1 \forall v_2 \ldots \forall v_p \neg F[a_1, a_2, \ldots, a_n, v_1, v_2, \ldots, v_p].$$

It is clear that $\mathcal{M} \models \exists v_1 \exists v_2 \ldots \exists v_p F[a_1, a_2, \ldots, a_n, v_1, v_2, \ldots, v_p]$ and this contradicts the fact that $\mathcal{A}_1^- \prec \mathcal{M}$. ∎

Obviously, we can replace T_1 by T_2 and obtain

Lemma 8.26 *Let $\mathcal{B}_1$ be a model of T_2, let $\mathcal{M}$ be a model of T, and assume that $\mathcal{B}_1^- \prec \mathcal{M}$. Then there exists a model $\mathcal{B}_2$ of T_2 such that $\mathcal{B}_1 \prec \mathcal{B}_2$ and $\mathcal{M} \prec \mathcal{B}_2^-$.*

We are now in a position to complete the proof of the consistency lemma by constructing a model of $T_1 \cup T_2$. We start with a model $\mathcal{A}_1$ of T_1; $\mathcal{A}_1^-$ is a model of T. So we may apply Lemma 8.24 to nd a model $\mathcal{B}_1$ of T_2 such that $\mathcal{A}_1^- \prec \mathcal{B}_1^-$. Then apply Lemma 8.25 to nd a model $\mathcal{A}_2$ of T_1 such that $\mathcal{A}_1 \prec \mathcal{A}_2$ and $\mathcal{B}_1^- \prec \mathcal{A}_2^-$. We continue by applying Lemmas 8.25 and 8.26 in alternation to produce structures $\mathcal{A}_n$ that are models of T_1 and $\mathcal{B}_n$ that are models of T_2 such that, for every integer n,

$$\mathcal{A}_n \prec \mathcal{A}_{n+1}, \qquad \mathcal{B}_n \prec \mathcal{B}_{n+1}, \qquad \mathcal{A}_n^- \prec \mathcal{B}_n^-, \quad \text{and} \quad \mathcal{B}_n^- \prec \mathcal{A}_{n+1}^-.$$

Set $\mathcal{A} = \bigcup_{n \geq 1} \mathcal{A}_n$ and $\mathcal{B} = \bigcup_{n \geq 1} \mathcal{B}_n$. By the union of chains theorem, $\mathcal{A}$ is an elementary extension of $\mathcal{A}_1$, hence is a model of T_1; in the same way, $\mathcal{B}$ is a model of T_2. Now these two structures have the same base set and same L-reduct, namely

$$\bigcup_{n \geq 1} \mathcal{A}_n^- = \bigcup_{n \geq 1} \mathcal{B}_n^-.$$

So we obtain a model of $T_1 \cup T_2$ by considering the $(L_1 \cup L_2)$-structure whose L_1-reduct is $\mathcal{A}$ and whose L_2-reduct is $\mathcal{B}$ (this is where the hypothesis that $L_1 \cap L_2 = L$ intervenes; it allows for an unambiguous interpretation of the symbols of $L_1 \cup L_2$ that do not belong to L). ■

The next theorem, called **Craig's interpolation theorem**, is a consequence of Robinson's consistency lemma and should be compared with Theorem 1.35.

Theorem 8.27 *Let F and G be two closed formulas and assume that $F \Rightarrow G$ is universally valid. Then there exists a formula H such that*

(1) $\vdash F \Rightarrow H$;

(2) $\vdash H \Rightarrow G$;

(3) *any symbol for a constant, a function, or a predicate that appears in H (with the exception of the equality predicate) must also appear in both F and G.*

A formula that satises conditions (1), (2), and (3) is called an **interpolant between F and G**.

We should note that this theorem does have a purely syntactical meaning in which only the notion of formal deduction intervenes and in which there is absolutely no mention of models. There are purely syntactical proofs of this theorem; but that is not the case for the proof which follows.

Proof Let L be the language consisting of the predicate for equality and of the symbols common to F and G. We need to nd a closed formula of L such that

$$\vdash (F \Rightarrow H) \wedge (H \Rightarrow G).$$

We will assume that this is not possible and arrive at a contradiction. The idea is to construct a complete theory T in L such that $T \cup \{F\}$ and $T \cup \{\neg G\}$ are both consistent. Robinson's consistency lemma would then imply that $T \cup \{F, \neg G\}$ is consistent, which is absurd since $\vdash F \Rightarrow G$.

The language L is countable, so we can nd an enumeration $\{J_n : n \in \mathbb{N}\}$ of all the closed formulas of L; assume that $J_0 = \exists v_0(v_0 \simeq v_0)$. We then construct, by induction on n, a sequence of formulas $(K_n : n \in \mathbb{N})$ of L in such a way that, for all $n \in \mathbb{N}$,

(i) $\vdash K_{n+1} \Rightarrow K_n$;

(ii) $\vdash K_n \Rightarrow J_n$ or $\vdash K_n \Rightarrow \neg J_n$;

(iii) there is no interpolant between the formulas $F \wedge K_n$ and $G \wedge K_n$.

We start by letting $K_0 = \exists v_0(v_0 \simeq v_0)$. Conditions (i) and (ii) are trivial and, because we assumed that F and G have no interpolant, condition (iii) is also satised. Next, we construct K_{n+1} from K_n.

The crucial observation is that at least one of the following two situations must arise:

- There is no interpolant between the formulas $F \wedge K_n \wedge J_{n+1}$ and $G \wedge K_n \wedge J_{n+1}$.
- There is no interpolant between the formulas $F \wedge K_n \wedge \neg J_{n+1}$ and $G \wedge K_n \wedge \neg J_{n+1}$.

Indeed, if we assume the contrary, there exist closed formulas H_0 and H_1 of L such that

$$\vdash (F \wedge K_n \wedge J_{n+1}) \Rightarrow H_0, \qquad \vdash H_0 \Rightarrow (G \wedge K_n \wedge J_{n+1}),$$
$$\vdash (F \wedge K_n \wedge \neg J_{n+1}) \Rightarrow H_1, \qquad \vdash H_1 \Rightarrow (G \wedge K_n \wedge \neg J_{n+1}).$$

This is not possible, for it would require that

$$\vdash (F \wedge K_n) \Rightarrow (H_0 \vee H_1) \quad \text{and} \quad \vdash (H_0 \vee H_1) \Rightarrow (G \wedge K_n),$$

which makes $(H_0 \vee H_1)$ an interpolant for $F \wedge K_n$ and $G \wedge K_n$.

So we may take K_{n+1} equal to $K_n \wedge J_{n+1}$ or $K_n \wedge \neg J_{n+1}$ according to whichever situation holds; as a result, $F \wedge K_{n+1}$ and $G \wedge K_{n+1}$ will have no interpolant and conditions (i), (ii), and (iii) will be satised. Set

$$T = \{K_n : n \in \mathbb{N}\}.$$

Observe that for every n, $F \wedge K_n$ is a consistent formula; otherwise, $\exists v_0 \neg v_0 \simeq v_0$ would be an interpolant between $F \wedge K_n$ and $G \wedge K_n$. Using condition (i) and the compactness theorem, we see that $T \cup \{F\}$ is a consistent theory. Similarly, $\neg G \wedge K_n$ is consistent, otherwise K_n would be an interpolant between $F \wedge K_n$

and $G \wedge K_n$. Therefore, $T \cup \{\neg G\}$ is also consistent. Finally, it is clear that T is a complete theory in L, for if H is a closed formula of L, then for some integer n, $H = J_n$ and we have arranged things so that either $\vdash K_n \Rightarrow J_n$ or $\vdash K_n \Rightarrow \neg J_n$. ■

Craig's interpolation theorem leads to another important result called Beth's denability theorem. We will begin with some remarks and notation.

Let L be a denumerable language, let P be a new n-ary predicate symbol and set $L' = L \cup \{P\}$. Let P_1 be another n-ary predicate symbol that does not appear in L. If G is a formula of L', we will let $G_{P_1/P}$ denote the formula obtained by substituting P_1 for all occurrences of P in G. Note that if F is a formula of L and if $\vdash F \Rightarrow G$, then $\vdash F \Rightarrow G_{P_1/P}$.

Let T be a theory in the enriched language L'. It is more or less clear that the following two conditions are equivalent:

- For every L-structure $\mathcal{M}$, there is at most one interpretation of P that results in an enrichment of $\mathcal{M}$ to a model of T.
- Suppose that P_1 is a new n-ary predicate symbol, that $L_1 = L \cup \{P_1\}$, and that T_1 is the theory obtained by replacing P with P_1 in T; then for every model $\mathcal{N}$ of $T \cup T_1$, we have

$$\mathcal{N} \models \forall v_1 \forall v_2 \ldots \forall v_n (P v_1 v_2 \ldots v_n \Leftrightarrow P_1 v_1 v_2 \ldots v_n).$$

The previous condition is also equivalent to

$$T \cup T_1 \vdash \forall v_1 \forall v_2 \ldots \forall v_n (P v_1 v_2 \ldots v_n \Leftrightarrow P_1 v_1 v_2 \ldots v_n).$$

If one of these conditions is satised, we say that P is **implicitly denable in** T. We will say that P is **explicitly denable in** T if there exists a formula $F[v_1, v_2, \ldots, v_n]$ of L such that

$$T \vdash \forall v_1 \forall v_2 \ldots \forall v_n (P v_1 v_2 \ldots v_n \Leftrightarrow F[v_1, v_2, \ldots, v_n]).$$

It is rather obvious that if P is explicitly denable in T, then it is implicitly denable. The converse of this is **Beth's denability theorem**.

Theorem 8.28 *If P is implicitly denable in T, then it is explicitly denable in T.*

Proof We retain the notations from the preceding paragraph. We introduce n constant symbols, $c_1, c_2, \ldots, c_n$. Because P is implicitly denable, the theory

$$T \cup \{P c_1 c_2 \ldots c_n\} \cup T_1 \cup \{\neg P_1 c_1 c_2 \ldots c_n\}$$

is contradictory. Using the compactness theorem, we can nd closed formulas F of L' and G of L_1 such that $F \wedge G \wedge P c_1 c_2 \ldots c_n \wedge \neg P_1 c_1 c_2 \ldots c_n$ is contradictory, while $T \vdash F$ and $T_1 \vdash$ G. The fact that

$$F \wedge G \wedge Pc_1c_2\ldots c_n \wedge \neg P_1c_1c_2\ldots c_n$$

is contradictory can be rephrased as

$$\vdash (F \wedge Pc_1c_2\ldots c_n) \Rightarrow (G \Rightarrow P_1c_1c_2\ldots c_n).$$

Now apply the interpolation theorem to obtain a formula $H[v_1, v_2, \ldots, v_n]$ of L (in which neither P nor P_1 occurs) such that $H[v_1, v_2, \ldots, v_n]$ is an interpolant between $F \wedge Pc_1c_2\ldots c_n$ and $G \Rightarrow P_1c_1c_2\ldots c_n$. Since $T \vdash F$, we have

$$T \vdash Pc_1c_2\ldots c_n \Rightarrow H[v_1, v_2, \ldots, v_n],$$

and since the constants, c_i, do not occur in T, we have

$$T \vdash \forall v_1 \forall v_2 \ldots \forall v_n (Pv_1v_2\ldots v_n \Rightarrow H[v_1, v_2, \ldots, v_n]).$$

Similarly, we have

$$T_1 \vdash \forall v_1 \forall v_2 \ldots \forall v_n (H[v_1, v_2, \ldots, v_n] \Rightarrow P_1v_1v_2\ldots v_n),$$

and when we substitute P for P_1, we have

$$T \vdash \forall v_1 \forall v_2 \ldots \forall v_n (H[v_1, v_2, \ldots, v_n] \Rightarrow Pv_1v_2\ldots v_n).$$ ■

8.4 Reduced products and ultraproducts

We are given a language L, a set I, and a family $(\mathcal{M}_i)_{i\in I}$ of L-structures. We are going to dene another L-structure, called the **product of the family** $(\mathcal{M}_i)_{i\in I}$ and denoted by $\prod_{i\in I} \mathcal{M}_i$, whose base set is the product set $\prod_{i\in I} M_i$. First, some notation: if a is an element of $\prod_{i\in I} M_i$ and $i \in I$, we let a^i denote the ith coordinate of a [in other words, a is equal to the sequence $(a^i : i \in I)$]; if X is a symbol (for a constant, a function or a predicate) of L, then X_i denotes the interpretation of this symbol in $\mathcal{M}_i$.

- If c is a constant symbol, then the interpretation of c in $\prod_{i\in I} \mathcal{M}_i$ is the sequence $(c_i : i \in I)$.
- If f is an n-ary function symbol, then the interpretation of f in $\prod_{i\in I} \mathcal{M}_i$ is the map which, with $(a_1, a_2, \ldots, a_n)$, associates

$$(f_i(a_1^i, a_2^i, \ldots, a_n^i) : i \in I).$$

- If R is an n-ary predicate symbol, then the interpretation of R in $\prod_{i\in I} \mathcal{M}_i$ is the set

$$\left\{(a_1, a_2, \ldots, a_n) \in \left(\prod_{i\in I} M_i\right)^n : \text{for every } i \in I,\ (a_1^i, a_2^i, \ldots, a_n^i) \in R_i\right\}.$$

To restate this in a more compact fashion, we could say that $\prod_{i\in I} \mathcal{M}_i$ is dened in such a way that, for every atomic formula $F[v_1, v_2, \ldots, v_n]$ and for all $a_1, a_2, \ldots, a_n$ in $\prod_{i \in I} M_i$,

$$\prod_{i \in I} \mathcal{M}_i \models F[a_1, a_2, \ldots, a_n]$$
$$\text{if and only if for all } i \in I,\ \mathcal{M}_i \models F[a_1^i, a_2^i, \ldots, a_n^i].$$

When I is nite, we refer to this as the **nite product**.

Example 8.29 If the $\mathcal{M}_i$ are groups, this is the familiar product of groups. Similarly for rings and for ordered sets. We should note that the product of elds is not, in general, a eld (see Exercise 14); also, the product of totally ordered sets is not, in general, totally ordered.

We will now generalize this denition. In addition to the set I and the structures $\mathcal{M}_i$, we are now also given a lter $\mathcal{F}$ on the Boolean algebra of subsets of I. Consider the binary relation $\approx_{\mathcal{F}}$ on $\prod_{i \in I} M_i$ dened by

$$(a^i : i \in I) \approx_{\mathcal{F}} (b^i : i \in I) \quad \text{if and only if} \quad \{i \in I\colon a^i = b^i\} \in \mathcal{F}.$$

Note that this is an equivalence relation. It is clearly symmetric and reexive. To see that it is transitive, suppose that $(a^i : i \in I)$, $(b^i : i \in I)$, and $(c^i : i \in I)$ are three elements of $\prod_{i \in I} M_i$ such that

$$(a^i : i \in I) \approx_{\mathcal{F}} (b^i : i \in I) \quad \text{and} \quad (b^i : i \in I) \approx_{\mathcal{F}} (c^i : i \in I);$$

this means that

$$\{i \in I : a^i = b^i\} \in \mathcal{F} \quad \text{and} \quad \{i \in I : b^i = c^i\} \in \mathcal{F}.$$

But $\{i \in I : a^i = b^i\} \cap \{i \in I : b^i = c^i\} \subseteq \{i \in I : a^i = c^i\}$ and, since $\mathcal{F}$ is a lter, it is clear that $\{i \in I : a^i = c^i\} \in \mathcal{F}$.

The set of equivalence classes of $\prod_{i \in I} M_i$ with respect to this relation will be denoted by $\prod_{i \in I} M_i / \mathcal{F}$. If $a \in \prod_{i \in I} M_i$, the equivalence class of a modulo $\approx_{\mathcal{F}}$ will be denoted by $\bar{a}$ provided this does not lead to any ambiguity.

We will now dene an L-structure on the set $\prod_{i \in I} M_i / \mathcal{F}$ which we will denote by $\prod_{i \in I} \mathcal{M}_i / \mathcal{F}$ and which we will call the **reduced product of the family** $(\mathcal{M}_i)_{i \in I}$ **modulo the lter** $\mathcal{F}$. This denition will be such that, for every atomic formula $F[v_1, v_2, \ldots, v_n]$ and for all $a_1, a_2, \ldots, a_n$ in $\prod_{i \in I} M_i$,

$$\prod_{i \in I} \mathcal{M}_i \Big/ \mathcal{F} \models F[\bar{a}_1, \bar{a}_2, \ldots, \bar{a}_n]$$

if and only if

$$\{i \in I : \mathcal{M}_i \models F[a_1^i, a_2^i, \ldots, a_n^i]\} \in \mathcal{F}.$$

- If c is a constant symbol, the interpretation of c in $\prod_{i \in I} \mathcal{M}_i / \mathcal{F}$ is the equivalence class of the sequence $(c_i : i \in I)$ modulo $\approx_{\mathcal{F}}$.

- Suppose R is an n-ary predicate symbol and that $\bar{a}_1, \bar{a}_2, \ldots, \bar{a}_n$ are elements of $\prod_{i \in I} M_i / \mathcal{F}$. Suppose, for all k from 1 to n inclusive, that $(b_k^i : i \in I)$ is some other representative of $\bar{a}_k$ in $\prod_{i \in I} M_i$. Then

$$\{i \in I : \text{for all } k \text{ from 1 to } n \text{ inclusive, } a_k^i = b_k^i\} \in \mathcal{F},$$

  and consequently

$$\{i \in I : \mathcal{M}_i \models Ra_1^i a_2^i \ldots a_n^i\} \in \mathcal{F}$$

  if and only if

$$\{i \in I : \mathcal{M}_i \models Rb_1^i b_2^i \ldots b_n^i\} \in \mathcal{F}.$$

  So it is natural and legitimate to decide that

$$\prod_{i \in I} \mathcal{M}_i \Big/ \mathcal{F} \models R\bar{a}_1\bar{a}_2 \ldots \bar{a}_n$$

  if and only if

$$\{i \in I : \mathcal{M}_i \models Ra_1^i a_2^i \ldots a_n^i\} \in \mathcal{F}.$$

- Suppose next that f is an n-ary function symbol and that $\bar{a}_1, \bar{a}_2, \ldots, \bar{a}_n$ are elements of $\prod_{i \in I} \mathcal{M}_i / \mathcal{F}$. Once again, we note that if for all k from 1 to n inclusive, $(b_k^i : i \in I)$ is some other representative of $\bar{a}_k$, then

$$(f_i(a_1^i, a_2^i, \ldots, a_n^i) : i \in I) \approx_{\mathcal{F}} (f_i(b_1^i, b_2^i, \ldots, b_n^i) : i \in I).$$

  The interpretation of f in $\prod_{i \in I} \mathcal{M}_i / \mathcal{F}$ is the mapping which, with $\bar{a}_1, \bar{a}_2, \ldots, \bar{a}_n$, associates the equivalence class of $(f_i(a_1^i, a_2^i, \ldots, a_n^i) : i \in I)$ relative to $\approx_{\mathcal{F}}$.

Example 8.30 The product dened earlier is a special case of the reduced product; it corresponds to the situation in which $\mathcal{F}$ is the lter $\mathcal{F}$ whose single element is the set I itself. We will show that the reduced product of groups, of rings, or of ordered sets is, respectively, a group, a ring, or an ordered set. We will also see (Exercise 14) that the reduced product of elds is only rarely a eld.

In case all the structures $\mathcal{M}_i$ are equal to the same structure $\mathcal{M}$, we will speak of the **reduced power of $\mathcal{M}$ modulo $\mathcal{F}$** and denote it by $\mathcal{M}^I / \mathcal{F}$.

If $\mathcal{F}$ is an ultralter, then $\prod_{i \in I} \mathcal{M}_i / \mathcal{F}$ is called the **ultraproduct of the family $(\mathcal{M}_i)_{i \in I}$ modulo the lter $\mathcal{F}$** and if all the structures $\mathcal{M}_i$ are equal to the same structure $\mathcal{M}$, it is called the **ultrapower of $\mathcal{M}$ modulo $\mathcal{F}$** (and, in agreement with the previous convention, we will denote it by $\mathcal{M}^I / \mathcal{F}$). What makes ultraproducts interesting is the next theorem, known as **Äos' theorem**.

Theorem 8.31 *Let $(\mathcal{M}_i : i \in I)$ be a family of L-structures and let $\mathcal{F}$ be an ultralter on I. Then for every formula $F[v_1, v_2, \ldots, v_n]$ of L and for all elements $a_1, a_2, \ldots, a_n$ of $\prod_{i\in I} M_i$,*

$$\prod_{i\in I} \mathcal{M}_i \Big/ \mathcal{F} \models F[\bar{a}_1, \bar{a}_2, \ldots, \bar{a}_n]$$

if and only if

$$\{i \in I : \mathcal{M}_i \models F[a_1^i, a_2^i, \ldots, a_n^i]\} \in \mathcal{F}.$$

Proof We may begin by assuming that the only propositional connectives occurring in F are $\neg$ and $\wedge$ and that the universal quantier does not appear. The proof is then by induction on the height of F. We already have the result if F is an atomic formula.

- Case in which $F[v_1, v_2, \ldots, v_n] = \neg G[v_1, v_2, \ldots, v_n]$. For all elements $a_1, a_2, \ldots, a_n$ of $\prod_{i\in I} M_i$,

$$\prod_{i\in I} \mathcal{M}_i \Big/ \mathcal{F} \models F[\bar{a}_1, \bar{a}_2, \ldots, \bar{a}_n]$$

if and only if

$$\prod_{i\in I} \mathcal{M}_i \Big/ \mathcal{F} \not\models G[\bar{a}_1, \bar{a}_2, \ldots, \bar{a}_n],$$

and thus, by the induction hypothesis, if and only if

$$\{i \in I : \mathcal{M}_i \models G[a_1^i, a_2^i, \ldots, a_n^i]\} \notin \mathcal{F}.$$

Because $\mathcal{F}$ is an ultralter, $\{i \in I : \mathcal{M}_i \models G[a_1^i, a_2^i, \ldots, a_n^i]\} \notin \mathcal{F}$ if and only if its complement does belong to $\mathcal{F}$. Therefore,

$$\prod_{i\in I} \mathcal{M}_i \Big/ \mathcal{F} \models F[\bar{a}_1, \bar{a}_2, \ldots, \bar{a}_n]$$

if and only if

$$\{i \in I : \mathcal{M}_i \models F[a_1^i, a_2^i, \ldots, a_n^i]\} \in \mathcal{F}.$$

- Case in which $F[v_1, v_2, \ldots, v_n] = G[v_1, v_2, \ldots, v_n] \wedge H[v_1, v_2, \ldots, v_n]$. For all elements $a_1, a_2, \ldots, a_n$ of $\prod_{i\in I} M_i$,

$$\prod_{i\in I} \mathcal{M}_i \Big/ \mathcal{F} \models F[\bar{a}_1, \bar{a}_2, \ldots, \bar{a}_n]$$

if and only if

$$\prod_{i\in I}\mathcal{M}_i\Big/\mathcal{F}\models G[\bar{a}_1,\bar{a}_2,\ldots,\bar{a}_n] \quad\text{and}$$

$$\prod_{i\in I}\mathcal{M}_i\Big/\mathcal{F}\models H[\bar{a}_1,\bar{a}_2,\ldots,\bar{a}_n].$$

By the induction hypothesis, this is equivalent to

$$\{i\in I:\mathcal{M}_i\models G[a_1^i,a_2^i,\ldots,a_n^i]\}\in\mathcal{F}\quad\text{and}$$
$$\{i\in I:\mathcal{M}_i\models H[a_1^i,a_2^i,\ldots,a_n^i]\}\in\mathcal{F}.$$

Because $\mathcal{F}$ is a lter, these two sets belong to $\mathcal{F}$ if and only if their intersection belongs to $\mathcal{F}$, and their intersection is equal to

$$\{i\in I:\mathcal{M}_i\models F[a_1^i,a_2^i,\ldots,a_n^i]\}.$$

- Case in which $F[v_1,v_2,\ldots,v_n]=\exists v_0G[v_0,v_1,\ldots,v_n]$. Let $a_1,a_2,\ldots,a_n$ be elements of $\prod_{i\in I}M_i$ and suppose that

$$\prod_{i\in I}\mathcal{M}_i\Big/\mathcal{F}\models F[\bar{a}_1,\bar{a}_2,\ldots,\bar{a}_n].$$

Then there exists $a_0\in\prod_{i\in I}M_i$ such that $\prod_{i\in I}\mathcal{M}_i/\mathcal{F}\models G[\bar{a}_0,\bar{a}_1,\ldots,\bar{a}_n]$. By the induction hypothesis, we see that the set

$$X=\{i\in I:\mathcal{M}_i\models G[a_0^i,a_1^i,\ldots,a_n^i]\}$$

belongs to $\mathcal{F}$. It is clear that

$$\text{if } i\in X,\quad\text{then } \mathcal{M}_i\models\exists v_0G[v_0,\bar{a}_1,\bar{a}_2,\ldots,\bar{a}_n];$$

thus, the set $\{i\in I:\mathcal{M}_i\models F[a_1^i,a_2^i,\ldots,a_n^i]\}$ includes X and belongs to $\mathcal{F}$. Conversely, suppose that

$$Y=\{i\in I:\mathcal{M}_i\models F[a_1^i,a_2^i,\ldots,a_n^i]\}\in\mathcal{F}.$$

We dene a sequence $a_0=(a_0^i:i\in I)$ as follows: if $i\in Y$, then a_0^i is a point in M_i such that $\mathcal{M}_i\models G[a_0^i,a_1^i,\ldots,a_n^i]$; if $i\notin Y$, then a_0^i can be chosen arbitrarily in M_i. It is then clear that

$$\{i\in I:\mathcal{M}_i\models G[a_0^i,a_1^i,\ldots,a_n^i]\}$$

includes Y, hence belongs to $\mathcal{F}$. So by the induction hypothesis,

$$\prod_{i\in I}\mathcal{M}_i\Big/\mathcal{F}\models G[\bar{a}_0,\bar{a}_1,\ldots,\bar{a}_n].$$ ∎

Observe that it is only for the step involving negation that we use the fact that $\mathcal{F}$ is an ultralter and not just a lter.

In the particular case in which F is a closed formula, Äos' theorem tells us that F is true in an ultraproduct of the family $(\mathcal{M}_i)_{i\in I}$ if and only if it is true in `almost all' the $\mathcal{M}_i$.

The construction of reduced products and ultraproducts is strictly algebraic and very little is required for the proof of Äos' theorem. The argument that follows provides a new proof of the compactness theorem. This proof is unencumbered by syntactic considerations, in opposition to the proof that we gave originally and which relied on the completeness theorem.

Corollary 8.32 *If every nite subset of a theory T has a model, then T has a model.*

Proof Let I be the set of nite subsets of T. For every $i \in I$, let $\mathcal{M}_i$ be a model of i (we know that one exists).

Now, for every formula F of T, consider the subset $X(F)$ of I dened by

$$X(F) = \{i \in I : F \in i\}.$$

We see that the intersection of any nite number of sets of the form $X(F)$ is never empty; for if $F_1, F_2, \ldots, F_n$ are in T, then $\{F_1, F_2, \ldots, F_n\}$ belongs to $\bigcap_{1\leq i\leq n} X(F_i)$. So the set $\{X(F) : F \in T\}$ is a lterbase and we can nd an ultralter $\mathcal{F}$ that includes this set (see Theorem 2.79). We may then use Äos' theorem to verify that $\prod_{i\in I} \mathcal{M}_i/\mathcal{F}$ is a model of T; indeed, if $F \in T$, then $\{i \in I : \mathcal{M}_i \models F\}$ includes $X(F)$ and hence belongs to $\mathcal{F}$. ■

The next proposition asserts that an ultrapower of $\mathcal{M}$ may be considered as an elementary extension of $\mathcal{M}$.

Proposition 8.33 *Let $\mathcal{M}$ be an L-structure, let I be a set and let $\mathcal{F}$ be an ultralter on I. For $a \in M$, let $c(a)$ denote the constant mapping from I into M whose value is a and let $h(a)$ denote the equivalence class of $c(a)$ modulo $\approx_{\mathcal{F}}$. Then h is an elementary mapping from $\mathcal{M}$ into $\mathcal{M}^I/\mathcal{F}$.*

Proof In light of Äos' theorem, this amounts to showing that, for every formula $F[v_1, v_2, \ldots, v_n]$ of L and for all elements $a_1, a_2, \ldots, a_n$ of M,

$$\mathcal{M} \models F[a_1, a_2, \ldots, a_n]$$

if and only if

$$\{i \in I : \mathcal{M} \models F[c(a_1)^i, c(a_2)^i, \ldots, c(a_n)^i]\} \in \mathcal{F}.$$

But this is obvious since, for all $i \in I$ and for all k from 1 to n inclusive, $c(a_k)^i$ is equal to a_k. ■

8.5 Preservation theorems

8.5.1 Preservation by substructures

Preservation theorems are results which relate the syntactical form of a theory with closure properties of its class of models. We begin with the easiest one; it concerns preservation of universal formulas. This problem has already been broached in Chapter 3 (Theorem 3.70). We recall a denition.

Denition 8.34 *A **universal formula** is a prenex formula in which the existential quantier does not appear. A **universal theory** is a theory that only contains universal formulas.*

For example, the theory of total orderings [axioms (i), (ii), and (iii) from Example 8.4] is a universal theory. The theory of groups is another interesting example. In the language that consists of one constant symbol, **1** and one binary function symbol, $\cdot$, its axioms are

(i) $\forall v_1 \forall v_2 \forall v_3\, (v_1 \cdot v_2) \cdot v_3 \simeq v_1 \cdot (v_2 \cdot v_3)$;

(ii) $\forall v_1 (v_1 \cdot \mathbf{1} \simeq v_1 \wedge \mathbf{1} \cdot v_1 \simeq v_1)$;

(iii) $\forall v_1 \exists v_2 (v_1 \cdot v_2 \simeq \mathbf{1} \wedge v_2 \cdot v_1 \simeq \mathbf{1})$.

In this form, we do not have a universal theory because of the third axiom. However, we may choose to include a unary function $^{-1}$ in the language to denote the inverse function. If we do this, we can replace the third axiom by

(iii)′ $\forall v_1 (v_1 \cdot v_1^{-1} \simeq \mathbf{1} \wedge v_1^{-1} \cdot v_1 \simeq \mathbf{1})$,

which, this time, provides a universal axiomatization.

Remark 8.35 The conjunction of two universal formulas is not, in general, a universal formula. It is, nonetheless, equivalent to a universal formula. Indeed,

$$\forall v_1 \ldots \forall v_n F_1 \wedge \forall w_1 \ldots \forall w_p F_2$$
$$\text{is equivalent to } \forall v_1 \ldots \forall v_n \forall w_1 \ldots \forall w_p (F_1 \wedge F_2)$$

provided that the v_i (for i from 1 to n inclusive) have no free occurrences in F_2 and that the w_j (for j from 1 to p inclusive) have no free occurrences in F_1. Subject to renaming the bound variables, we can always guarantee that this condition is satised. The conjunction of a nite number of universal formulas is also equivalent to a universal formula. The same applies to the disjunction of universal formulas:

$$\forall v_1 \ldots \forall v_n F_1 \vee \forall w_1 \ldots \forall w_p F_2$$
$$\text{is equivalent to } \forall v_1 \ldots \forall v_n \forall w_1 \ldots \forall w_p (F_1 \vee F_2)$$

provided the condition just mentioned is also satised.

Denition 8.36 *A theory T is said to be* **preserved under substructures** *if, for every model $\mathcal{M}$ of T and for every substructure $\mathcal{N}$ of $\mathcal{M}$, $\mathcal{N}$ is a model of T. A closed formula F is* **preserved under substructures** *if the theory $\{F\}$ is.*

Here is the promised **preservation theorem for universal formulas**.

Theorem 8.37 *Let T be a theory in a language L. The following two conditions are equivalent:*

(i) *There exists a universal theory Ψ in L which is equivalent to T.*

(ii) *The theory T is preserved under substructures.*

Proof The direction (i) $\Rightarrow$ (ii) was proved in Chapter 3 (Theorem 3.70). For the converse, we resurrect an idea that was developed in Exercise 20 from Chapter 3. Suppose that (ii) is satised. Set

$$\Psi = \{G : G \text{ is a closed universal formula and } T \vdash G\}.$$

It is clear that every formula in Ψ is a consequence of T. So let $\mathcal{M}$ be a model of Ψ. Consider the simple diagram of $\mathcal{M}$:

$$\begin{aligned}\Delta(\mathcal{M}) = \{H[\underline{a}_1, \underline{a}_2, \ldots, \underline{a}_n] &: H \text{ is a quantier-free formula} \\ &\text{and } \mathcal{M} \models H[a_1, a_2, \ldots, a_n]\}.\end{aligned}$$

It sufces to show that $\Delta(\mathcal{M}) \cup T$ has a model; for we have seen that if this is the case, then $\Delta(\mathcal{M}) \cup T$ has a model that is an extension of $\mathcal{M}$ (Lemma 8.13). Together with hypothesis (ii), this implies that $\mathcal{M}$ is a model of T.

Suppose that $\Delta(\mathcal{M}) \cup T$ is not consistent; then there exist quantier-free formulas $H_1[v_1, v_2, \ldots, v_n]$, $H_2[v_1, v_2, \ldots, v_n], \ldots, H_p[v_1, v_2, \ldots, v_n]$ and points a_1, $a_2, \ldots, a_n$ in M such that

$$\begin{aligned}T \vdash \neg(H_1[\underline{a}_1, \underline{a}_2, \ldots, \underline{a}_n] &\wedge H_2[\underline{a}_1, \underline{a}_2, \ldots, \underline{a}_n] \\ &\wedge \cdots \wedge H_p[\underline{a}_1, \underline{a}_2, \ldots, \underline{a}_n])\end{aligned}$$

and

$$\mathcal{M} \models H_1[a_1, a_2, \ldots, a_n] \wedge H_2[a_1, a_2, \ldots, a_n] \wedge \cdots \wedge H_p[a_1, a_2, \ldots, a_n].$$

But since the symbols $\underline{a}_i$, for $1 \leq i \leq n$, do not appear in T, we have

$$\begin{aligned}T \vdash \forall v_1 \forall v_2 \ldots \forall v_n \neg(H_1[v_1, v_2, \ldots, v_n] &\wedge H_2[v_1, v_2, \ldots, v_n] \\ &\wedge \cdots \wedge H_p[v_1, v_2, \ldots, v_n]),\end{aligned}$$

which proves that

$$\begin{aligned}&\forall v_1 \forall v_2 \ldots \forall v_n \\ &\quad \neg(H_1[v_1, v_2, \ldots, v_n] \wedge H_2[v_1, v_2, \ldots, v_n] \wedge \cdots \wedge H_p[v_1, v_2, \ldots, v_n])\end{aligned}$$

is a formula in Ψ. This contradicts the fact that $\mathcal{M}$ is a model of Ψ. ∎

Let us return to the example of groups. If the language is $(\mathbf{1}, \cdot)$, then it is false that every substructure of a group is a group; for example, $\mathbb{N}$ is a substructure of $\mathbb{Z}$ but is not a subgroup of $\mathbb{Z}$. This shows that in this language, the theory of groups is not equivalent to a universal theory. But if we add the inverse function to the language, the notion of substructure changes (a substructure has to be closed under all the functions in the language) and a substructure of a group is now necessarily a group (which is not surprising since we do have a universal theory in this case).

Corollary 8.38 *Let F be a closed formula in a language L. Then the following two conditions are equivalent:*

(i) *There exists a closed universal formula G of L which is equivalent to F.*

(ii) *The formula F is preserved under substructures.*

Proof The direction (i) $\Rightarrow$ (ii) is part of the preceding theorem. So assume (ii). The preceding theorem also tells us that there is a universal theory Ψ equivalent to F. By the compactness theorem, there exists a nite subset Ψ_0 of Ψ which is equivalent to F; F is therefore equivalent to the conjunction of the formulas in Ψ_0 which, in turn, is equivalent to a universal formula by Remark 8.35. ∎

There are dual versions of these theorems. A closed formula F (or a theory T) is **preserved under extensions** if, for every model $\mathcal{M}$ of F (of T) and for every extension $\mathcal{N}$ of $\mathcal{M}$, $\mathcal{N}$ is a model of F (of T). Also, a prenex formula in which the universal quantier does not occur is called an **existential formula**; an **existential theory** is a theory that consists only of existential formulas. We note that a formula is equivalent to an existential formula if and only if its negation is equivalent to a universal formula; also, a formula is preserved under extensions if and only if its negation is preserved under substructures. From all this, we obtain the following **preservation theorem for existential formulas**.

Theorem 8.39 *A closed formula is preserved under extensions if and only if it is equivalent to an existential formula.*

The previous theorem also extends to existential theories.

Theorem 8.40 *A theory T is preserved under extensions if and only if it is equivalent to an existential theory.*

Proof If T is equivalent to an existential theory, it is clear that any extension of a model of T is also a model of T.

Conversely, suppose that T is preserved under extensions. If F is a closed formula, let

$$U(F) = \{G : G \text{ is universal and } \vdash F \Rightarrow G\}.$$

Using the method of diagrams, as above, we see that any model of $U(F)$ can be embedded in a model of F. Consequently, if $F \in T$, then $U(\neg F) \cup T$ is contradictory: to see this, observe that if $\mathcal{M}$ were a model of $U(F) \cup T$, $\mathcal{M}$ would embed in some model $\mathcal{M}'$ of $\neg F$ which would also be a model of T since T is preserved under extensions. Now, by compactness, there exists $G_F \in U(\neg F)$ such that $T \cup G_F$ is contradictory; in other words, $T \vdash \neg G_F$. Set $T' = \{\neg G_F : F \in T\}$. Clearly, T' is equivalent to an existential theory and is a consequence of T. Since $\vdash \neg G_F \Rightarrow F$ is true for every $F \in T$, T' is equivalent to T. ∎

8.5.2 Preservation by unions of chains

We now proceed to something more complicated.

Denition 8.41 *An* $\forall\exists$ ***formula*** *is a prenex formula of the form*

$$\forall v_1 \forall v_2 \ldots \forall v_n \exists v_{n+1} \exists v_{n+2} \ldots \exists v_{n+p} G,$$

where G is a formula without quantiers. An $\forall\exists$ ***theory*** *is a theory consisting of* $\forall\exists$ *formulas.*

So an $\forall\exists$ formula begins with a certain number (possibly zero) of universal quantiers followed by a certain number of existential quantiers followed, nally, by a quantier-free formula. Universal and existential formulas are special cases of $\forall\exists$ formulas.

Note that the conjunction or disjunction of a nite number of $\forall\exists$ formulas is equivalent to an $\forall\exists$ formula (this is easy to prove, as before, simply by renaming the bound variables).

Denition 8.42 *A theory T is* ***preserved under unions of chains*** *if every union of a chain of models of T is again a model of T. A closed formula F is* ***preserved under unions of chains*** *if the theory $\{F\}$ is.*

To paraphrase, T is preserved under unions of chains if and only if the following condition is satised:

If $(I, <)$ is a totally ordered set, if $(\mathcal{M}_i : i \in I)$ is a family of models of T, and for all i and j in I, if $i < j$, then $\mathcal{M}_i \subseteq \mathcal{M}_j$, then $\bigcup_{i \in I} \mathcal{M}_i$ is a model of T.

Theorem 8.43 *A theory is preserved under unions of chains if and only if it is equivalent to an* $\forall\exists$ *theory.*

Proof We will rst prove that a closed $\forall\exists$ formula is preserved under unions of chains. This will clearly imply that a closed formula which is equivalent to a closed $\forall\exists$ formula, as well as a theory which is equivalent to an $\forall\exists$ theory, are also preserved under unions of chains.

So consider the formula

$$F = \forall v_1 \forall v_2 \ldots \forall v_n \exists v_{n+1} \exists v_{n+2} \ldots \exists v_{n+p} H[v_1, v_2, \ldots, v_{n+p}],$$

where H is a quantier-free formula. Let $(I, <)$ be a totally ordered set and let $(\mathcal{M}_i : i \in I)$ be a chain of models of F. Set $\mathcal{M} = \bigcup_{i \in I} \mathcal{M}_i$. To show that $\mathcal{M}$ is a model of F, we will consider arbitrary elements $a_1, a_2, \ldots, a_n$ of M and will show that

$$\mathcal{M} \models \exists v_{n+1} \exists v_{n+2} \ldots \exists v_{n+p} H[a_1, a_2, \ldots, a_n, v_1, v_2, \ldots, v_{n+p}].$$

Since the family $(\mathcal{M}_i : i \in I)$ is totally ordered by inclusion, there exists an index $i \in I$ such that all the a_k, for k from 1 to n inclusive, are contained in M_i. Because $\mathcal{M}_i$ is a model of F, there exist points $a_{n+1}, a_{n+2}, \ldots, a_{n+p}$ in M_i such that $\mathcal{M}_i \models H[a_1, a_2, \ldots, a_{n+p}]$. Since H is quantier-free and $\mathcal{M}$ is an extension of $\mathcal{M}_i$, it follows that $\mathcal{M} \models H[a_1, a_2, \ldots, a_{n+p}]$. Thus, $\mathcal{M} \models F$.

For the converse, we will prove that a theory T which is preserved under unions of chains is equivalent to an $\forall\exists$ theory. Set

$$\Psi = \{G : G \text{ is a closed } \forall\exists \text{ formula and } T \vdash G\}.$$

It is clear that every formula of Ψ is a consequence of T; we will now prove that the converse is also true. We will need a denition and two lemmas.

Denition 8.44 *Suppose that $\mathcal{M}$ and $\mathcal{N}$ are two L-structures and that $\mathcal{M} \subseteq \mathcal{N}$. We say that $\mathcal{M}$ is a* **1-elementary substructure** *of $\mathcal{N}$ (and write $\mathcal{M} \prec_1 \mathcal{N}$) if for every universal formula $F[v_1, v_2, \ldots, v_n]$ of L and for all elements $a_1, a_2, \ldots, a_n$ of M,*

$$\text{if } \mathcal{M} \models F[a_1, a_2, \ldots, a_n], \quad \text{then} \quad \mathcal{N} \models F[a_1, a_2, \ldots, a_n].$$

It is easy to see that if $\mathcal{M}$ is a 1-elementary substructure of $\mathcal{N}$, then for every universal or existential formula $F[v_1, v_2, \ldots, v_n]$ of L and for all $a_1, a_2, \ldots, a_n$ of M,

$$\mathcal{M} \models F[a_1, a_2, \ldots, a_n] \quad \text{if and only if} \quad \mathcal{N} \models F[a_1, a_2, \ldots, a_n].$$

This is what justies the expression `1-elementary' (elementary for formulas which have only a single block of the same quantier).

Lemma 8.45 *Assume that $\mathcal{M} \prec_1 \mathcal{N}$. Then there exists $\mathcal{M}'$ such that $\mathcal{M} \prec \mathcal{M}'$ and $\mathcal{N} \subseteq \mathcal{M}'$.*

Proof We use the method of diagrams. Consider the following theories in L_N (the simple diagram of $\mathcal{N}$ and the complete diagram of $\mathcal{M}$):

$$\Delta(\mathcal{N}) = \{H[\underline{a}_1, \underline{a}_2, \ldots, \underline{a}_n] : H \text{ is a formula of } L \text{ without quantiers,}$$
$$a_1, a_2, \ldots, a_n \text{ are points of } N \text{ and } \mathcal{N} \models H[a_1, a_2, \ldots, a_n]\}$$

and

$$D(\mathcal{M}) = \{F[\underline{a}_1, \underline{a}_2, \ldots, \underline{a}_n] : F \text{ is a formula of } L,$$
$$a_1, a_2, \ldots, a_n \text{ are points of } M \text{ and } \mathcal{M} \models F[a_1, a_2, \ldots, a_n]\}.$$

We are going to prove that $\Delta(\mathcal{N}) \cup D(\mathcal{M})$ is a consistent theory; we will then be nished since, thanks to Lemma 8.13, we will have a (simple) extension of $\mathcal{N}$ that is a model of $D(\mathcal{M})$, i.e. an elementary extension of $\mathcal{M}$.

The set $\Delta(\mathcal{N})$ is closed under conjunction. If we suppose that $\Delta(\mathcal{N}) \cup D(\mathcal{M})$ is not consistent, then, invoking the compactness theorem, we obtain a formula of $\Delta(\mathcal{N})$ of the form $H[\underline{a}_1, \underline{a}_2, \ldots, \underline{a}_n, \underline{a}_{n+1}, \underline{a}_{n+2}, \ldots, \underline{a}_{n+p}]$, where H is quantier-free, $a_1, a_2, \ldots, a_n$ are points of M and $a_{n+1}, a_{n+2}, \ldots, a_{n+p}$ are points of $N - M$ such that

$$D(\mathcal{M}) \vdash \neg H[\underline{a}_1, \underline{a}_2, \ldots, \underline{a}_n, \underline{a}_{n+1}, \ldots, \underline{a}_{n+p}].$$

Since the a_i, for i from $n+1$ to $n+p$ inclusive, do not occur in $D(\mathcal{M})$, we have

$$D(\mathcal{M}) \vdash \forall v_1 \forall v_2 \ldots \forall v_p \neg H[\underline{a}_1, \underline{a}_2, \ldots, \underline{a}_n, v_1, v_2, \ldots, v_p]$$

and hence

$$\mathcal{M} \models \forall v_1 \forall v_2 \ldots \forall v_p \neg H[a_1, a_2, \ldots, a_n, v_1, v_2, \ldots, v_p].$$

But this formula is clearly not satised in $\mathcal{N}$, which satises

$$H[a_1, a_2, \ldots, a_n, a_{n+1}, a_{n+2}, \ldots, a_{n+p}],$$

so this contradicts the fact that $\mathcal{M} \prec_1 \mathcal{N}$. ■

Lemma 8.46 *Let $\mathcal{N}$ be a model of Ψ. Then there exists a* 1*-elementary extension of $\mathcal{N}$ that is a model of T.*

Proof Consider the theory

$$\Delta_1(\mathcal{N}) = \{H[\underline{a}_1, \underline{a}_2, \ldots, \underline{a}_n] : H \text{ is a conjunction of universal formulas of } L,$$
$$a_1, a_2, \ldots, a_n \text{ are points of } N \text{ and } \mathcal{N} \models H[a_1, a_2, \ldots, a_n]\}.$$

We will begin by showing that $\Delta_1(\mathcal{N}) \cup T$ is a consistent theory. If it is not, then there exists a formula $H[\underline{a}_1, \underline{a}_2, \ldots, \underline{a}_n]$ of $\Delta_1(\mathcal{N})$ (which is closed under conjunction) such that

$$T \vdash \neg H[\underline{a}_1, \underline{a}_2, \ldots, \underline{a}_n],$$

and since the a_i, for i from 1 to n inclusive, do not appear in T,

$$T \vdash \forall v_1 \forall v_2 \ldots \forall v_n \neg H[v_1, v_2, \ldots, v_n].$$

Since H is equivalent to a universal formula, $\neg H[v_1, v_2, \ldots, v_n]$ is equivalent to an existential formula; so $\forall v_1 \forall v_2 \ldots \forall v_n \neg H[v_1, v_2, \ldots, v_n]$ is equivalent to an $\forall\exists$ formula which belongs to Ψ, according to the denition of this latter set. This contradicts the fact that $\mathcal{N}$ is a model of Ψ.

Thus we can nd a model $\mathcal{M}$ of $\Delta_1(\mathcal{N}) \cup T$ which we may assume, again thanks to Lemma 8.13, is an extension of $\mathcal{N}$. The fact that it is a model of $\Delta_1(\mathcal{N})$ shows immediately that $\mathcal{N} \prec_1 \mathcal{M}$. ■

We are now in a position to prove that every model of Ψ is a model of T. Starting with a model $\mathcal{M}_0$ of Ψ, we use Lemma 8.46 to obtain a model $\mathcal{M}_1$ of T such that $\mathcal{M}_0 \prec_1 \mathcal{M}_1$. Then use Lemma 8.45 to obtain an elementary extension $\mathcal{M}_2$ of $\mathcal{M}_0$ (which is therefore a model of Ψ) and which is also an extension of $\mathcal{M}_1$. Alternately invoking the two lemmas, we construct a chain $(\mathcal{M}_k : k \in \mathbb{N})$ of L-structures such that, for all k, $\mathcal{M}_{2k}$ is a model of Ψ, $\mathcal{M}_{2k+1}$ is a model of T, $\mathcal{M}_{2k} \prec_1 \mathcal{M}_{2k+1}$ and $\mathcal{M}_{2k} \prec \mathcal{M}_{2k+2}$.

Set $\mathcal{M} = \bigcup_{k\in\mathbb{N}} \mathcal{M}_k$. Note that $\mathcal{M}$ is also equal to both $\bigcup_{k\in\mathbb{N}} \mathcal{M}_{2k}$ and $\bigcup_{k\in\mathbb{N}} \mathcal{M}_{2k+1}$. Because T is preserved by unions of chains, $\mathcal{M}$ is a model of T. According to Tarski's unions of chains theorem (Theorem 8.21), since the chain $(\mathcal{M}_{2k} : k \in \mathbb{N})$ is elementary, $\mathcal{M}$ is an elementary extension of $\mathcal{M}_0$, thus $\mathcal{M}_0$ is also a model of T. ■

Remark 8.47 In the preceding argument, we only used the fact that T was preserved by unions of chains that are indexed by the integers. In other words, we have also proved that

> if, for every increasing chain $(\mathcal{M}_n : n \in \mathbb{N})$ of models of T, $\bigcup_{n\in\mathbb{N}} \mathcal{M}_n$ is a model of T, then T is equivalent to an $\forall\exists$ theory.

As examples of $\forall\exists$ theories (that are therefore preserved under unions of chains), we have the theory of groups, of elds, and of dense orderings without endpoints. By contrast, the theory of dense orderings with a rst and last element is not preserved under unions of chains. Here is an axiomatization of that theory:

(i) $\forall v_0(\neg v_0 < v_0)$;

(ii) $\forall v_0 \forall v_1((v_0 < v_1 \Leftrightarrow \neg v_1 < v_0) \vee v_0 \simeq v_1)$;

(iii) $\forall v_0 \forall v_1 \forall v_2((v_0 < v_1 \wedge v_1 < v_2) \Rightarrow v_0 < v_2)$;

(iv) $\neg\forall v_0 \exists v_1\, v_0 < v_1$;

(v) $\neg\forall v_0 \exists v_1\, v_1 < v_0$;

(vi) $\forall v_0 \forall v_1 \exists v_2(v_0 < v_1 \Rightarrow (v_0 < v_2 \wedge v_2 < v_1))$.

For every positive integer i, let $\mathcal{M}_i$ be the interval of reals $[-i, +i]$; each $\mathcal{M}_i$ is clearly a model of the theory above, but the union of the $\mathcal{M}_i$ is the whole of $\mathbb{R}$ which does not have a rst or a last element. This shows that it is impossible to nd an $\forall\exists$ axiomatization for the notion of a dense ordering with rst and last element.

Using an argument by compactness which is analogous to the one given for Corollary 8.38, we obtain the next theorem.

Theorem 8.48 *A closed formula is preserved under unions of chains if and only if it is equivalent to a closed $\forall\exists$ formula.*

8.5.3 Preservation by reduced products

At this point, we will take a new leap forward and deal with preservation under reduced products.

Denition 8.49 *An **elementary Horn formula** is a formula of the form*

$$H_1 \vee H_2 \vee \cdots \vee H_n$$

in which at most one of the H_i (for i from 1 to n inclusive) is an atomic formula, while the others are negations of atomic formulas.

*A **Horn formula** is one that can be obtained from elementary Horn formulas using conjunction together with the universal and existential quantiers.*

Elementary Horn formulas are clauses (see Denition 4.65). Therefore, up to logical equivalence, an elementary Horn formula has one of the following two forms:

- $(F_1 \wedge F_2 \wedge \cdots \wedge F_{n-1}) \Rightarrow F_n$ where, for k from 1 to n inclusive, the F_k are atomic formulas; this is the case in which there is in fact one atomic formula among the H_k in the denition;
- $H_1 \vee H_2 \vee \cdots \vee H_n$ where, for k from 1 to n inclusive, the H_k are negations of atomic formulas. These could also be written in the previous form if we accept the formula that is always false as an atomic formula.

By denition, the conjunction of a nite number of Horn formulas is a Horn formula. So it is easy to see that a Horn formula is equivalent to a formula of the form

$$Q_1 v_1 Q_2 v_2 \ldots Q_n v_n (G_1 \wedge G_2 \wedge \cdots \wedge G_n),$$

where each Q_i, for i from 1 to n inclusive, represents either the quantier $\forall$ or the quantier $\exists$ and where the G_j, for j from 1 to n inclusive, are elementary Horn formulas.

Example 8.50 The theory of groups and that of rings can be axiomatized with Horn formulas. By contrast, we will see (Exercise 14) that it is impossible to axiomatize the theory of elds by Horn formulas. It is the axiom

$$\forall v_1 \exists v_2 (\neg v_1 \simeq \mathbf{0} \Rightarrow v_1 \cdot v_2 \simeq \mathbf{1})$$

that is not equivalent to a Horn formula.

Denition 8.51 *A closed formula F in a language L is **preserved under reduced products** if for every set I, for every lter $\mathcal{F}$ on I, and for every family $(\mathcal{M}_i : i \in I)$ of L-structures, if $\{i \in I : \mathcal{M}_i \models F\} \in \mathcal{F}$, then*

$$\prod_{i \in I} \mathcal{M}_i \Big/ \mathcal{F} \models F.$$

Proposition 8.52 *Horn formulas are preserved under reduced products.*

Proof The proposition is a consequence of the following property:

> for any Horn formula $G[v_1, v_2, \ldots, v_n]$ of L, for any family $(\mathcal{M}_i : i \in I)$ of L-structures, for any lter $\mathcal{F}$ on I and for all $a_1, a_2, \ldots, a_n$ in $\prod_{i \in I} M_i$,
> if $\{i \in I : \mathcal{M}_i \models G[a_1^i, a_2^i, \ldots, a_n^i]\} \in \mathcal{F}$,
> then $\prod_{i \in I} \mathcal{M}_i / \mathcal{F} \models G[\bar{a}_1, \bar{a}_2, \ldots, \bar{a}_n]$. $(*)$

(Here, we are reusing the notations from Section 8.4: if $a \in \prod_{i \in I} M_i$, then a^i is the ith coordinate of a and $\bar{a}$ is the equivalence class of a with respect to $\approx_{\mathcal{F}}$.)

We will prove $(*)$ by induction on the number of steps required to obtain $G[v_1, v_2, \ldots, v_n]$ starting from elementary Horn formulas.

- Case in which $G[v_1, v_2, \ldots, v_n]$ is an elementary Horn formula. We can then rewrite $G[v_1, v_2, \ldots, v_n]$ in the form

$$H_1[v_1, v_2, \ldots, v_n] \vee H_2[v_1, v_2, \ldots, v_n] \vee \cdots \vee H_k[v_1, v_2, \ldots, v_n],$$

where H_1 is either an atomic formula or a formula that is always false and where, for j from 2 to k inclusive, $H_j = \neg J_j$, where J_j is an atomic formula.

For each j from 1 to k inclusive, set

$$X_j = \{i \in I : \mathcal{M}_i \models H_j[a_1^i, a_2^i, \ldots, a_n^i]\}$$

and let Y_j be the complement of X_j in I. By hypothesis,

$$\bigcup_{1 \leq j \leq k} X_j = \{i \in I : \mathcal{M}_i \models G[a_1^i, a_2^i, \ldots, a_n^i]\} \in \mathcal{F}.$$

We distinguish two possibilities.

(a) First possibility: for some integer j with $2 \leq j \leq k$, Y_j does not belong to $\mathcal{F}$. This means that $\{i \in I : \mathcal{M}_i \models J_j[a_1^i, a_2^i, \ldots, a_n^i]\} \notin \mathcal{F}$. By denition of the reduced product and because J_j is atomic, this implies that

$$\prod_{i \in I} \mathcal{M}_i \Big/ \mathcal{F} \not\models J_j[\bar{a}_1, \bar{a}_2, \ldots, \bar{a}_n],$$

hence that

$$\prod_{i \in I} \mathcal{M}_i \Big/ \mathcal{F} \models G[\bar{a}_1, \bar{a}_2, \ldots, \bar{a}_n].$$

(b) In the opposite case, $\bigcap_{2 \leq j \leq k} Y_j \in \mathcal{F}$. But

$$\left[\bigcup_{1 \leq j \leq k} X_j\right] \cap \left[\bigcap_{2 \leq j \leq k} Y_j\right] \subseteq X_1,$$

and this implies that $X_1 = \{i \in I : \mathcal{M}_i \models H_1[a_1^i, a_2^i, \ldots, a_n^i]\} \in \mathcal{F}$; in this case, H_1 is an atomic formula (otherwise, the empty set would belong to $\mathcal{F}$) and, according to the denition of reduced product,

$$\prod_{i \in I} \mathcal{M}_i \Big/ \mathcal{F} \models H_1[\bar{a}_1, \bar{a}_2, \ldots, \bar{a}_n],$$

so

$$\prod_{i \in I} \mathcal{M}_i \Big/ \mathcal{F} \models G[\bar{a}_1, \bar{a}_2, \ldots, \bar{a}_n].$$

- Case in which $G = G_1 \wedge G_2$. This presents no problem. If

$$\{i \in I : \mathcal{M}_i \models G[a_1^i, a_2^i, \ldots, a_n^i]\} \in \mathcal{F},$$

then

$$\{i \in I : \mathcal{M}_i \models G_1[a_1^i, a_2^i, \ldots, a_n^i]\} \in \mathcal{F} \quad \text{and}$$
$$\{i \in I : \mathcal{M}_i \models G_2[a_1^i, a_2^i, \ldots, a_n^i]\} \in \mathcal{F},$$

and, by the induction hypothesis,

$$\prod_{i \in I} \mathcal{M}_i \Big/ \mathcal{F} \models G_1[\bar{a}_1, \bar{a}_2, \ldots, \bar{a}_n] \quad \text{and}$$
$$\prod_{i \in I} \mathcal{M}_i \Big/ \mathcal{F} \models G_2[\bar{a}_1, \bar{a}_2, \ldots, \bar{a}_n];$$

so

$$\prod_{i \in I} \mathcal{M}_i \Big/ \mathcal{F} \models G_1[\bar{a}_1, \bar{a}_2, \ldots, \bar{a}_n] \wedge G_2[\bar{a}_1, \bar{a}_2, \ldots, \bar{a}_n].$$

- Case in which $G = \exists v_0 G_1[v_0, v_1, \ldots, v_n]$. If the set

$$X = \{i \in I : \mathcal{M}_i \models G[a_1^i, a_2^i, \ldots, a_n^i]\}$$

belongs to $\mathcal{F}$, we can dene an element $a_0 \in \prod_{i \in I} M_i$ in the following way: if $i \in X$, a_0^i is a point in M_i such that

$$\mathcal{M}_i \models G_1[a_0^i, a_1^i, \ldots, a_n^i];$$

if $i \notin X$, we can choose a_0^i to be any element of M_i. We then observe that

$$\{i \in I : \mathcal{M}_i \models G_1[a_0^i, a_1^i, \ldots, a_n^i]\} = X,$$

and, by the induction hypothesis,

$$\prod_{i\in I} \mathcal{M}_i \Big/ \mathcal{F} \models G_1[\bar{a}_0, \bar{a}_1, \ldots, \bar{a}_n]$$

and hence that

$$\prod_{i\in I} \mathcal{M}_i \Big/ \mathcal{F} \models G[\bar{a}_1, \ldots, \bar{a}_n].$$

- Case in which $G = \forall v_0 G_1[v_0, v_1, v_2, \ldots, v_n]$. Our hypothesis is that

$$X = \{i \in I : \mathcal{M}_i \models \forall v_0 G_1[v_0, a_1^i, \ldots, a_n^i]\} \in \mathcal{F};$$

let $a_0 \in \prod_{i\in I} M_i$; we wish to show that $\prod_{i\in I} \mathcal{M}_i / \mathcal{F} \models G_1[\bar{a}_0, \bar{a}_1, \ldots, \bar{a}_n]$. Now, for all $i \in X$, $\mathcal{M}_i \models G_1[a_0^i, a_1^i, \ldots, a_n^i]$; hence the set

$$\{i \in I : \mathcal{M}_i \models G_1[a_0^i, a_1^i, \ldots, a_n^i]\}$$

includes X and therefore belongs to $\mathcal{F}$. By the induction hypothesis, this implies that

$$\prod_{i\in I} \mathcal{M}_i \Big/ \mathcal{F} \models G_1[\bar{a}_0, \bar{a}_1, \ldots, \bar{a}_n]. \qquad \blacksquare$$

The converse of this proposition is true: every formula that is preserved by reduced products is equivalent to a Horn formula. But the proof of this requires techniques that we have not included in this brief introduction and we will not give it here. We will be satised to consider universal Horn formulas.

Proposition 8.53 *Let F be a closed formula. The following three conditions are equivalent:*

(i) *F is equivalent to a universal Horn formula.*

(ii) *F is preserved under reduced products and substructures.*

(iii) *F is preserved under nite products and substructures.*

Proof The implication (i) $\Rightarrow$ (ii) follows from Proposition 8.52 and Corollary 8.38. The implication (ii) $\Rightarrow$ (iii) is obvious. Let us deal with the implication (iii) $\Rightarrow$ (i). Because F is preserved under substructures, it is equivalent to a universal formula,

$$G = \forall v_1 \forall v_2 \ldots \forall v_n H[v_1, v_2, \ldots, v_n].$$

We may assume, in addition, that H is written in conjunctive normal form (see Chapter 3, Section 3.4), in other words,

$$H_1[v_1, v_2, \ldots, v_n] \wedge H_2[v_1, v_2, \ldots, v_n] \wedge \cdots \wedge H_k[v_1, v_2, \ldots, v_n],$$

where, for i from 1 to k inclusive, H_i is a disjunction of atomic formulas and negations of atomic formulas. Because universal quantiers commute with conjunction, the formula G is equivalent to the following set of formulas:

$$\{\forall v_1 \forall v_2 \ldots \forall v_n H_i[v_1, v_2, \ldots, v_n] : 1 \leq i \leq k\}.$$

We will now replace each of the formulas $\forall v_1 \forall v_2 \ldots \forall v_n H_i[v_1, v_2, \ldots, v_n]$ by a Horn formula thanks to the following lemma:

Lemma 8.54 *Let T be a theory and let K be a closed formula of the form*

$$\forall v_1 \forall v_2 \ldots \forall v_n (A_1 \vee A_2 \vee \cdots \vee A_u \vee \neg B_1 \vee \neg B_2 \vee \cdots \vee \neg B_t),$$

where, for $1 \leq i \leq u$ and for $1 \leq j \leq t$, A_i and B_j are atomic formulas. Assume that $T \cup \{K\}$ is preserved under nite products. Then there is a universal Horn formula J such that

$$T \vdash K \Leftrightarrow J.$$

Proof For every s from 1 to u inclusive, consider the formula

$$K_s = \forall v_1 \forall v_2 \ldots \forall v_n (A_s \vee \neg B_1 \vee \neg B_2 \vee \cdots \vee \neg B_t).$$

This is clearly a universal Horn formula. It is just as clear that $K_s \Rightarrow K$. We will prove that there exists an integer s between 1 and u such that

$$T \vdash K \Rightarrow K_s.$$

For an argument by contradiction, assume the contrary; so for $1 \leq s \leq u$, we obtain a structure $\mathcal{M}_s$ which is a model of T, of K and of $\neg K_s$. So there exist points $a_1^s, a_2^s, \ldots, a_n^s$ in M_s such that $\mathcal{M}_s \models \neg A_s[a_1^s, a_2^s, \ldots, a_n^s]$ and $\mathcal{M}_s \models B_j[a_1^s, a_2^s, \ldots, a_n^s]$ for $1 \leq j \leq t$.

Then consider the product $\mathcal{M} = \prod_{1 \leq i \leq u} \mathcal{M}_i$ and, in this product, the points $a_k = (a_k^s : 1 \leq s \leq u)$ for $1 \leq k \leq n$. By denition of the product, we have

$$\mathcal{M} \models \neg A_s[a_1, a_2, \ldots, a_n] \quad \text{for } 1 \leq s \leq u$$

and

$$\mathcal{M} \models B_j[a_1, a_2, \ldots, a_n] \quad \text{for } 1 \leq j \leq t.$$

We conclude from this that K is not satisable in $\mathcal{M}$; but this contradicts the fact that $T \cup \{K\}$ is preserved under nite products. ∎

Returning now to the proof of the proposition, we apply this lemma to the formula $K = \forall v_1 \forall v_2 \ldots \forall v_n H_1[v_1, v_2, \ldots, v_n]$ and the theory

$$T = \{\forall v_1 \forall v_2 \ldots \forall v_n H_i[v_1, v_2, \ldots, v_n] : 2 \leq i \leq k\}.$$

We obtain a universal Horn formula J_1 with the property that G is equivalent to the set

$$\{J_1\} \cup \{\forall v_1 \forall v_2 \ldots \forall v_n H_i[v_1, v_2, \ldots, v_n] : 2 \leq i \leq k\}.$$

We repeat this process and replace $\forall v_1 \forall v_2 \ldots \forall v_n H_2[v_1, v_2, \ldots, v_n]$ by an equivalent universal Horn formula, and so on, until we obtain a set of universal Horn formula, which, in turn, is equivalent to a single universal Horn formula. ■

8.6 $\aleph_0$-categorical theories

8.6.1 The omitting types theorem

Recall that a theory is $\aleph_0$-categorical if it has a denumerable model and all of its denumerable models are isomorphic. The denumerable models of such a theory have a certain number of very nice properties which we will now display. Throughout this section, T will denote a complete theory in a denumerable language L and the word `model' will always mean `model of T'. We begin by introducing the notions of type and of isolated type which are essential in this area.

Denition 8.55 (1) *Let n be an integer. An n-**type** is a set p of formulas of L that is closed under conjunction and is such that any free variables in a formula of p must be among $v_0, v_1, \ldots, v_{n-1}$. We will use the word **type** when we do not need to specify the integer n.*

(2) *Let p be an n-type, let $\mathcal{M}$ be a model and let $a_0, a_1, \ldots, a_{n-1}$ be points of $\mathcal{M}$. We say that the sequence $(a_0, a_1, \ldots, a_{n-1})$ **realizes** p if for every formula $F[v_0, v_1, \ldots, v_{n-1}]$ of p, we have*

$$\mathcal{M} \models F[a_0, a_1, \ldots, a_{n-1}].$$

(3) *We say that a model $\mathcal{M}$ **realizes** an n-type p (or that p is **realized in** $\mathcal{M}$) if there exists a sequence from $\mathcal{M}$ that realizes p. In the opposite case, we say that $\mathcal{M}$ **omits** p.*

(4) *Let $\bar{a} = (a_0, a_1, \ldots, a_{n-1})$ be a sequence of points of $\mathcal{M}$. The **type of** $\bar{a}$ **in** $\mathcal{M}$, denoted by $t(\bar{a}/\mathcal{M})$, is the n-type*

$$\{F[v_0, v_1, \ldots, v_{n-1}] : \mathcal{M} \models F[a_0, a_1, \ldots, a_{n-1}]\}.$$

(5) *Let p be an n-type and let $G[v_0, v_1, \ldots, v_{n-1}]$ be a formula. We say that $G[v_0, v_1, \ldots, v_{n-1}]$ **isolates** p if*

$$T \vdash \exists v_0 \exists v_1 \ldots \exists v_{n-1} G[v_0, v_1, \ldots, v_{n-1}]$$

and, for every formula $F[v_0, v_1, \ldots, v_{n-1}]$ of p,

$$T \vdash \forall v_0 \forall v_1 \ldots \forall v_{n-1}(G[v_0, v_1, \ldots, v_{n-1}] \Rightarrow F[v_0, v_1, \ldots, v_{n-1}]).$$

*We say that p is **isolated** if there exists a formula that isolates it.*

Before approaching the key theorem (the omitting types theorem), we will need a few lemmas and remarks.

Lemma 8.56 *If p is an n-type, the following three conditions are equivalent:*

(i) *there exists a denumerable model that realizes p;*

(ii) *there exists a model that realizes p;*

(iii) *for every formula $F[v_0, v_1, \ldots, v_{n-1}]$ of p,*

$$T \vdash \exists v_0 \exists v_1 \ldots \exists v_{n-1} F[v_0, v_1, \ldots, v_{n-1}].$$

Proof It is obvious that condition (i) implies condition (ii). To prove that (ii) implies (iii), suppose that $\mathcal{M}$ is a model of T, that $(a_0, a_1, \ldots, a_{n-1})$ is a sequence of points of M that realizes p and that $F[v_0, v_1, \ldots, v_{n-1}]$ belongs to p; then

$$\mathcal{M} \models F[a_0, a_1, \ldots, a_{n-1}]$$

and hence

$$\mathcal{M} \models \exists v_0 \exists v_1 \ldots \exists v_{n-1} F[v_0, v_1, \ldots, v_{n-1}].$$

Since T is a complete theory, any other model of T is elementarily equivalent to $\mathcal{M}$; consequently,

$$T \vdash \exists v_0 \exists v_1 \ldots \exists v_{n-1} F[v_0, v_1, \ldots, v_{n-1}].$$

Finally, suppose that (iii) is satised. Let us add constant symbols $c_0, c_1, \ldots, c_{n-1}$ to the language L and consider the following theory T':

$$T' = T \cup \{F[c_0, c_1, \ldots, c_{n-1}] : F[v_0, v_1, \ldots, v_{n-1}] \in p\}.$$

This theory is consistent; otherwise, by the compactness theorem, there exists a nite subset p_0 of p such that

$$T \cup \{F[c_0, c_1, \ldots, c_{n-1}] : F[v_0, v_1, \ldots, v_{n-1}] \in p_0\}$$

is inconsistent. Let $G[v_0, v_1, \ldots, v_{n-1}]$ denote the conjunction of the formulas of p_0. The formula $G[v_0, v_1, \ldots, v_{n-1}]$ belongs to p because p is closed under conjunction; so we have

$$T \vdash \neg G[c_0, c_1, \ldots, c_{n-1}].$$

Since the symbols $c_0, c_1, \ldots, c_{n-1}$ do not appear in T, it follows, as usual, that

$$T \vdash \neg \exists v_0 \exists v_1 \ldots \exists v_{n-1} G[v_0, v_1, \ldots, v_{n-1}],$$

which contradicts (iii).

By the LowenheimñSkolem theorem, we can nd a denumerable model $\mathcal{M}'$ of T'. If, for i from 0 to $n-1$ inclusive, we let a_i denote the point in this model that interprets the symbol c_i, and if we let $\mathcal{M}$ denote the reduct of $\mathcal{M}'$ to the original language L, then the sequence $(a_0, a_1, \ldots, a_{n-1})$ realizes the type p in $\mathcal{M}$. ∎

Denition 8.57 *A type that satises the conditions of Lemma 8.56 is called a* ***consistent type****.*

The next two remarks are evident.

Remark 8.58 If $\mathcal{M}' \prec \mathcal{M}$ and if the sequence $(a_0, a_1, \ldots, a_{n-1})$ realizes the type p in $\mathcal{M}'$, then this same sequence realizes the type p in $\mathcal{M}$.

Remark 8.59 If the models $\mathcal{M}$ and $\mathcal{M}'$ are isomorphic and if p is realized in one of the models, then it is realized in the other one.

Lemma 8.60 *Let $\mathcal{M}$ be a model and let p be a consistent type. Then there exists an elementary extension $\mathcal{M}_1$ of $\mathcal{M}$ in which p is realized.*

Proof We know there exists a model $\mathcal{M}'$ in which p is realized. Because $\mathcal{M}$ and $\mathcal{M}'$ are elementarily equivalent, there exists an elementary extension $\mathcal{M}_1$ of $\mathcal{M}$ and an elementary embedding from $\mathcal{M}'$ into $\mathcal{M}_1$ (Theorem 8.22), and hence an elementary substructure $\mathcal{M}_2$ of $\mathcal{M}_1$ that is isomorphic to $\mathcal{M}'$. The two previous remarks show that p is realized in $\mathcal{M}_1$. ∎

Remark 8.61 An isolated type is realized in every model of T (and is therefore a consistent type).

To see this, suppose that p is an isolated n-type and let $G[v_0, v_1, \ldots, v_{n-1}]$ be a formula that isolates it. If $\mathcal{M}$ is a model of T, there exist points $a_0, a_1, \ldots, a_{n-1}$ in M such that

$$\mathcal{M} \models G[a_0, a_1, \ldots, a_{n-1}].$$

It is then clear that the sequence $(a_0, a_1, \ldots, a_{n-1})$ realizes p.

The **omitting types theorem** is a converse of Remark 8.61.

Theorem 8.62 *Let p be an n-type that is not isolated. Then there exists a denumerable model of T that omits p.*

Proof will construct a model of T using Henkin's method which we employed in Chapter 4 to prove the completeness theorem. To produce a model that omits p, we must play our cards right.

Let $C = \{c_i : i \in \omega\}$ be an innite set of new constant symbols that we adjoin to L to produce L'. We will construct a theory T' in L' that has the following properties:

(1) $T \subseteq T'$;

(2) T' is complete in L';

(3) T' admits Henkin witnesses, i.e. if $F[v_0]$ is a formula of L', then there exists an integer i such that the formula $\exists v_0 F[v_0] \Rightarrow F[c_i]$ belongs to T';

(4) if $(d_0, d_1, \ldots, d_{n-1})$ is a sequence of length n of elements of C, then there exists a formula $F[v_0, v_1, \ldots, v_{n-1}] \in p$ such that

$$\neg F[d_0, d_1, \ldots, d_{n-1}] \in T'.$$

The reader is advised to review the proofs of Propositions A and B from Section 4.2 of Chapter 4. Conditions (1), (2), and (3) above appeared there also; using them, we prove the following:

- The binary relation R on C dened for all i and j in ω by

  $$R(c_i, c_j) \quad \text{if and only if} \quad T' \vdash c_i \simeq c_j$$

  is an equivalence relation. If $d \in C$, $\bar{d}$ will denote the equivalence class of d modulo R.
- If M denotes the set of equivalence classes for this relation, we can dene an L'-structure $\mathcal{M}'$ whose base set is M and is such that

  for every integer n, for all symbols $d_0, d_1, \ldots, d_{n-1}$ of C,
  and for every formula $F[v_0, v_1, \ldots, v_{n-1}]$ of L',
  $\mathcal{M}' \models F[\bar{d_0}, \bar{d_1}, \ldots, \bar{d}_{n-1},]$ if and only if $F[d_0, d_1, \ldots, d_{n-1}] \in T'$. (∗)

We then use condition (4) to show that $\mathcal{M}$, the reduct of $\mathcal{M}'$ to L, omits the type p; to do this, let $(a_0, a_1, \ldots, a_{n-1})$ be a sequence of points of M and, for all i from 0 to $n-1$ inclusive, let d_i be a point of C whose equivalence class is a_i. From (4), we know that there exists a formula $F[v_0, v_1, \ldots, v_{n-1}] \in p$ such that $\neg F[d_0, d_1, \ldots, d_{n-1}] \in T'$; we can then deduce from (∗) that $\mathcal{M} \models \neg F[a_0, a_1, \ldots, a_{n-1}]$. In other words, no sequence from M realizes p.

A brief digression: for the reader who is not keen to dive back into the proof of the completeness theorem, the argument above (for constructing from T' a model that omits p) can be replaced by the following argument. T' is a consistent theory, so it has a model $\mathcal{M}'$; if we let N denote the set of interpretations in $\mathcal{M}'$ of the symbols in C, then we can see, using condition (3) together with the test from the TarskiñVaught theorem (Theorem 8.7) and its accompanying remark, that N is the base set of an elementary substructure $\mathcal{N}$ of $\mathcal{M}'$; the reduct of $\mathcal{N}$ to the language L is therefore a model of T and we can verify, using condition (4), that it omits p.

It remains to construct T'. To do this, we will need

- an enumeration $(K_i : i \in \mathbb{N})$ of all the closed formulas of L';
- an enumeration $(G_i[v_0] : i \in \mathbb{N})$ of all formulas of L' with one free variable;
- an enumeration $(\gamma_i : i \in \mathbb{N})$ of all sequences from C of length n.

By induction on the integer k, we will dene a sequence of theories $(T_k : k \in \mathbb{N})$ which will, among others, satisfy the following properties:

- for all $k \in \mathbb{N}$, T_k is the union of T and a nite set of closed formulas of L';
- for all $k \in \mathbb{N}$, T_k is a consistent theory;
- for k and m in $\mathbb{N}$, if $k \leq m$, then $T_k \subseteq T_m$.

The theory T' will be the union of the theories T_k for $k \in \mathbb{N}$; we can already see (by compactness) that the result will be a consistent theory that includes T.

The induction begins with $T_0 = T$.

Let k be an integer greater than or equal to 0 and assume that T_m has already been dened for all $m \leq k$. The denition of T_k splits into three cases depending on whether k is congruent to 0, to 1, or to 2 modulo 3.

- Case in which $k = 3i$ for some integer i. If $T_k \cup \{K_i\}$ is a consistent theory, then set $T_{k+1} = T_k \cup \{K_i\}$; if not, then since T_k is consistent by the induction hypothesis, the reason is that $T_k \cup \{\neg K_i\}$ is consistent, so we set $T_{k+1} = T_k \cup \{\neg K_i\}$.
- Case in which $k = 3i + 1$ for some integer i. We choose an integer j such that c_j does not occur in T_k or in G_i (this is possible since T_k is the union of T, in which no constants from C occur, and a nite number of formulas). Set

$$T_{k+1} = T_k \cup \{\exists v_0 G_i[v_0] \Rightarrow G_i[c_j]\}.$$

The theory T_{k+1} is consistent by Lemma 4.26.

So far, we have merely copied the proof of the completeness theorem, while leaving some freedom for a third stage of the construction. We are already assured that the theory T' will be complete and will admit Henkin witnesses.

- Case in which $k = 3i + 2$ for some integer i. Suppose that $d_0, d_1, \ldots, d_{n-1}$ are the symbols from C such that $\gamma_i = (d_0, d_1, \ldots, d_{n-1})$. We know from the induction hypothesis that there exists a closed formula H of L' such that T_k is equivalent to $T \cup \{H\}$. Also, H can be written in the form

$$H = D[d_0, d_1, \ldots, d_{n-1}, e_0, e_1, \ldots, e_{m-1}],$$

where $D[v_0, v_1, v_2, \ldots, v_{n-1}, v_n, v_{n+1}, \ldots, v_{n+m-1}]$ is a formula of L and where, for i and j satisfying $0 \leq i \leq n-1$ and $0 \leq j \leq m-1$, $e_j \in C$ and $d_i \neq e_j$. Set

$$E[v_0, \ldots, v_{n-1}] = \exists v_n \exists v_{n+1} \ldots \exists v_{n+m-1} D[v_0, \ldots, v_{n-1}, v_n, \ldots, v_{n+m-1}].$$

The formula $\exists v_0 \exists v_1 \ldots \exists v_{n-1} E[v_0, v_1, \ldots, v_{n-1}]$ is a consequence of T. Since this is a closed formula and T is complete, we have

$$T \vdash \exists v_0 \exists v_1 \ldots \exists v_{n-1} E[v_0, v_1, \ldots, v_{n-1}].$$

Because p is not an isolated type, it contains a formula $F[v_0, v_1, \ldots, v_{n-1}]$ such that

$$T \cup \{\neg(\forall v_0 \forall v_1 \ldots \forall v_{n-1}(E[v_0, v_1, \ldots, v_{n-1}] \Rightarrow F[v_0, v_1, \ldots, v_{n-1}]))\}$$

is consistent. This amounts to saying that

$$T \cup \{\exists v_0 \ldots \exists v_{n-1}(\exists v_n \ldots \exists v_{n+m-1} D[v_0, \ldots, v_{n-1}, v_n, \ldots, v_{n+m-1}] \wedge \neg F[v_0, \ldots, v_{n-1}])\}$$

is consistent or, equivalently, that the theory

$$T \cup \{\exists v_0 \exists v_1 \ldots \exists v_{n+m-1} (D[v_0, \ldots, v_{n-1}, v_n, \ldots, v_{n+m-1}] \wedge \neg F[v_0, \ldots, v_{n-1}])\}$$

is consistent. Because the constants from C do not appear in this theory, it easily follows that

$$T \cup \{D[d_0, d_1, \ldots, d_{n-1}, e_0, e_1, \ldots, e_{m-1}] \wedge \neg F[d_0, d_1, \ldots, d_{n-1}]\}$$

is a consistent theory. So it sufces to set

$$T_{k+1} = T \cup \{\neg F[d_0, d_1, \ldots, d_{n-1}]\}.$$

■

Remark 8.63 The preceding proof can be modied to obtain the following stronger result: let $\{p_j : j \in \mathbb{N}\}$ be a denumerable set of non-isolated types (to be precise, suppose that p_j is an n_j-type); then there exists a denumerable model of T that omits all the types p_j for $j \in \mathbb{N}$.

To achieve this, we simply have to do more at stages of the form $3i + 2$ in the preceding proof. Enumerate the set of pairs (γ, j), where γ is a nite sequence from C, j is an integer, and the length of γ is equal to n_j. Using the same strategy of sabotage as above (which allowed us to prevent the sequence γ_i from realizing the type p by requiring this sequence to satisfy the negation of some formula in p), we ensure at stage $3i + 2$ that if (γ, j) is the ith element in the enumeration, then γ does not realize p_j.

8.6.2 $\aleph_0$-categorical structures

Corollary 8.64 *Suppose that T is an $\aleph_0$-categorical theory; then every consistent type is isolated.*

Proof Suppose there exists a consistent type p that is not isolated. According to Denition 8.57, there exists a denumerable model that realizes p. Theorem 8.62, for its part, guarantees the existence of a denumerable model that omits p. These two models cannot be isomorphic (Remark 8.59) so T is not $\aleph_0$-categorical. ■

Denition 8.65 *An n-type p is* ***complete*** *if, rst, it is consistent and, second, if for every formula $F[v_0, v_1, \ldots, v_{n-1}]$ of L,*

$$\text{either } F[v_0, v_1, \ldots, v_{n-1}] \in p \quad \text{or } \neg F[v_0, v_1, \ldots, v_{n-1}] \in p.$$

The set of complete n-types will be denoted by S_n.

We note that if p and q are two complete n-types and if $p \subseteq q$, then $p = q$; for if $F[v_0, v_1, \ldots, v_{n-1}] \notin p$, then $\neg F[v_0, v_1, \ldots, v_{n-1}] \in p$, so

$$\neg F[v_0, v_1, \ldots, v_{n-1}] \in q$$

and, since q is consistent, $F[v_0, v_1, \ldots, v_{n-1}] \notin q$.

For example, if $\bar{a}$ is a sequence of length n from a model $\mathcal{M}$, then $t(\bar{a}/\mathcal{M})$ is a complete type. Conversely, if p is a complete n-type, then there exists a model $\mathcal{M}$, which we may assume is denumerable, and a sequence $\bar{a}$ from $\mathcal{M}$ that realizes p and it is easy to see that $t(\bar{a}/\mathcal{M}) = p$.

Remark 8.66 If a formula $F[v_0, v_1, \ldots, v_{n-1}]$ isolates a complete n-type p, then it must belong to p. To see this, note that because

$$T \vdash \exists v_0 \exists v_1 \ldots \exists v_{n-1} F[v_0, v_1, \ldots, v_{n-1}],$$

it is false that

$$T \vdash \forall v_0 \forall v_1 \ldots \forall v_{n-1} (F[v_0, v_1, \ldots, v_{n-1}] \Rightarrow \neg F[v_0, v_1, \ldots, v_{n-1}]);$$

thus $\neg F[v_0, v_1, \ldots, v_{n-1}]$ does not belong to p. Then, since p is a complete type, $F[v_0, v_1, \ldots, v_{n-1}]$ does belong to p.

Corollary 8.67 *Suppose that T is an $\aleph_0$-categorical theory; then, for every integer n, the set S_n is nite.*

Proof If p is a complete n-type, it is isolated (Corollary 8.64) so we may choose a formula $F_p[v_0, v_1, \ldots, v_{n-1}]$ that isolates it. We have just observed that $F_p[v_0, v_1, \ldots, v_{n-1}]$ belongs to p.

If p and q are distinct complete n-types, then $\neg F_p \in q$; for if not, we obtain a contradiction as follows. Because q is complete, we would have $F_p \in q$. If $F[v_0, v_1, \ldots, v_{n-1}] \in p$, then by the choice of F_p,

$$\neg F[v_0, v_1, \ldots, v_{n-1}] \wedge F_p[v_0, v_1, \ldots, v_{n-1}]$$

is not consistent, hence $\neg F[v_0, v_1, \ldots, v_{n-1}] \notin q$. Because q is complete, it follows that $F[v_0, v_1, \ldots, v_{n-1}] \in q$. So we have proved that $p \subseteq q$; but we have already remarked that this implies $p = q$.

We continue the proof, arguing again by contradiction; assume that, for some integer n, S_n is innite. Add to the language a set $\{c_i : 0 \leq i \leq n-1\}$ of new, pairwise distinct constant symbols and consider the theory

$$T' = T \cup \{\neg F_p[c_0, c_1, \ldots, c_{n-1}] : p \in S_n\}.$$

This theory is consistent. To prove this, it is sufcient, by the compactness theorem, to show that for every nite subset X of S_n, the theory

$$T_X = T \cup \{\neg F_p[c_0, c_1, \ldots, c_{n-1}] : p \in X\}$$

is consistent. Choose a complete n-type q that does not belong to X (this is possible because X is nite and S_n is innite), a model $\mathcal{M}$, and a sequence of points $(a_0, a_1, \ldots, a_{n-1})$ of M such that $t((a_0, a_1, \ldots, a_{n-1})/\mathcal{M}) = q$. We saw above that, for all $p \in X$, $\neg F_p \in q$; hence

$$\mathcal{M} \models \neg F_p[a_0, a_1, \ldots, a_{n-1}].$$

So to obtain a model of T_X, it sufces to interpret c_i by a_i for i between 0 and $n-1$ inclusive.

Consequently, there exists a model $\mathcal{M}$ of T' containing points $b_0, b_1, \ldots, b_{n-1}$ such that, for all $p \in S_n$, $\mathcal{M} \models \neg F_p[b_0, b_1, \ldots, b_{n-1}]$. Thus, $t((b_0, b_1, \ldots, b_{n-1})/\mathcal{M})$ cannot belong to S_n; this is a contradiction. ■

We will prove a converse to Corollary 8.64.

Theorem 8.68 *Suppose that, for every n, every complete n-type is isolated; then T is* $\aleph_0$*-categorical.*

Proof The proof of the theorem is preceded by two lemmas.

Lemma 8.69 *Suppose that, for every n, every complete n-type is isolated. Let* $\mathcal{M}$ *and* $\mathcal{N}$ *be two models of T and let*

$$(a_0, a_1, \ldots, a_{n-1}) \quad \textit{and} \quad (b_0, b_1, \ldots, b_{n-1})$$

be two sequences from $\mathcal{M}$ *and* $\mathcal{N}$*, respectively, that satisfy*

$$t((a_0, a_1, \ldots, a_{n-1})/\mathcal{M}) = t((b_0, b_1, \ldots, b_{n-1})/\mathcal{N}).$$

Then for every $a \in M$*, there exists* $b \in N$ *such that*

$$t((a_0, a_1, \ldots, a_{n-1}, a)/\mathcal{M}) = t((b_0, b_1, \ldots, b_{n-1}, b)/\mathcal{N}).$$

Proof Set $p = t((a_0, a_1, \ldots, a_{n-1})/\mathcal{M})$ and $q = t((a_0, a_1, \ldots, a_{n-1}, a)/\mathcal{M})$. Let $F_1[v_0, v_1, \ldots, v_{n-1}]$ and $F_2[v_0, v_1, \ldots, v_{n-1}, v_n]$ be formulas that isolate p and q, respectively. We then see that

$$\exists v_n F_2[v_0, v_1, \ldots, v_{n-1}, v_n] \in p$$

and, since $t((b_0, b_1, \ldots, b_{n-1})/\mathcal{N}) = p$,

$$\mathcal{N} \models \exists v_n F_2[b_0, b_1, \ldots, b_{n-1}, v_n].$$

So let b be a point of N such that $\mathcal{N} \models F_2[b_0, b_1, \ldots, b_{n-1}, b]$; then $t((b_0, b_1, \ldots, b_{n-1}, b)/\mathcal{N}) = q$. ■

Lemma 8.70 *Suppose that, for every n, every complete n-type is isolated. Let* $\mathcal{M}$ *and* $\mathcal{N}$ *be two denumerable models of T and let*

$$(a_0, a_1, \ldots, a_{n-1}) \quad \textit{and} \quad (b_0, b_1, \ldots, b_{n-1})$$

be two sequences from $\mathcal{M}$ and $\mathcal{N}$, respectively, that satisfy

$$t((a_0, a_1, \ldots, a_{n-1})/\mathcal{M}) = t((b_0, b_1, \ldots, b_{n-1})/\mathcal{N}).$$

Then there exists an isomorphism f from $\mathcal{M}$ onto $\mathcal{N}$ such that, for all i from 1 to $n-1$ inclusive, $f(a_i) = b_i$.

Proof We will construct f using the back-and-forth technique that we used in Example 8.19. Let $(c_i : i \in \mathbb{N})$ be an enumeration of M and let $(d_i : i \in \mathbb{N})$ be an enumeration of N. The mapping f is already dened on the set $\{a_0, a_1, \ldots, a_{n-1}\}$. We will extend it by dening, with an induction on k, a point a_{n+k} and a point b_{n+k} in such a way that, for every integer k,

$$t((a_0, a_1, \ldots, a_{n+k})/\mathcal{M}) = t((b_0, b_1, \ldots, b_{n+k})/\mathcal{N}). \qquad (*)$$

To dene a_{n+k} and b_{n+k} assuming that the points a_i and b_i for $i < n+k$ have already been dened in such a way that $(*)$ holds, we will distinguish two cases:

- If k is even, say $k = 2i$, then set $a_{n+k} = c_i$; by the induction hypothesis, we have

  $$t((a_0, a_1, \ldots, a_{n+k-1})/\mathcal{M}) = t((b_0, b_1, \ldots, b_{n+k-1})/\mathcal{N}),$$

  and Lemma 8.69 allows us to nd a point b_{n+k} such that

  $$t((a_0, a_1, \ldots, a_{n+k})/\mathcal{M}) = t((b_0, b_1, \ldots, b_{n+k})/\mathcal{N}).$$

- If k is odd, say $k = 2i + 1$, then set $b_{n+k} = d_i$ and use this same lemma to nd a point a_{n+k} such that

  $$t((a_0, a_1, \ldots, a_{n+k})/\mathcal{M}) = t((b_0, b_1, \ldots, b_{n+k})/\mathcal{N}).$$

Observe that for all integers m and p, $a_m = a_p$ if and only if $b_m = b_p$; to see this, note for example that if $p \geq m$,

$$\begin{aligned} a_m = a_p \quad & \text{if and only if} \quad v_m \simeq v_p \in t((a_0, a_1, \ldots, a_p)/\mathcal{M}) \\ & \text{if and only if} \quad v_m \simeq v_p \in t((b_0, b_1, \ldots, b_p)/\mathcal{N}) \\ & \text{if and only if} \quad b_m = b_p. \end{aligned}$$

So we can dene a bijection f from $\{a_m : m \in \mathbb{N}\}$ onto $\{b_m : m \in \mathbb{N}\}$ by setting $f(a_m) = b_m$ for every $m \in \mathbb{N}$. But the choice of a_{n+k} for even k guarantees that $\{a_m : m \in \mathbb{N}\} = M$ and the choice of b_{n+k} for odd k guarantees that $\{b_m : m \in \mathbb{N}\} = N$; thus f is a bijection from M onto N.

The fact that this bijection is an isomorphism from $\mathcal{M}$ onto $\mathcal{N}$ is a consequence of the fact that, for every formula $F[v_0, v_1, \ldots, v_{m-1}]$, the following conditions are equivalent:

(1) $\mathcal{M} \models F[a_0, a_1, \ldots, a_{m-1}]$;

(2) $F[v_0, v_1, \ldots, v_{m-1}] \in t((a_0, a_1, \ldots, a_{m-1})/\mathcal{M})$;

(3) $F[v_0, v_1, \ldots, v_{m-1}] \in t((b_0, b_1, \ldots, b_{m-1})/\mathcal{N})$;

(4) $\mathcal{N} \models F[b_0, b_1, \ldots, b_{m-1}]$. ■

The proof of the theorem can now be completed. Suppose that $\mathcal{M}$ and $\mathcal{N}$ are two denumerable models of T. The complete type realized in $\mathcal{M}$ by the empty sequence is the theory of $\mathcal{M}$ and it is equal to the complete type realized in $\mathcal{N}$ by the empty sequence [because, since T is complete, we have $\mathrm{Th}(\mathcal{M}) = \mathrm{Th}(\mathcal{N})$]. So Lemma 8.70 can be applied to conclude that $\mathcal{M}$ and $\mathcal{N}$ are isomorphic. ■

There is another interesting consequence of Lemma 8.70 in case the models $\mathcal{M}$ and $\mathcal{N}$ are equal. We obtain:

Proposition 8.71 *Let $\mathcal{M}$ be a denumerable model of an $\aleph_0$-categorical theory and suppose that $(a_0, a_1, \ldots, a_{n-1})$ and $(b_0, b_1, \ldots, b_{n-1})$ are two sequences from M such that $t((a_0, a_1, \ldots, a_{n-1})/\mathcal{M}) = t((b_0, b_1, \ldots, b_{n-1})/\mathcal{M})$. Then there exists an automorphism f of $\mathcal{M}$ such that, for all i between 0 and $n-1$ inclusive, $f(a_i) = b_i$.*

Let n be an integer. It is possible to dene, on the set of formulas of L whose only free variables are among $v_0, v_1, \ldots, v_n$, a relation $\equiv$ that is obviously an equivalence relation: for all formulas $F[v_0, v_1, \ldots, v_{n-1}]$ and $G[v_0, v_1, \ldots, v_{n-1}]$,

$$F[v_0, v_1, \ldots, v_{n-1}] \equiv G[v_0, v_1, \ldots, v_{n-1}]$$

if and only if

$$T \vdash \forall v_0 \forall v_1 \ldots \forall v_{n-1}(F[v_0, v_1, \ldots, v_{n-1}] \Leftrightarrow G[v_0, v_1, \ldots, v_{n-1}]).$$

In other words, $F[v_0, v_1, \ldots, v_{n-1}] \equiv G[v_0, v_1, \ldots, v_{n-1}]$ if and only if, for every model $\mathcal{M}$ of T and for every sequence $(a_0, a_1, \ldots, a_{n-1})$ from M,

$$\mathcal{M} \models F[a_0, a_1, \ldots, a_{n-1}] \Leftrightarrow G[a_0, a_1, \ldots, a_{n-1}].$$

The set of equivalence classes for this relation is denoted by Lind_n (the **Lindenbaum algebra**); it is, of course, a Boolean algebra.

If $F = F[v_0, v_1, \ldots, v_{n-1}]$ is a formula of L, let

$$S_n(F) = \{p \in S_n : F \in p\}.$$

It is clear, rst of all, that if two formulas

$$F[v_0, v_1, \ldots, v_{n-1}] \quad \text{and} \quad G[v_0, v_1, \ldots, v_{n-1}]$$

are equivalent modulo $\equiv$, then $S_n(F) = S_n(G)$. Conversely, if F and G are not equivalent modulo $\equiv$, then there exists a model $\mathcal{M}$ of T and a sequence

$(a_0, a_1, \ldots, a_{n-1})$ of points of M that satises one of the formulas, F for example, and not the other. Then $t((a_0, a_1, \ldots, a_{n-1})/\mathcal{M})$ is in $S_n(F)$ but not in $S_n(G)$; as a consequence, there cannot be more elements in Lind_n than there are subsets of S_n; thus, if S_n is nite, so is Lind_n.

The next theorem summarizes and completes the results which we have obtained concerning $\aleph_0$-categorical theories.

Theorem 8.72 *The ve properties which follow are equivalent.*

(i) *T is $\aleph_0$-categorical;*

(ii) *for every integer n, every consistent n-type is isolated;*

(iii) *for every integer n, every complete n-type is isolated;*

(iv) *for every integer n, the set S_n is nite;*

(v) *for every integer n, the set Lind_n is nite.*

Proof The implication (i) $\Rightarrow$ (ii) is Corollary 8.64; (ii) $\Rightarrow$ (iii) is obvious; (iii) $\Rightarrow$ (i) is Theorem 8.68. In addition, (i) $\Rightarrow$ (iv) is Corollary 8.67 and we have just proved the implication (iv) $\Rightarrow$ (v). So to nish the proof, we will show that $\neg$(iii) $\Rightarrow$ $\neg$(v).

By hypothesis, therefore, there exists an integer n and a complete n-type p that is not isolated. By induction on k, we construct a sequence $(F_k[v_0, v_1, \ldots, v_{n-1}]: k \in \mathbb{N})$ of formulas belonging to p in such a way that for all k,

$$T \vdash \forall v_0 \forall v_1 \ldots \forall v_{n-1}(F_{k+1}[v_0, v_1, \ldots, v_{n-1}] \Rightarrow F_k[v_0, v_1, \ldots, v_{n-1}])$$

and

$$T \vdash \neg \forall v_0 \forall v_1 \ldots \forall v_{n-1}(F_k[v_0, v_1, \ldots, v_{n-1}] \Rightarrow F_{k+1}[v_0, v_1, \ldots, v_{n-1}]).$$

We start with an arbitrary formula $F_0[v_0, v_1, \ldots, v_{n-1}]$ of p. Assume that $F_k[v_0, v_1, \ldots, v_{n-1}]$ has already been dened. Because p is not isolated, we know there exists a formula $G[v_0, v_1, \ldots, v_{n-1}] \in p$ such that

$$T \vdash \neg \forall v_0 \forall v_1 \ldots \forall v_{n-1}(F_k[v_0, v_1, \ldots, v_{n-1}] \Rightarrow G[v_0, v_1, \ldots, v_{n-1}]).$$

Set $F_{k+1}[v_0, v_1, \ldots, v_{n-1}] = F_k[v_0, v_1, \ldots, v_{n-1}] \wedge G[v_0, v_1, \ldots, v_{n-1}]$.

It is then clear that the formulas $F_k[v_0, v_1, \ldots, v_{n-1}]$ are pairwise inequivalent modulo $\equiv$. ■

A structure will be called $\aleph_0$-categorical if its complete theory is $\aleph_0$-categorical. Let $\mathcal{M}$ be a denumerable structure that is $\aleph_0$-categorical. Consider, for every integer n, the following binary relation R_n on M^n: for all sequences $(a_0, a_1, \ldots, a_{n-1})$

and $(b_0, b_1, \ldots, b_{n-1})$ from M,

$$R_n((a_0, a_1, \ldots, a_{n-1}), (b_0, b_1, \ldots, b_{n-1}))$$

if and only if

there exists an automorphism f of $\mathcal{M}$ such that,
for all i from 0 to $n-1$ inclusive, $f(a_i) = b_i$.

It is easy to see that this relation is an equivalence relation. We should note that the set of automorphisms of $\mathcal{M}$ is a subgroup of the group of permutations of M and that the equivalence classes modulo R_n are none other than the orbits under the action of this group on M^n.

We have seen (Proposition 8.71) that if two sequences of the same length from M realize the same complete type, then they are equivalent modulo R_n. It follows from Theorem 8.72 that if $\mathcal{M}$ is $\aleph_0$-categorical, then the relations R_n each have only nitely many equivalence classes.

The converse is also true; this allows us to give a purely algebraic characterization (with no reference to the notion of formula) of $\aleph_0$-categorical structures.

Theorem 8.73 *Let $\mathcal{M}$ be a denumerable structure. Then the following conditions are equivalent:*

(i) *$\mathcal{M}$ is $\aleph_0$-categorical;*

(ii) *for every integer n, the relation R_n dened above has only a nite number of equivalence classes.*

Proof Assume that for every integer n, the relation R_n has only a nite number of equivalence classes. We need to prove that $T = \mathrm{Th}(\mathcal{M})$ is $\aleph_0$-categorical.

Let n be xed. If two sequences $(a_0, a_1, \ldots, a_{n-1})$ and $(b_0, b_1, \ldots, b_{n-1})$ are equivalent modulo R_n, they realize the same complete type. So there are only a nite number of complete n-types realized in $\mathcal{M}$. Let $\{p_1, p_2, \ldots, p_k\}$ be the set of these complete n-types; we will prove that this set is equal to $S_n(T)$. Suppose the contrary and let $q \in S_n(T)$ be different from all the p_i. Then q contains a formula $F[v_0, v_1, \ldots, v_{n-1}]$ such that, for all i, $\neg F_k[v_0, v_1, \ldots, v_{n-1}] \in p_i$. We do have $\exists v_0 \exists v_1 \ldots \exists v_{n-1} F[v_0, v_1, \ldots, v_{n-1}] \in T$, so there must exist points $a_0, a_1, \ldots, a_{n-1}$ such that $\mathcal{M} \models F[a_0, a_1, \ldots, a_{n-1}]$; but this is not possible since, for some i with $1 \leq i \leq k$, the sequence $(a_0, a_1, \ldots, a_{n-1})$ must realize one of the types p_i.

This shows that for every integer n, $S_n(T)$ is nite; so T is $\aleph_0$-categorical by Theorem 8.72. ∎

EXERCISES FOR CHAPTER 8

1. Consider the rst-order language L that has two unary predicate symbols E and P and one binary predicate symbol A. Let T be the theory of L consisting of the following formulas:

$$H_0 : \forall v_0(Ev_0 \Leftrightarrow \neg Pv_0)$$
$$H_1 : \forall v_0 \forall v_1(Av_0v_1 \Rightarrow (Ev_0 \wedge Pv_1))$$
$$H_2 : \forall v_1 \forall v_2((Pv_1 \wedge Pv_2 \wedge \forall v_0(Av_0v_1 \Leftrightarrow Av_0v_2)) \Rightarrow v_1 \simeq v_2)$$
$$H_3 : \exists v_0(Pv_0 \wedge \forall v_1 \neg Av_1v_0)$$
$$H_4 : \forall v_1(Pv_1 \Rightarrow \exists v_2(Pv_2 \wedge \forall v_0(Ev_0 \Rightarrow (Av_0v_1 \Leftrightarrow \neg Av_0v_2))))$$
$$H_5 : \forall v_1 \forall v_2 \exists v_3((Pv_1 \wedge Pv_2) \Rightarrow \forall v_0(Av_0v_3 \Leftrightarrow (Av_0v_1 \vee Av_0v_2)))$$

and, for every integer $n \geq 1$, the formula

$$F_n = \forall v_1 \forall v_2 \ldots \forall v_n \Bigg(\Bigg(\bigwedge_{1 \leq i \leq n} Ev_i \Bigg)$$
$$\Rightarrow \exists w_1 \forall w_0 \Bigg(Aw_0w_1 \Leftrightarrow \Bigg(\bigvee_{1 \leq i \leq n} w_0 \simeq v_i \Bigg) \Bigg) \Bigg).$$

(a) Let X be a non-empty set and $\wp(X)$ be the set of subsets of X, which we assume to be disjoint from X. We dene an L-structure $\mathcal{M}$ as follows:

- the base set is $M_X = X \cup \wp(X)$;
- the interpretation of E is X;
- the interpretation of P is $\wp(X)$;
- the interpretation of A is the set

$$\bar{A} = \{(x, y) \in M_X^2 : x \in X, \ y \in \wp(X) \text{ and } x \in y\}.$$

Show that $\mathcal{M}$ is a model of T.

(b) Does T have a denumerable model?

(c) For which integers n, does T have a model whose base set has cardinality n?

(d) Show that T is equivalent to $\{H_0, H_1, H_2, H_3, H_4, H_5, F_1\}$.

(e) Show that T is not $\aleph_0$-categorical.

2. This is a follow-up to Exercise 15 from Chapter 3.

(a) Let $\mathcal{M} = \langle M, \bar{d}, \bar{g} \rangle$ be an arbitrary model of T. Dene a binary relation $\approx$ on M by

$$a \approx b \quad \text{if and only if} \quad \text{there exist integers } m, n, p, \text{ and } q \text{ such that } \bar{d}^m(\bar{g}^n(a)) = \bar{d}^p(\bar{g}^q(b)).$$

Show that $\approx$ is an equivalence relation on M. An equivalence class under $\approx$ will be called a **grill**. Show that every grill is stable for $\bar{d}$ and for $\bar{g}$. Show that every grill, together with the restrictions of the mappings $\bar{d}$ and $\bar{g}$, is a substructure of $\mathcal{M}$ that is a model of T.

(b) Let L' be the language obtained by adding two new constant symbols λ and μ to L. For every four-tuple (m, n, p, q) of natural numbers, let G_{mnpq} denote the following closed formula of L':

$$\neg d^m g^n \lambda \simeq d^p g^q \mu.$$

Using this family of formulas, prove the existence of a non-standard model of T (i.e. a model of T that is not isomorphic to the standard model).

(c) Let A be a non-empty set. Construct a model of T whose set of grills is equipotent with A.

(d) Show that two models of T whose sets of grills are equipotent are isomorphic.

(e) Show that T is not $\aleph_0$-categorical. Consider a set $\mathcal{X}$ of L-structures that has the following properties:

- the elements of $\mathcal{X}$ are denumerable models of T;
- if $\mathcal{M} \in \mathcal{X}$, $\mathcal{N} \in \mathcal{X}$, and $\mathcal{M} \neq \mathcal{N}$, then $\mathcal{M}$ and $\mathcal{N}$ are not isomorphic;
- every denumerable model of T is isomorphic to one of the elements of $\mathcal{X}$.

What is the cardinality of $\mathcal{X}$?

(f) Let κ be an uncountable cardinal. Show that T is κ-categorical.

3. Let $\langle G, \cdot, e \rangle$ be a group. With this group, we associate a rst-order language L_G that has a unary function symbol f_α for every element $\alpha \in G$. Let T denote the following theory of L_G:

$$\begin{aligned}&\{\forall v_0 f_e v_0 \simeq v_0\}\\ &\quad \cup \{\forall v_0 f_\alpha f_\beta v_0 \simeq f_{\alpha\beta} v_0 : \alpha \in G \text{ and } \beta \in G\}\\ &\quad \cup \{\forall v_0 \neg f_\alpha v_0 \simeq v_0 : \alpha \in G \text{ and } \alpha \neq e\}.\end{aligned}$$

(a) Show that for every term t of L_G, there exists an element $\alpha \in G$ and a symbol x for a variable such that

$$T \vdash \forall x\, t \simeq f_\alpha x.$$

(b) After observing that any atomic formula of L_G can involve at most two variables, show that, for every atomic formula $F = F[v_0, v_1]$ of L_G, one of the following three possibilities holds:
- $T \vdash \forall v_0 \forall v_1 F$;
- $T \vdash \forall v_0 \forall v_1 \neg F$;
- there exists an element $\alpha \in G$ such that $T \vdash \forall v_0 \forall v_1 (F \Leftrightarrow v_0 \simeq f_\alpha v_1)$.

(c) Let $\mathcal{G}$ be the L-structure whose base set is G and in which, for every $\alpha \in G$, the symbol f_α is interpreted by the map $\beta \mapsto \alpha \cdot \beta$ from G into G (i.e. left multiplication by α). Show that $\mathcal{G}$ is a model of T.

(d) Let $\mathcal{M} = \langle M, (\phi_\alpha)_{\alpha \in G} \rangle$ be a model of T and let a be an element of M. Consider the set

$$O(a) = \{x \in M : \text{ there exists } \alpha \in G \text{ such that } x = \phi_\alpha(a)\}.$$

Show that $O(a)$ is a substructure of $\mathcal{M}$ that is isomorphic to $\mathcal{G}$.
Show that $X_M = \{O(a) : a \in M\}$ is a partition of M.
Show that if $\mathcal{M}$ and $\mathcal{N}$ are two models of T and if X_M and X_N are equipotent, then $\mathcal{M}$ and $\mathcal{N}$ are isomorphic.

(e) Show that if G is innite, then the theory T is complete.

(f) Suppose that G is nite. Does there exist an innite cardinal λ such that T is λ-categorical? Is the theory T complete?

4. Consider the language L that consists of a single binary relation symbol R. Let L_∞ denote the language obtained by adding denumerably many new constant symbols $c_0, c_1, \ldots, c_n, \ldots$ to L.

For each integer n, let L_n denote the language $L \cup \{c_0, c_1, \ldots, c_n\}$.

Given an L_∞-structure $\mathcal{M}$ and an integer n, let $\mathcal{M}_n$ denote the reduct of $\mathcal{M}$ to the language L_n.

Consider the theory in the language L expressing that the interpretation of R is an equivalence relation that has innitely many equivalence classes, each of which is innite.

(a) Write down axioms for the theory T and give an example of a model of T.

(b) Show that T is not equivalent to any nite theory in L.

(c) For which innite cardinals λ is the theory T λ-categorical? Find two models $\mathcal{M}_1$ and $\mathcal{M}_2$ of T such that there is no elementary embedding from $\mathcal{M}_1$ into $\mathcal{M}_2$ or from $\mathcal{M}_2$ into $\mathcal{M}_1$.

(d) Is the theory T complete?

(e) Let T_+ be the following theory in the language L_∞:

$$T_+ = T \cup \{\neg R c_n c_m : n \in \mathbb{N},\ m \in \mathbb{N} \text{ and } n \neq m\}.$$

Give an example of a model of T_+.
Show that T_+ is not equivalent to any nite theory in the language L_∞.

(f) For which innite cardinals λ is the theory T_+ λ-categorical?

(g) Let $\mathcal{M}_1$ and $\mathcal{M}_2$ be two denumerable models of T_+. Show that for every integer n, the reducts of $\mathcal{M}_1$ and of $\mathcal{M}_2$ to the language L_n, which we will denote by $\mathcal{M}_1 \restriction L_n$ and $\mathcal{M}_2 \restriction L_n$, are isomorphic. Conclude from this that T_+ is a complete theory in L_∞.

5. Consider a denumerable rst-order language L and let $\mathcal{F}_1$ denote the formulas of L that have at most one free variable.

Given a formula $F[x] \in \mathcal{F}_1$ and an L-structure $\mathcal{M} = \langle M, \dots \rangle$, the **value of** F **in** $\mathcal{M}$, denoted by $Val(F, \mathcal{M})$, is the subset of M dened by the formula F, in other words, the set

$$Val(F, \mathcal{M}) = \{a \in M : \mathcal{M} \models F[a]\}.$$

For any innite cardinal λ, a λ-**structure** is an innite L-structure $\mathcal{M}$ with the property that, for every formula $F \in \mathcal{F}_1$, the value of F in $\mathcal{M}$ is either a nite set or a set of cardinality λ.

A model of a formula or of a theory that is a λ-structure will be called a λ-**model**.

(a) Show that if λ is an innite cardinal, then every λ-structure has cardinality λ.

(b) Show that every structure of cardinality $\aleph_0$ is an $\aleph_0$-structure.

(c) Let T be a theory in L and let $F[x]$ be a formula in $\mathcal{F}_1$. Suppose that for every integer n, T has a model in which the value of the formula F is a set that contains at least n elements.

Show that for every innite cardinal λ, T has at least one model in which the value of F is a set whose cardinality is equal to λ.

(*Hint*: Add to the language a set of constant symbols of cardinality λ.)

(d) Let T be a theory in L that has at least one innite model. Prove that for every innite cardinal λ, T has at least one λ-model.

(*Hint*: Choose an innite model $\mathcal{M}_0$ of T and, with each formula $G \in \mathcal{F}_1$ whose value in $\mathcal{M}_0$ is an innite set, associate a set C_G of constant symbols of cardinality λ.)

(e) Let S be a consistent theory in L that only has innite models. Assume that for some innite cardinal λ, all λ-models of T are isomorphic. Show that S is complete.

6. Consider the languages $L_1 = \{f\}$ and $L_2 = \{f, P\}$, where f is a unary function symbol and P is a unary relation symbol. Let T_1 denote the following theory in L_1:

$$\{\forall x \forall y (fx \simeq fy \Rightarrow x \simeq y)\} \cup \{\forall x \exists y\, fy \simeq x\} \\ \cup \{\forall x \neg f^n x \simeq x : n \in \mathbb{N}^*\}.$$

[The term $f^n x$ is dened as usual: $f^0 x = x$ and, for all $n \in \mathbb{N}$, $f^{n+1} x = f(f^n x)$.]

Let T_2 denote the following theory in L_2:

$$T_1 \cup \{\exists x Px, \quad \exists x \neg Px, \quad \forall x(Px \Leftrightarrow Pfx)\}.$$

(a) Show that T_1 is a complete theory in L_1.

(b) Show that T_2 is not categorical in any innite cardinal.

(c) Use the results from the preceding exercise to show that T_2 is a complete theory in L_2.

7. (a) Let L_0 be the language consisting of the single binary predicate R. Let T_0 be the theory expressing that the interpretation of R is a total ordering, together with the following two additional formulas:

$$\forall v_1 \exists v_2 (R v_1 v_2 \wedge \neg v_1 \simeq v_2 \wedge \forall v_3((R v_1 v_3 \wedge \neg v_1 v_3) \Rightarrow R v_2 v_3));$$
$$\forall v_1 \exists v_2 (R v_2 v_1 \wedge \neg v_1 \simeq v_2 \wedge ((R v_3 v_1 \wedge \neg v_1 \simeq v_3) \Rightarrow R v_3 v_2)).$$

Show that we can nd two models $\mathcal{M}_0$ and $\mathcal{M}_1$ of T_0 such that $\mathcal{M}_0$ is a substructure of $\mathcal{M}_1$, but is not an elementary substructure.

(b) Show that T_0 is not equivalent to an $\forall\exists$ theory in L_0.

8. Let L be a denumerable language and let T be a consistent theory in L that only has innite models.

T is called **model-complete** if and only if for all models $\mathcal{M}$ and $\mathcal{N}$ of T, if $\mathcal{N}$ is an extension of $\mathcal{M}$, then it is an elementary extension.

A model $\mathcal{M}$ of T is called a **prime model of** T if and only if every model of T is isomorphic to a (simple) extension of $\mathcal{M}$.

(a) Show that a model-complete theory that has a prime model is complete.

(b) Show that the following four conditions are equivalent:

(1) T is model-complete;

(2) for every model $\mathcal{M}$ of T, every formula of $D(\mathcal{M})$ is a consequence of $\Delta(\mathcal{M}) \cup T$ (for the notations, see Section 8.2 of Chapter 8);

(3) for every denumerable model $\mathcal{M}$ of T, every formula of $D(\mathcal{M})$ is a consequence of $\Delta(\mathcal{M}) \cup T$;

(4) for all denumerable models $\mathcal{M}$ and $\mathcal{M}'$ of T, if $\mathcal{M} \subseteq \mathcal{M}'$, then $\mathcal{M} \prec \mathcal{M}'$.

(c) Show that if T is model-complete, T is equivalent to an $\forall\exists$ theory. Is the converse true?

(d) Let $F[v_0, v_1, \ldots, v_n]$ be a formula of L. Consider the following property of T and $F[v_0, v_1, \ldots, v_n]$:

for all models $\mathcal{M}$ and $\mathcal{M}'$ of T such that $\mathcal{M} \subseteq \mathcal{M}'$,
for all elements $a_0, a_1, \ldots, a_n$ of M,
if $\mathcal{M} \models F[a_0, a_1, a_2, \ldots, a_n]$, then $\mathcal{M}' \models F[a_0, a_1, a_2, \ldots, a_n]$. (*)

Prove that (∗) holds if and only if there is an existential formula $G[v_0, v_1, v_2, \ldots, v_n]$ of L such that

$$T \vdash \forall v_0 \forall v_1 \ldots \forall v_n (F[v_0, v_1, \ldots, v_n] \Leftrightarrow G[v_0, v_1, \ldots, v_n]).$$

(*Hint*: Adjoin constant symbols $c_0, c_1, \ldots, c_n$ and reread the proof of Theorem 8.37 for inspiration.)

(e) Show that T is model-complete if and only if for every universal formula $F[v_0, v_1, \ldots, v_n]$ of L, there exists an existential formula $G[v_0, v_1, \ldots, v_n]$ of L such that

$$T \vdash \forall v_0 \forall v_1 \ldots \forall v_n (F[v_0, v_1, \ldots, v_n] \Leftrightarrow G[v_0, v_1, \ldots, v_n]).$$

9. The purpose of this exercise is to prove **Lindstrom's theorem**: *let T be an $\forall\exists$ theory in a denumerable language; if T has no nite models and is categorical in some innite cardinal, then T is model-complete.*

We resume with the notations and results from the preceding exercise.

(a) Let λ be an innite cardinal and let $F[v_0, v_1, \ldots, v_n]$ be a formula of L. Show that the following two conditions are equivalent:

(1) For all models $\mathcal{M}$ and $\mathcal{M}'$ of T of cardinality λ such that $\mathcal{M} \subseteq \mathcal{M}'$ and for all elements $a_0, a_1, \ldots, a_n$ of M, if $\mathcal{M} \models F[a_0, a_1, a_2, \ldots, a_n]$, then $\mathcal{M}' \models F[a_0, a_1, a_2, \ldots, a_n]$.

(2) There is an existential formula $G[v_0, v_1, \ldots, v_n]$ of L such that

$$T \vdash \forall v_0 \forall v_1 \ldots \forall v_n (F[v_0, v_1, \ldots, v_n] \Leftrightarrow G[v_0, v_1, \ldots, v_n]).$$

(b) Assume that T is an $\forall\exists$ theory. Let $F[v_0, v_1, \ldots, v_n]$ be a universal formula and let λ be an innite cardinal. Prove that T has a model $\mathcal{M}$ of cardinality λ which has the following property:

for every model $\mathcal{M}'$ of T such that $\mathcal{M} \subseteq \mathcal{M}'$,
for all elements $a_0, a_1, \ldots, a_n$ of M,
if $\mathcal{M} \models F[a_0, a_1, a_2, \ldots, a_n]$, then $\mathcal{M}' \models F[a_0, a_1, a_2, \ldots, a_n]$. (∗∗)

(c) Prove Lindstrom's theorem.

(d) Assume that the language consists of a single unary function symbol f and set

$$T_0 = \{\forall v_0 \forall v_1 (f v_0 \simeq f v_1 \Rightarrow v_0 \simeq v_1)\} \cup \{\forall v_0 \neg f^n v_0 \simeq v_0 : n \in \mathbb{N}\}.$$

Show that T_0 is an $\forall\exists$ theory that is neither complete nor model-complete.

10. The language L consists of a binary predicate symbol R and a denumerably innite set of constant symbols $\{c_0, c_1, \ldots, c_n, \ldots\}$.

Let A be a closed formula of L expressing that the interpretation of R is a strict, dense, total ordering without endpoints. For every $n \in \mathbb{N}$, F_n is the formula Rc_nc_{n+1}.

Consider the theory

$$\{A\} \cup \{F_n : n \in \mathbb{N}\}.$$

We let $\mathcal{A}$, $\mathcal{B}$, and $\mathcal{C}$ denote the three L-structures whose base set is $\mathbb{Q}$, in which the interpretation of R is the usual strict ordering, and in which the sequence of constant symbols $(c_n)_{n\in\mathbb{N}}$ is interpreted, respectively, by the following sequences of rationals:

$$\alpha = (\alpha_n)_{n\in\mathbb{N}}, \qquad \beta = (\beta_n)_{n\in\mathbb{N}}, \qquad \gamma = (\gamma_n)_{n\in\mathbb{N}}$$

where

$$\alpha_n = n, \qquad \beta_n = -\frac{1}{n+1}, \qquad \gamma_n = \sum_{k=0}^{n} \frac{1}{k!}.$$

(a) Show that T is complete in L.

(b) Show that every denumerable model of T is isomorphic to one of the three structures $\mathcal{A}$, $\mathcal{B}$, or $\mathcal{C}$.

(c) Show that the theory T is model-complete (see Exercise 8).

(d) Show that every denumerable model of T has an elementary extension that is isomorphic to $\mathcal{B}$ and an elementary extension that is isomorphic to $\mathcal{C}$.

11. Let L be the rst-order language consisting of a unary function symbol f and a binary relation symbol R. Let A denote the conjunction of the following seven formulas:

$$\forall v_0 R v_0 v_0;$$
$$\forall v_0 \forall v_1 ((R v_0 v_1 \Leftrightarrow R v_1 v_0) \Rightarrow v_0 \simeq v_1);$$
$$\forall v_0 \forall v_1 \forall v_2 ((R v_0 v_1 \wedge R v_1 v_2) \Rightarrow R v_0 v_2);$$
$$\forall v_0 \exists v_1 (f v_1 \simeq v_0 \wedge \forall v_2 (f v_2 \simeq v_0 \Rightarrow v_2 \simeq v_1));$$
$$\forall v_0 \forall v_1 (R v_0 v_1 \Leftrightarrow R f v_0 f v_1);$$
$$\forall v_0 (R v_0 f v_0 \wedge \neg v_0 \simeq f v_0);$$
$$\forall v_0 \forall v_1 ((\neg v_0 \simeq v_1 \wedge R v_0 v_1) \Rightarrow R f v_0 v_1).$$

(a) Show that in every model of the formula A, the interpretation of the symbol R is a total ordering of the base set of the model, with no least or greatest element, such that every element has a successor, i.e. a strict least upper bound.

(b) Show that $\mathbb{Z}$ with its usual ordering and the successor function is a model of A.

Let $X = \langle B, \leq \rangle$ be an arbitrary totally ordered set. Consider the following L-structure $\mathcal{M}_X$:

- the base set of $\mathcal{M}_X$ is the set $B \times \mathbb{Z}$;
- the interpretation of R in $\mathcal{M}_X$ is the set

$$\{((x, n), (y, m)) \in (B \times \mathbb{Z})^2 : x < y \text{ or } (x = y \text{ and } n \leq m)\};$$

- the interpretation of f in $\mathcal{M}_X$ is the mapping that, with $(x, n) \in (B \times \mathbb{Z})$, associates $(x, n + 1)$.

Show that $\mathcal{M}_X$ is a model of A.

(c) Let $\mathcal{M} = \langle M, \bar{f}, \bar{R} \rangle$ be a model of A. We wish to prove that there exists a totally ordered set X such that $\mathcal{M}$ is isomorphic to $\mathcal{M}_X$.

On the base set M of $\mathcal{M}$, we dene two binary relations $\ll$ and $\approx$ as follows: for all a and b in M,

$$a \ll b \quad \text{if and only if} \quad \text{for all } n \in \mathbb{N},\ \mathcal{M} \models Rf^n ab$$

and

$$a \approx b \quad \text{if and only if} \quad \text{there exist integers } n \text{ and } p \text{ such that } \mathcal{M} \models f^n a \simeq f^p b.$$

Show that $\ll$ is irreexive and transitive, that $\approx$ is an equivalence relation, and that

$$a \approx b \quad \text{if and only if} \quad a \ll b \text{ and } b \ll a \text{ are both false.}$$

Show that each equivalence class modulo $\approx$ is a substructure of $\mathcal{M}$ that is isomorphic to $\mathbb{Z}$.

Show that the relation $\ll$ allows us to dene a total ordering on the set $M/\approx$ of equivalence classes.

Show that if $X = \langle C, \lhd \rangle$ is the ordered set obtained in this way, then $\mathcal{M}$ is isomorphic to $\mathcal{M}_X$.

(d) Show that if X and Y are two totally ordered sets, then $\mathcal{M}_X$ and $\mathcal{M}_Y$ are isomorphic if and only if X and Y are isomorphic.

Show that A only has innite models and is not categorical in any innite cardinal.

(e) We wish to show that $\{A\}$ is a complete theory.

(1) Show that if a and b are two points in a model $\mathcal{M}$ of A which satisfy $a \ll b$, then there exists an elementary extension $\mathcal{M}_1$ of $\mathcal{M}$ and a point c in $\mathcal{M}_1$ such that

$$a \ll c \quad \text{and} \quad c \ll b.$$

(2) Show also that if a is a point of M, then there exists an elementary extension $\mathcal{M}_1$ of $\mathcal{M}$ and points b and c of M_1 such that

$$b \ll a \quad \text{and} \quad a \ll c.$$

(3) Let $\mathcal{M}$ and $\mathcal{N}$ be two models of A and let $(a_1, a_2, \ldots, a_n)$ and $(b_1, b_2, \ldots, b_n)$ be two nite sequences of the same length from $\mathcal{M}$ and $\mathcal{N}$, respectively.

Consider the following condition

$$P((\mathcal{M}, a_1, a_2, \ldots, a_n), (\mathcal{N}, b_1, b_2, \ldots, b_n)) :$$

for every atomic formula $F[v_1, v_2, \ldots, v_n]$ of L,
$\mathcal{M} \models F[a_1, a_2, \ldots, a_n]$ if and only if $\mathcal{N} \models F[b_1, b_2, \ldots, b_n]$.

Show that this condition is equivalent to

for all integers i and j such that $1 \leq i, j \leq n$ and for all $k \in \mathbb{N}$,
$\mathcal{M} \models a_i \simeq f^k a_j$ if and only if $\mathcal{N} \models b_i \simeq f^k b_j$, and
$\mathcal{M} \models Ra_i a_j$ if and only if $\mathcal{N} \models Rb_i b_j$.

(4) Assume that the condition

$$P((\mathcal{M}, a_1, a_2, \ldots, a_n), (\mathcal{N}, b_1, b_2, \ldots, b_n))$$

is satised.

Show that if c is an element of M, then

- if $c \approx a_i$ for some index i between 1 and n inclusive, then there exists a point $d \in N$ such that

$$P((\mathcal{M}, a_1, a_2, \ldots, a_n, c), (\mathcal{N}, b_1, b_2, \ldots, b_n, d));$$

- if not, there exists an elementary extension $\mathcal{N}'$ of $\mathcal{N}$ and a point d of N' such that $P((\mathcal{M}, a_1, a_2, \ldots, a_n, c), (\mathcal{N}', b_1, b_2, \ldots, b_n, d))$.

(5) Use induction on the height of the formula $G[v_1, v_2, \ldots, v_n]$ to prove the following assertion:

If $\mathcal{M}$ and $\mathcal{N}$ are two models of A and if $(a_1, a_2, \ldots, a_n)$ and $(b_1, b_2, \ldots, b_n)$ are two sequences from $\mathcal{M}$ and $\mathcal{N}$, respectively, then $P((\mathcal{M}, a_1, a_2, \ldots, a_n), (\mathcal{N}, b_1, b_2, \ldots, b_n))$ implies
$\mathcal{M} \models G[a_1, a_2, \ldots, a_n]$ if and only if $\mathcal{N} \models G[b_1, b_2, \ldots, b_n]$.

(6) Conclude from this that $\{A\}$ is a complete theory.

12. Let L be the language consisting of a single binary predicate symbol $\leq$ and let T be the theory of dense linear orderings with no rst or last element. Show that for every formula $F[v_0, v_1, v_2, \ldots, v_n]$, there exists a quantier-free formula $H[v_0, v_1, v_2, \ldots, v_n]$ such that

$$T \vdash \forall v_0 \forall v_1 \ldots \forall v_n (F[v_0, v_1, v_2, \ldots, v_n] \Leftrightarrow H[v_0, v_1, v_2, \ldots, v_n]).$$

13. Let L be a language. A class $\mathcal{C}$ of L-structures is **closed under ultraproducts** if for every set I, for every ultralter $\mathcal{U}$ on I, and for every sequence $(M_i : i \in I)$ of structures from $\mathcal{C}$, $\prod_{i \in I} \mathcal{M}_i/\mathcal{U}$ belongs to $\mathcal{C}$.

Let T be a theory in L. Show that the class of L-structures that are not models of T is closed under ultraproducts if and only if T is nitely axiomatizable.

14. Let K be a eld (the language is $\{\mathbf{0}, \mathbf{1}, +, \times\}$). Let I be a set and let $\mathcal{F}$ be a lter on I.

(a) Show that $K^I/\mathcal{F}$ is a ring and that it is a eld if and only if $\mathcal{F}$ is an ultralter on I.

(b) Let $\mathcal{J}$ be the subset of K^I dened by

$$\mathcal{J} = \{(k_i : i \in I) \in K^I : \{i \in I : k_i = \mathbf{0}\} \in \mathcal{F}\}.$$

Show that $\mathcal{J}$ is an ideal in the ring K^I and that the quotient ring $K^I/\mathcal{J}$ is equal to $K^I/\mathcal{F}$.

15. The language is that of ordered sets: $L = \{\leq\}$.

(a) Let α be an innite ordinal. Show that there exists an ordered set that is elementarily equivalent to $\langle \alpha, \leq \rangle$ but is not a well-ordering.

(b) Show that there exists a denumerable ordinal α such that

$$\langle \alpha, \leq \rangle \prec \langle \aleph_1, \leq \rangle.$$

($\aleph_1$ denotes the least uncountable ordinal.)

(c) Show that there exist two distinct denumerable ordinals α and β such that

$$\langle \alpha, \leq \rangle \prec \langle \beta, \leq \rangle.$$

16. Consider the uncountable language L that contains:

- for every integer n a constant symbol $\underline{n}$;
- for every subset A of $\mathbb{N}$, a unary predicate symbol $\underline{A}$;
- for every mapping f from $\mathbb{N}$ into $\mathbb{N}$, a unary function symbol $\underline{f}$.

Let $\mathcal{N}$ be the L-structure whose base set is $\mathbb{N}$ and in which each symbol $\underline{X}$ of L is interpreted by X. Let T be the theory of $\mathcal{N}$.

(a) Show that every model of T is isomorphic to an elementary extension of $\mathcal{N}$.

(b) Let $\mathcal{M}$ be a proper elementary extension of $\mathcal{N}$ and let a be a point of $\mathcal{M}$ that does not belong to $\mathbb{N}$. Show that the set

$$\mathcal{F}_a = \{A \subseteq \mathbb{N} : \mathcal{M} \models \underline{A}a\}$$

is a non-trivial ultralter on $\mathbb{N}$.

(c) Let α be a bijection from $\mathbb{N}^2$ onto $\mathbb{N}$. For every positive real number r, choose two sequences of natural numbers $(p_r(n) : n \in \mathbb{N})$ and $(q_r(n) : n \in \mathbb{N})$

such that the sequence $(p_r(n)/q_r(n) : n \in \mathbb{N})$ converges to the limit r. Dene the mapping f_r from $\mathbb{N}$ into $\mathbb{N}^2$ by setting

$$f_r(n) = \alpha(p_r(n), q_r(n))$$

for all $n \in \mathbb{N}$.

Show that if r and s are distinct positive reals, then the set

$$\{n \in \mathbb{N} : f_r(n) = f_s(n)\}$$

is nite.

(d) Show that every model of T that is not isomorphic to $\mathcal{N}$ must have cardinality greater than or equal to $2^{\aleph_0}$.

Show that T is $\aleph_0$-categorical.

(e) Let L' be the language obtained by adding a new unary predicate symbol X to L and let T' be the theory $T \cup \{X\underline{n} : n \in \mathbb{N}\}$. Show that T' has no nite model, is $\aleph_0$-categorical, and is not complete.

Solutions

Solutions to the exercises for Chapter 5

1. Since the set of primitive recursive subsets of $\mathbb{N}$ is closed under nite unions, it is sufcient to show that singleton subsets are primitive recursive. Note that the characteristic function of $\{n\}$ is equal to

$$\lambda x.((x+1)\dot{-}n)\,((n+1)\dot{-}x).$$

2. If we set $g(n) = \alpha_2(f(n), f(n+1))$, we have

$$\begin{aligned} g(0) &= \alpha_2(1, 1), \\ g(n+1) &= \alpha_2(\beta_2^2(g(n)), \beta_2^2(g(n)) + \beta_2^1(g(n))); \end{aligned}$$

this shows that g is primitive recursive; and so is f which is equal to $\beta_2^1 \circ g$.

3. (a) If $\alpha(\sigma) = \alpha(\sigma') = n$, then σ and σ' have the same length $p = \beta_2^1(n)$ and are therefore equal since $\alpha_p(\sigma) = \alpha_p(\sigma') = \beta_2^2(n)$. The image of α is the set $\{x : \beta_2^1(x) \neq 0\}$ which is primitive recursive.

(b) It is easy to verify that

$$\alpha_2(x_1, x_2) < (x_1 + x_2 + 1)^2,$$

and, by induction on p, that

$$\alpha_p(x_1, x_2, \ldots, x_p) < (x_1 + x_2 + \cdots + x_p + 1)^{2^{p-1}}.$$

So it sufces to take $g(x) = (x+1)^{2^x}$.

(c) Let us rst show that the function $\psi = \lambda px.\phi(p, p, x)$ is primitive recursive. When we refer to the denition of the functions β_i^j (in Proposition 5.4), we see that ψ can be dened by recursion:

$$\begin{aligned} \psi(0, x) &= 0; \\ \psi(1, x) &= x; \\ \psi(p+1, x) &= \beta_2^2(\psi(p, x)). \end{aligned}$$

Next, the function ϕ itself is denable by recursion:

$$\begin{aligned}
\phi(0, i, x) &= 0; \\
\phi(1, i, x) &= x \quad \text{if } i = 1 \text{ and } 0 \text{ otherwise}; \\
\phi(p+1, i, x) &= 0 \quad \text{if } i = 0 \text{ or if } i > p+1; \\
\phi(p+1, i, x) &= \phi(p, i, x) \quad \text{if } 0 < i < p; \\
\phi(p+1, i, x) &= \beta_2^1(\phi(p, i, x)) \quad \text{if } p > 0 \text{ and } i = p; \\
\phi(p+1, i, x) &= \psi(p+1, x) \quad \text{if } p > 0 \text{ and } i = p+1.
\end{aligned}$$

(d) The fact that the function γ is injective is easily proved using the theorem that the decomposition of a number into prime factors is unique. The image of γ is the set

$$\{x : \text{ for all } p \text{ less than } x \text{ and for all } q \text{ less than } p, \\ \text{if } p \text{ and } q \text{ are prime and if } p \text{ divides } x, \text{ then } q \text{ divides } x\};$$

this set is dened from primitive recursive sets using bounded quantications and Boolean operations; it is therefore primitive recursive.

(e) Let σ be a nite sequence and suppose that $\alpha(\sigma) = x$; we may then calculate the length of σ which is equal to $p = \beta_2^1(x)$ and we see that

$$\gamma(\sigma) = \prod_{i=0}^{i=p-1} \pi(i)^{\phi(p,i+1,x)+1};$$

so we may set

$$f(x) = \prod_{i=0}^{i=p-1} \pi(i)^{\phi(p,i+1,x)+1},$$

where $p = \beta_2^1(x)$ [and $f(x) = 1$ in case $\beta_2^1(x) = 0$].

For the function h, we begin by doing the same thing: it is easy to dene two primitive recursive functions $p(x)$ and $\theta(i, x)$ which behave as follows: if $\sigma = (x_1, x_2, \ldots, x_p)$ is a nite sequence and $\gamma(\sigma) = x$, then $p(x)$ is equal to the length p of x and, for all i from 1 to p inclusive, $x_i = \theta(i, x)$. We may then use the function g described in part (b) and dene h by

$$h(x) = \mu y \leq g(x) \quad (\beta_2^1(y) = p(x) \text{ and}, \\ \text{for all } i \text{ from 1 to } p \text{ inclusive}, \theta(i, x) = \phi(p(x), i, \beta_2^2(y))).$$

4. The number e is the sum of the series

$$1 + \frac{1}{1!} + \frac{1}{2!} + \cdots + \frac{1}{n!} + \cdots.$$

Set $e_n = 1 + \frac{1}{1!} + \frac{1}{2!} + \cdots + \frac{1}{n!} = \alpha_n/n!$, where α_n is an integer. A simple calculation shows that

$$\frac{1}{(n+1)!} < e - e_n < \frac{1}{n \cdot n!},$$

and we may then observe that $e \cdot n! - \alpha_n$ lies strictly between $1/(n+1)$ and $1/n$; thus if $n > 2$, then α_n is the integer part of $e \cdot n!$. If p is an arbitrary integer, we have

$$|e \cdot n! - p| > \frac{1}{n+1} \quad \text{and} \quad \left|e - \frac{p}{n!}\right| > \frac{1}{(n+1)!}.$$

Now let p/q be any positive rational. It can be written in the form $p'/q!$; therefore

$$\left|e - \frac{p}{q}\right| > \frac{1}{(q+1)!}. \qquad (*)$$

Fix n and set $q = 10^n + 1$. We are going to prove that the nth digit in the decimal expansion of e_q is equal to the nth digit in the decimal expansion of e: let β be the integer part of $10^n \cdot e_q$. It is clear that $\beta < 10^n \cdot e$, so it sufces to show that $10^n \cdot e$ is less than $\beta + 1$.

If we assume the contrary, we obtain, from the inequality $(*)$, that

$$e - \frac{\beta+1}{10^n} > \frac{1}{q!},$$

but we have already seen that

$$0 < e - e_q < \frac{1}{q \cdot q!} < \frac{1}{q!};$$

this contradicts the denition of β.

It is now easy to see that the function $\alpha = \lambda n.\alpha_n$ is primitive recursive and that the function which sends n into the nth digit in the decimal expansion of $\alpha(10^n + 1)/(10^n + 1)!$ is also. This concludes the exercise.

5. (a) Let us get rid of the case where a_0 is zero [there is then at least one integer root; this remark should be repeated for part (b)]. In all other cases, it is clear that some root of $P = a_0 + a_1 X + \cdots + a_p X^p$ must be negative and, if it is an integer, it must divide a_0. Assume, for example, that p is even (the other case is treated the same way); we then see that P has a root in $\mathbb{Z}$ if and only if there exists $y \in \mathbb{N}$ less than or equal to a_0 such that

$$a_0 + a_2 y^2 + \cdots + a_p y^p = a_1 y + a_3 y^3 + \cdots + a_{p-1} y^{p-1}.$$

The set E is therefore dened by applying a bounded quantication to a primitive recursive set.

(b) We may again assume that a_0 is not zero and, arguing by induction on p, that a_p is also different from zero. Suppose that $y = -r/s$ is a root of P, where r and s belong to $\mathbb{N}$, s is non-zero, and r and s are relatively prime. It is then easy to see that r must divide a_0 and that s must divide a_p. Assuming, once again, that p is even, we see that P has a root in $\mathbb{Q}$ if and only if there exists an integer r less than or equal to a_0 and an integer s less than or equal to a_p such that

$$\begin{aligned} a_0 s^p + a_2 s^{p-2} r^2 + \cdots + a_p r^p \\ = a_1 s^{p-1} r + a_3 s^{p-3} r^3 + \cdots + a_{p-1} s r^{p-1}. \end{aligned}$$

(c) We begin by constructing two primitive recursive functions $\theta_1(x, y)$ and $\theta_2(x, y)$ such that, if x codes the sequence $(a_0, a_1, \ldots, a_p)$ [in other words, if $\Omega(a_0, a_1, \ldots, a_p) = x$], then

$$\theta_1(x, y) = \sum_{2i \leq x} \delta(2i, x) \cdot y^{2i}, \quad \text{and}$$
$$\theta_2(x, y) = \sum_{2i+1 \leq x} \delta(2i+1, x) \cdot y^{2i+1}$$

(see Denition 5.5 for the denition of δ).

These functions are primitive recursive. We may then use the same argument as in part (a):

$$\begin{aligned} x \in F \quad & \text{if and only if} \\ & \text{there exists } y \leq \delta(0, x) \text{ such that } \theta_1(x, y) = \theta_2(x, y). \end{aligned}$$

6. The formula F has a model of cardinality n if and only if it has a model $\mathcal{M}$ whose base set is $A_n = \{0, 1, \ldots, n-1\}$. This model will then be characterized by the interpretation $R \subseteq A_n^2$ of the binary predicate. We code the model by the pair $(n, u(R))$, where $u(R)$ is the integer dened by

$$u(R) = \prod_{(i,j) \in R} \pi(\alpha_2(i, j)).$$

It is easy to see that $u(R)$ is bounded by a primitive recursive function of n [$\alpha_2(n, n)!$, for example]. Without too much difculty, we also see that the set of codes of nite L-structures,

$$M = \{(n, u) : \text{ there exists } R \subseteq A_n^2 \text{ such that } u = u(R)\},$$

is primitive recursive. We will show that the set

$$U(F) = \{(n, u(R)) : \ R \subseteq A_n^2 \text{ and } (A_n, R) \text{ is a model of } F\}$$

is primitive recursive. It will follow that $n \in Sp(F)$ if and only if there exists an integer u less than $\alpha_2(n, n)!$ such that $(n, u) \in U(F)$ and, hence, that $Sp(F)$ is primitive recursive.

The formula F is equivalent to a formula of the form

$$Q_1 Q_2 \ldots Q_p B[v_1, v_2 \ldots, v_p],$$

where p is an integer, where for i between 1 and n inclusive Q_i represents either the quantier $\exists v_i$ or the quantier $\forall v_i$, and where $B[v_1, v_2 \ldots, v_p]$ is a formula without quantiers.

We begin by showing that if $C[v_1, v_2, \ldots, v_p]$ is a formula of L without quantiers and whose free variables are among $v_1, v_2, \ldots, v_p$, then the set

$$X(C) = \{(n, u(R), a_1, a_2, \ldots, a_p)\colon (A_n, R) \models C[a_1, a_2, \ldots, a_p]\}$$

is primitive recursive. To do this by induction on the complexity of C is straightforward. If C is atomic, i.e. of the form $Rv_i v_j$ where i and j are integers between 1 and p inclusive, then

$$(n, u, a_1, a_2, \ldots, a_p) \in X(C)$$

if and only if

$$(n, u) \in M \quad \text{and } a_1, a_2, \ldots, a_p \text{ are all between 0 and } n-1$$
$$\text{and } \pi(\alpha_2(a_i, a_j)) \text{ divides } u.$$

Then, we note that

$$X(C_1 \wedge C_2) = X(C_1) \cap X(C_2),$$
$$X(C_1 \vee C_2) = X(C_1) \cup X(C_2),$$

and that $(n, u, a_1, a_2, \ldots, a_p) \in X(\neg C)$ if and only if $(n, u) \in M$, $a_1, a_2, \ldots, a_p$ are all between 0 and $n-1$, and $(n, u, a_1, a_2, \ldots, a_p) \notin X(C)$.

As a consequence, $X(B)$ is primitive recursive. Now, the set $U(F)$ can be dened by

$$(n, q) \in U(F) \quad \text{if and only if}$$
$$T_1 T_2 \ldots T_p((n, q, x_1, x_2, \ldots, x_p) \in X(B)),$$

where, for all i from 1 to p inclusive, T_i is equal to $\exists x_i \leq n-1$ if Q_i is the quantier $\exists v_i$ and T_i is equal to $\forall x_i \leq n-1$ if Q_i is the quantier $\forall v_i$. This shows that $U(F)$ is primitive recursive.

7. This exercise is left to the reader.

8. The machine can have as many bands as one wishes; it is only the rst that matters, so we will neglect the others. The machine has three states: e_0, e_1, and e_f.

Here is its table:

$$M(e_0, d) = (e_0, d, +1); \qquad M(e_0, |) = (e_1, b, +1);$$
$$M(e_1, |) = (e_0, b, +1); \qquad M(e_1, b) = (e_1, b, 0);$$
$$M(e_0, b) = (e_f, b, 0).$$

9. (a) Let $\mathcal{M}$ be a machine with n bands that computes f; we will simulate the computation performed by $\mathcal{M}$ using a machine $\mathcal{N}$ that has three bands in the following way; the computation will, in reality, take place on the third band. Cells numbered $1, n+1, 2n+1$, etc. of this band will play the role of the rst band of $\mathcal{M}$; cells numbered $2, n+2, 2n+2$, etc. will play the role of the second band of $\mathcal{M}$; and so on. Machine $\mathcal{N}$ must begin by copying the contents of the rst band of $\mathcal{M}$ onto the third band using only one out of every n cells; it must then simulate the computation of $\mathcal{M}$. After that, it must recopy the contents of cells 2, $n+2$, $2n+2$, etc. onto the second band; nally, it must erase the contents of the third band. We leave it to the reader, if desired, to specify the exact number of states required by $\mathcal{N}$ and to write down its table.

(b) The set $\mathcal{M}_n$ is nite!

(c) It sufces to add $p+1$ new states $f_0, f_1, \ldots, f_p$ to the set of states of $\mathcal{M}$. The initial state of $\mathcal{N}_p$ is f_0; when the machine is in state f_i $(0 \leq i \leq p-1)$, it adds a stroke to the rst band and enters state f_{i+1}; when it is in state f_p, it returns its head to the beginning of the tape and enters the initial state of $\mathcal{M}$. Thus, $\mathcal{N}_p$ has $n+p+1$ states.

(d) Suppose that the function Σ is T-computable; then so is the function $\lambda x.\Sigma(2x+1)+1$ and we may assume that it is computable by a machine that has three bands and n states. Then the machine $\mathcal{N}_n$ constructed in part (c) has $2n+1$ states and, when stated with a blank tape, will halt with $\Sigma(2n+1)$ strokes on its second band. This contradicts the denition of Σ.

10. If f is recursive, then the characteristic function of its graph G is

$$\chi_G = \lambda xy.(1\dot{-}[(y\dot{-}f(x)) + (f(x)\dot{-}y)]),$$

which is obviously recursive. Conversely,

$$f(x) = \mu y\, (x, y) \in G,$$

so if G is recursive, then so is f.

11. (a) We leave it to the reader to verify that the relation $\ll$ is transitive, reexive, and antisymmetric. If $(a, b, c) \in \mathbb{N}^3$ and $(x, y, z) \ll (a, b, c)$, then all three of x, y, and z are less than $\sup(a, b, c)$; this shows that the set

$$\{(x, y, z) \in \mathbb{N}^3 : \ (x, y, z) \ll (a, b, c)\}$$

has at most $(\sup(a, b, c)+1)^3$ elements.

Let $(a, b, c) \in \mathbb{N}^3$. We will dene, by distinguishing several cases, another element (a', b', c') of $\mathbb{N}^3$ and the reader should verify that it is the immediate successor of (a, b, c). Let $\sup(a, b, c) = k$;

if $k > c$,	then $a' = a, b' = b, c' = c + 1$;
if $k = c, k > b + 1,$ and $k > a,$	then $a' = a, b' = b + 1, c' = c$;
if $k = c, k > b + 1,$ and $k = a,$	then $a' = a, b' = b + 1, c' = 0$;
if $k = c = b + 1,$	then $a' = a, b' = b + 1, c' = 0$;
if $k = c = b$ and $k > a + 1,$	then $a' = a + 1, b' = 0, c' = c$;
if $k = c = b = a + 1,$	then $a' = a + 1, b' = 0, c' = 0$;
if $k = c = b = a,$	then $a' = 0, b' = 0, c' = c + 1.$

(b) The functions γ_1, γ_2, and γ_3 are dened simultaneously by induction as in the double recursion which follows Denition 5.5. $\gamma_1(0) = \gamma_2(0) = \gamma_3(0) = 0$ and $\gamma_1(n + 1), \gamma_2(n + 1), \gamma_3(n + 1)$ are dened from $\gamma_1(n), \gamma_2(n), \gamma_3(n)$ as follows:

if $\sup(\gamma_1(n), \gamma_2(n), \gamma_3(n)) > \gamma_3(n)$,
then $\gamma_1(n + 1) = \gamma_1(n)$, $\gamma_2(n + 1) = \gamma_2(n)$,
and $\gamma_3(n + 1) = \gamma_3(n) + 1$;

if $\sup(\gamma_1(n), \gamma_2(n), \gamma_3(n)) = \gamma_3(n)$,
$\gamma_2(n) + 1 < \sup(\gamma_1(n), \gamma_2(n), \gamma_3(n))$,
and $\gamma_1(n) < \sup(\gamma_1(n), \gamma_2(n), \gamma_3(n))$,
then $\gamma_1(n + 1) = \gamma_1(n)$, $\gamma_2(n + 1) = \gamma_2(n) + 1$,
and $\gamma_3(n + 1) = \gamma_3(n)$;

if $\sup(\gamma_1(n), \gamma_2(n), \gamma_3(n)) = \gamma_3(n)$,
$\gamma_2(n) + 1 < \sup(\gamma_1(n), \gamma_2(n), \gamma_3(n))$,
and $\gamma_1(n) = \sup(\gamma_1(n), \gamma_2(n), \gamma_3(n))$,
then $\gamma_1(n + 1) = \gamma_1(n)$, $\gamma_2(n + 1) = \gamma_2(n) + 1$,
and $\gamma_3(n + 1) = 0$;

if $\sup(\gamma_1(n), \gamma_2(n), \gamma_3(n)) = \gamma_3(n) = \gamma_2(n) + 1$,
then $\gamma_1(n + 1) = \gamma_1(n)$, $\gamma_2(n + 1) = \gamma_2(n) + 1$,
and $\gamma_3(n + 1) = 0$;

if $\sup(\gamma_1(n), \gamma_2(n), \gamma_3(n)) = \gamma_3(n) = \gamma_2(n)$
and $\gamma_1(n) + 1 < \sup(\gamma_1(n), \gamma_2(n), \gamma_3(n))$,
then $\gamma_1(n + 1) = \gamma_1(n) + 1$, $\gamma_2(n + 1) = 0$,
and $\gamma_3(n + 1) = \gamma_3(n)$;

if $\sup(\gamma_1(n), \gamma_2(n), \gamma_3(n)) = \gamma_3(n) = \gamma_2(n) = \gamma_1(n) + 1$,
then $\gamma_1(n + 1) = \gamma_1(n) + 1$ and $\gamma_2(n + 1) = \gamma_3(n + 1) = 0$;

if $\sup(\gamma_1(n), \gamma_2(n), \gamma_3(n)) = \gamma_3(n) = \gamma_2(n) = \gamma_1(n)$,
then $\gamma_1(n + 1) = \gamma_2(n + 1) = 0$ and $\gamma_3(n + 1) = \gamma_3(n) + 1$.

It is clear that $(0, 0, 0)$ is the minimum element of $\mathbb{N}^3$ for the relation $\ll$. Also, by comparing this with the results in part (a), we see that, for every n, $(\gamma_1(n+1), \gamma_2(n+1), \gamma_3(n+1))$ is the immediate successor of $(\gamma_1(n), \gamma_2(n), \gamma_3(n))$ for the relation $\ll$.

For every integer n, we can see by induction on $p \geq 0$ that

$$(\gamma_1(n+p), \gamma_2(n+p), \gamma_3(n+p)) \gg (\gamma_1(n), \gamma_2(n), \gamma_3(n))$$

and that the inequality is strict for $p > 0$. This establishes that, for all m and n,

$$(\gamma_1(m), \gamma_2(m), \gamma_3(m)) \ll (\gamma_1(n), \gamma_2(n), \gamma_3(n)) \\ \text{if and only if } m \leq n,$$

and that if $m < n$, then the inequality

$$(\gamma_1(m), \gamma_2(m), \gamma_3(m)) \ll (\gamma_1(n), \gamma_2(n), \gamma_3(n))$$

is strict. So the map $\Gamma = \lambda n.(\gamma_1(n), \gamma_2(n), \gamma_3(n))$ is injective.

Let us next deal with surjectivity. Suppose $(a, b, c) \in \mathbb{N}^3$ and let $d = \sup(a, b, c)$. We will argue by contradiction, so assume that, for all $n < (d+1)^3$, $\Gamma(n) \neq (a, b, c)$. We will then prove, by induction, that, for all $n \leq (d+1)^3$,

$$(\gamma_1(n), \gamma_2(n), \gamma_3(n)) \ll (a, b, c).$$

This is true for 0 since $(\gamma_1(0), \gamma_2(0), \gamma_3(0)) = (0, 0, 0)$ is the minimum element for $\ll$. Assuming it is true for n, together with our assumption that $(\gamma_1(n), \gamma_2(n), \gamma_3(n)) \neq (a, b, c)$, we may conclude that $(\gamma_1(n), \gamma_2(n), \gamma_3(n))$ is strictly less than (a, b, c) for the ordering $\ll$; and since

$$(\gamma_1(n+1), \gamma_2(n+1), \gamma_3(n+1))$$

is the immediate successor of $(\gamma_1(n), \gamma_2(n), \gamma_3(n))$, it follows that

$$(\gamma_1(n+1), \gamma_2(n+1), \gamma_3(n+1)) \ll (a, b, c).$$

So we see that the set $\{(x, y, z) \in \mathbb{N}^3 : (x, y, z) \ll (a, b, c)\}$ has at least $(d+1)^3 + 1$ elements; this is impossible since it contradicts the fact, from part (a), that this set has at most $(d+1)^3$ elements.

(c) The fact that H is a primitive recursive set is not entirely obvious: to compute $\chi_H(n)$, we need to have at our disposal all the values $\chi_H(p)$ for $p < n$ [not only $\chi_H(n-1)$, as in a standard induction]. The procedure to follow in this situation is explained in the solution of Exercise 13.

Let us turn to proving the equivalence

$$n \in H \quad \text{if and only if} \quad \gamma_1(n) = \xi(\gamma_2(n), \gamma_3(n)).$$

This is proved by induction on n. For $n = 0$, it is true since $n \notin H$ but $\gamma_1(0) = \gamma_2(0) = \gamma_3(0) = 0$ and $\xi(\gamma_2(n), \gamma_3(n)) = 1$. For $n \neq 0$, we must distinguish several cases:

- if $\gamma_2(n) = 0$, then the equivalence follows without any difculty from the denitions;
- the same holds in case $\gamma_3(n) = 0$;
- in the remaining case, suppose $z = \gamma_1(n)$, $y = \gamma_2(n)$, and $x = \gamma_3(n)$. Assume rst that $z = \xi(x, y)$ and that we wish to conclude $n \in H$. From the denition of Ackerman's function,

$$z = \xi(y - 1, \xi(y, x - 1)).$$

Since Γ is a bijection, there exist two integers p and q such that

$$\gamma_1(p) = \xi(y, x - 1), \qquad \gamma_2(p) = y, \quad \text{and} \quad \gamma_3(p) = x - 1$$

and

$$\gamma_1(q) = z, \qquad \gamma_2(q) = y, \quad \text{and} \quad \gamma_3(q) = \gamma_1(p).$$

It follows easily from the properties of Ackerman's function that

$$\begin{aligned} &(\gamma_1(p), \gamma_2(p), \gamma_3(p)) \ll (\gamma_1(n), \gamma_2(n), \gamma_3(n)), \quad \text{and} \\ &(\gamma_1(q), \gamma_2(q), \gamma_3(q)) \ll (\gamma_1(n), \gamma_2(n), \gamma_3(n)), \quad \text{and} \\ &(\gamma_1(p), \gamma_2(p), \gamma_3(p)) \neq (\gamma_1(n), \gamma_2(n), \gamma_3(n)), \quad \text{and} \\ &(\gamma_1(q), \gamma_2(q), \gamma_3(q)) \neq (\gamma_1(n), \gamma_2(n), \gamma_3(n)), \end{aligned}$$

and thus $p < n$ and $q < n$; so by the induction hypothesis, p and q belong to H. The recursive denition of H then shows that n also belongs to H.

Conversely, suppose that $n \in H$ and let us consider the integers p and q that are involved in the recursive denition of H. We then see, by the induction hypothesis, that $\gamma_1(p) = \xi(y, x - 1)$ and $z = \xi(y - 1, \gamma_1(p))$; this shows that $z = \xi(x, y)$.

(d) From what we have seen, $(y, x, z) \in G$ if and only if there exists $n \leq (\sup(x, y, z) + 1)^3$ such that $n \in H$ and $\gamma_1(n) = z$, $\gamma_2(n) = y$ and $\gamma_3(n) = x$; so Ackerman's function is recursive (see Exercise 10).

12. Let $f \in \mathcal{F}_1$ be an increasing recursive function; if f is bounded, its image is nite, hence recursive. If not, dene the function g by

$$g(x) = \mu y \;\; f(y) \geq x;$$

then g is a total recursive function and $x \in \mathsf{Im}(f)$ if and only if $f(g(x)) = x$.

Now let $A \subseteq \mathbb{N}$ be an innite recursive set and dene the function f by induction:

$$f(0) = \mu y \; y \in A;$$
$$f(n+1) = \mu y \; (y \in A \; \text{ and } \; y > f(n)).$$

Then f is recursive, increasing and its image is A.

13. Dene the function g as follows:

$$g(0) = f(0);$$
$$g(n+1) = f(p) \text{ where } p \text{ is the least integer}$$
$$\text{such that } f(p) \notin \{g(0), g(1), \ldots, g(n)\}.$$

It is more or less clear that g is a total injective function and that its image is equal to the image of f. It is less obvious that it is recursive, for to compute $g(n+1)$, one needs to know not only $g(n)$, as in a classical induction, but the values of $g(i)$ for all $i \leq n$. So we will begin by dening the function

$$h(x) = \prod_{t=0}^{t=x} \pi(g(t))$$

by the following induction, which, by contrast, is entirely orthodox:

$$h(0) = \pi(f(0));$$
$$h(n+1) = h(n) \cdot \pi(f(\mu y(\pi(f(y)) \text{ does not divide } h(n)))).$$

The function g can then be dened by $g(0) = f(0)$ and

$$g(n+1) = \mu y \; (\pi(y) \text{ divides } h(n+1) \text{ but does not divide } h(n)).$$

In the body of Chapter 5, we saw that there exist recursive functions whose image is not recursive. Now we know that there exist such functions which are, moreover, injective.

14. Let A be an innite recursively enumerable subset of $\mathbb{N}^p$. We wish to show that it includes an innite recursive set. By replacing A with its image under the map α_p (see Proposition 5.4),we may restrict our attention to the case where $A \subseteq \mathbb{N}$. We then know that A is the range of a primitive recursive function $f \in \mathcal{F}_1$. Dene the function g by

$$g(0) = f(0);$$
$$g(n+1) = \sup(g(n), f(n+1)).$$

g is then an increasing primitive recursive function whose range is innite and is included in A. According to Exercise 12, this range is recursive.

15. (a) The set B is recursively enumerable since it is the projection of a recursive set. We claim that, for every $x_0 \in \mathbb{N}$, there exists $x_1 \in \mathbb{N}$ such that $x_1 > x_0$

and $x_1 \notin B$; to see this, it sufces to choose $x_1 > x_0$ such that $\alpha(x_1)$ is the least element of $\{\alpha(y) : y > x_0\}$.

(b) It is clear that A is recursively enumerable; so it sufces, by Theorem 5.38, to show, under the assumptions of part (b), that the complement of A is recursively enumerable. Since C is included in the complement of B, if $x \in C$ and $y > x$, then $\alpha(y) > \alpha(x)$; thus α is strictly increasing on C. Since C is innite, the set $\{\alpha(x) : x \in C\}$ is unbounded. Let t be an integer and let x_0 be an element of C such that $\alpha(x_0) > t$; then t is in A if and only if there exists $y < x_0$ such that $\alpha(y) = t$. In other words,

$$t \notin A \quad \text{if and only if} \quad \text{there exists } x \in C \text{ such that}$$
$$\alpha(x) > t \text{ and, for all } y < x, \alpha(y) \neq t;$$

this shows that the complement of A is the projection of a recursively enumerable set.

(c) Let A be an innite subset of $\mathbb{N}$ that is recursively enumerable but not recursive. It is the range of a total recursive function, so, according to Exercise 13, it is also the range of an injective recursive function that we will call α. If we set

$$B = \{x : \text{ there exists } y > x \text{ such that } \alpha(y) < \alpha(x)\}$$

and set $D = \mathbb{N} - B$, we see that B is recursively enumerable and that D is innite [see part (a)]. But D cannot include an innite recursively enumerable subset, otherwise A would be recursive by part (b). The conclusion is that every innite recursively enumerable set has non-empty intersection with B.

16. (a) The set of bijections together with the operation of composition forms a group, so it sufces to show that the set of recursive bijections is a subgroup. To do this, we must show that the identity is recursive (which is obvious), that the composition of two recursive bijections is recursive (which is also obvious), and that the inverse of a recursive bijection is recursive; this last fact is true because if f is a bijection, then f^{-1} is dened by

$$f^{-1}(x) = \mu y\,(f(y) = x).$$

(b) Let us recall the denition of $C^1 \subseteq \mathbb{N}^4$: $(i, t, x, y) \in C^1$ if and only if the machine whose index is i, when started with x on its rst band, completes its computation at time t with y strokes on its second band. We have seen that C^1 is primitive recursive. If we suppose that f is primitive recursive and that, for all x, $f(x) \geq T(x)$, then

$$\phi(x) = \mu y \leq ST^1(e, f(x), x)\,((e, f(x), x, y) \in C^1)$$

is primitive recursive; this contradicts our hypotheses.

The graph G of T is dened by

$$(x, y) \in G \quad \text{if and only if}$$
$$(e, y, x) \in B^1 \text{ and, for all } z < x,\ (e, z, x) \notin B^1;$$

this shows that it is primitive recursive.

(c) The fact that g is recursive and strictly increasing is more or less obvious. Since $g(x) \geq T(x)$ for all x, it is not primitive recursive by part (a). Its graph G_1 and its image I are primitive recursive since:

- $(x, y) \in G_1$ if and only if there exists $i \leq x$ such that $(i, y - 2x) \in G$ and, for all $j \leq x$, there exists $z \leq y - 2x$ such that $(j, z) \in G$;
- $y \in I$ if and only if there exists $x \leq y$ such that $(x, y) \in G_1$.

(d) We have no choice: $g'(n)$ must be the $(n + 1)$st element of $\mathbb{N} - I$. Since 0 clearly does not belong to I (a calculation requires at least one step), we must set $g'(0) = 0$. On the other hand, for every n, the set $I \cap \{y : y \leq 2n\}$ has at most n elements; hence the set $(\mathbb{N} - I) \cap \{y : y \leq 2n\}$ has at least $n + 1$ elements, which proves that $g'(n) \leq 2n$; this allows us to dene g' by recursion, setting

$$g'(n + 1) = \mu y \leq 2n + 2 \ \ (y \notin I \text{ and } y > g'(n)).$$

(e) It is clear from its denition that the function h is recursive, injective, and surjective. We also see that it cannot be primitive recursive, otherwise g would be as well. Now we may dene h^{-1} in the following way:

$$h^{-1}(x) = \begin{cases} 2(\mu y \leq x\ ((y, x) \in G_1)) & \text{if } x \in I, \\ 2(\mu y \leq x\ (g'(y) = x)) + 1 & \text{otherwise,} \end{cases}$$

which shows that h^{-1} is primitive recursive. By contrast, we have just seen that its inverse, h, is not primitive recursive.

17. Let us take a set $B' \subseteq \mathbb{N}$ that is recursively enumerable but not recursive [for example, the domain of the partial function $\lambda x.\phi^1(x, x)$]. B' must be the projection of some recursive set C;

$$B' = \{x : \text{ there exists } y \in \mathbb{N} \text{ such that } (x, y) \in C\}.$$

The complement A of C is also recursive and

$$B = \mathbb{N} - B' = \{x : \ \text{ for all } y \in \mathbb{N},\ (x, y) \in A\}$$

is not recursively enumerable, otherwise B' would be recursive (Theorem 5.38).

18. Consider the partial function $g \in \mathcal{F}_2^*$ dened by

$$g(x, t) = \mu y\ (\phi^1(x, y) = t).$$

It is recursive and, if ϕ_x^1 is a bijection from $\mathbb{N}$ into $\mathbb{N}$, then $\lambda t.g(x,t)$ is the inverse bijection. Let i be an index for g; so we have

$$\text{for all } x \text{ and for all } t, \quad g(x,t) = \phi^2(i,x,t).$$

Now, by applying the *smn* theorem, we obtain

$$g(x,t) = \phi^1(s_1^1(i,x),t),$$

and we observe that $s_1^1(i,x)$ is an index for the inverse bijection. Thus, for α, we may take the function $\lambda x.s_1^1(i,x)$, which is primitive recursive.

19. Let us dene

$$f_0 = g$$

and, by recursion on x,

$$f_{x+1}(y) = h(f_x(\alpha(y)), y, x).$$

It is then clear that the partial function $\lambda xy.f_x(y)$ is the unique partial function that satises the stated conditions. It is also clear that each f_x is recursive, but it is not clear, *a priori*, that f itself is. To prove this, we will imitate the proof that Ackerman's function is recursive.

Consider the map that, with each partial function $k \in \mathcal{F}_2^*$, assigns $k^* \in \mathcal{F}_2^*$ dened by

$$k^*(0,y) = g(y);$$
$$k^*(x+1,y) = h(k(x,\alpha(y)), y, x).$$

Observe that f is the unique partial function that satises $f^* = f$. Also, as was the case for Ackerman's function, we can nd, using the *smn* theorem, a primitive recursive function β such that if $k = \phi_x^2$, then $k^* = \phi_{\beta(x)}^2$. The xed point theorem then tells us that there exists an integer i such that $\phi_i^2 = \phi_{\beta(i)}^2$, so f is equal to ϕ_i^2, which is recursive.

20. If the function $\lambda x.T^1(i,x)$ can be extended to a total recursive function h, then A is recursive; to decide whether $n \in A$, we observe whether the machine whose index is i has completed its computation after $h(n)$ steps.

21. (a) To prove that a primitive recursive function is computable within a time that is primitive recursive, it is sufcient to repeat the proof that every partial recursive function is T-computable. If the μ-scheme is not used, as is the case when we are dealing with a primitive recursive function, we note that the time taken by the computation can be bounded by a primitive recursive function.

The converse of this is precisely the remark that follows Theorem 5.29.

(b) This follows from Corollary 5.14.

(c) When we x n (and A and i), the function $\lambda x.\xi(n, x)$ is primitive recursive and so is the function $\lambda x.g(i, A, n, x)$. Conversely, suppose that $f \in \mathcal{F}_1$ is primitive recursive; then according to (a) and (b), there exist integers i, n, and A such that $f(x)$ is computed by the machine whose index is i within a time that is bounded by $\sup(A, \xi(n, x))$. In other words,

$$f = \lambda x.g(i, A, n, x).$$

(d) Thus we see that the set of functions $\lambda x.g(i, A, n, x)$, where i, A, and n are integers, is equal to the set of all primitive recursive functions of one variable. The desired function is obtained by setting

$$\psi(x, y) = g(\beta_3^1(x), \beta_3^2(x), \beta_3^3(x), y).$$

(e) We employ a diagonal argument. The set

$$X = \{x : \psi(x, x) = 0\}$$

is obviously recursive. It is not primitive recursive, for if it were, there would exist an integer y such that its characteristic function would equal $\lambda x.\psi(y, x)$ and it would follow that

$$y \in X \quad \text{if and only if} \quad y \notin X,$$

which is absurd.

22. (a) The same diagonal argument is applicable. If we suppose that the set of total recursive functions of one variable is listed by the function $F(x, y)$, we obtain a contradiction when we consider the function $\lambda x.F(x, x) + 1$.

(b) Let $F(x, y)$ be a recursive function that enumerates the primitive recursive functions of one variable (see Exercise 21). We dene $G(x, y)$ by

$$\begin{aligned} G(x, 0) &= F(x, 0); \\ G(x, y + 1) &= \sup(G(x, y) + 1,\ F(x, y + 1)). \end{aligned}$$

Verify that, for all x, the function $G_x = \lambda y.G(x, y)$ is primitive recursive and strictly increasing and that, moreover, if F_x is strictly increasing, then $G_x = F_x$. The set $\{G_x : x \in \mathbb{N}\}$ is therefore equal to the set of all strictly increasing primitive recursive functions of one variable.

(c) We use the same technique. Dene the function H by

$$\begin{aligned} &H(x, 0) = F(x, 0); \\ &H(x, y + 1) \\ &\quad = \begin{cases} F(x, y + 1) & \text{if } F(x, y + 1) \notin \{H(x, t) : 0 \le t \le y\}, \\ \sup\{H(x, i) + 1 : 0 \le i \le y\} & \text{otherwise.} \end{cases} \end{aligned}$$

(To prove that the function H is recursive and that the functions H_x are all primitive recursive, one must use the technique outlined in the solution to Exercise 13.)

(d) We will construct a strictly increasing recursive function g whose range, B, does not include any of the sets A_x. The set B will be recursive according to Exercise 12; so this will answer the question. We dene g by

$$g(0) = 0;$$
$$g(x+1) = \beta_2^2(\mu t\,[\beta_2^2(t) = F(x, \beta_2^1(t)) \text{ and } \beta_2^2(t) > g(x)]) + 1.$$

The function g is clearly recursive and the fact that the range of the function $\lambda y.F(x, y)$ is innite guarantees that g is total.

For every integer x, set $a = \mu t\,[\beta_2^2(t) = F(x, \beta_2^1(t)) \text{ and } \beta_2^2(t) > g(x)]$, $b = \beta_2^2(a)$, and $c = \beta_2^1(a)$. We then have

$$g(x+1) = b+1, \quad b > g(x) \text{ and } b = F(x, c).$$

This shows that g is strictly increasing and that b, which belongs to the range of the function $\lambda y.F(x, y)$ (i.e. to A_x), lies strictly between $g(x)$ and $g(x+1)$ and hence does not belong to the range of g.

If the set of strictly increasing recursive functions (or the set of injective recursive functions) were listed by some recursive F, it would be listed by some $F \in \mathcal{F}_2$ with the property that, for any integer x, the range of $\lambda y.F(x, y)$ is innite. But we have just constructed a strictly increasing recursive function g (which is therefore injective) which cannot be equal to any of the functions $\lambda y.F(x, y)$.

23. (a) This follows immediately from the fact that the set of total recursive functions contains the identity and is closed under composition.

(b) Suppose that B is a recursively enumerable set. It is therefore the domain of some partial recursive function h. If A is reducible to B, it is because there exists a recursive function f such that

if $x \in A$, then $f(x) \in B$, so $(h \circ f)(x)$ is dened;
if $x \notin A$, then $f(x) \notin B$, so $(h \circ f)(x)$ is not dened.

This shows that A is the domain of $h \circ f$, so it is recursively enumerable.

It is certainly clear that $A \leq B$ if and only if $\mathbb{N} - A \leq \mathbb{N} - B$. So if we assume that B is recursive, then A and $\mathbb{N} - A$ are both recursively enumerable; thus A is recursive.

(c) We know that Y is recursively enumerable; thus, using part (b), we see that if $A \leq Y$, then A is recursively enumerable. Conversely, suppose that A is the domain of the partial recursive function whose index is e.

Then

$$x \in A \quad \text{if and only if} \quad \phi^1(e, x) \text{ is dened}$$
$$\text{if and only if} \quad \alpha_2(e, x) \in Y;$$

thus A is reducible to Y.

(d) First of all, it is clear that A and B are both reducible to C since $x \in A$ if and only if $2x \in C$ and $x \in B$ if and only if $2x + 1 \in C$.

Let $D \subseteq \mathbb{N}$ and let f and g be two functions that satisfy

$$x \in A \quad \text{if and only if} \quad f(x) \in D, \quad \text{and}$$
$$x \in B \quad \text{if and only if} \quad g(x) \in D.$$

We must prove that C is reducible to D. It sufces to consider the function h dened by

$$h(x) = \begin{cases} f(x/2) & \text{if } x \text{ is even;} \\ g((x-1)/2) & \text{if } x \text{ is odd.} \end{cases}$$

It is then easy to see that $x \in C$ if and only if $h(x) \in D$.

(e) Let $B \subseteq \mathbb{N}$. Let C be the set obtained by applying the construction from part (d) to the sets B and $\mathbb{N} - B$. We will prove that C is self-dual. Since B and $\mathbb{N} - B$ are both reducible to C, we see that $\mathbb{N} - B$ and B are reducible to $\mathbb{N} - C$. It follows from the minimality property of C [proved in part (d)] that C is reducible to $\mathbb{N} - C$.

(f) (i) Let f be a partial recursive function that does not belong to $\mathcal{T}$. Consider the function $\theta(x, y) = f(y) + \phi^1(x, x) - \phi^1(x, x)$ and, for every integer n, set

$$\theta_n = \lambda y.\theta(n, y).$$

Thus θ_n is the partial function whose domain is empty if $n \in \mathbb{N} - X$ and is equal to f otherwise. If e is an index for θ, the *smn* theorem tells us that $s_1^1(e, n)$ is an index for θ_n. We see that $n \in \mathbb{N} - X$ if and only if $s_1^1(e, n) \in A$; this shows that $\mathbb{N} - X$ is reducible to A and, consequently, X is reducible to $\mathbb{N} - A$.

(ii) This time, choose a partial recursive function f that does belong to $\mathcal{T}$ and, once again, set $\theta(x, y) = f(y) + \phi^1(x, x) - \phi^1(x, x)$ and, for every integer n, $\theta_n = \lambda y.\theta(n, y)$. If e is an index for θ, $s_1^1(e, n)$ is an index for θ_n. Thus $n \in X$ if and only if $s_1^1(e, n) \in A$, so X is reducible to A.

(iii) We will argue by contradiction and assume that there exists a recursive function f such that for every integer x, $x \in A$ if and only if $f(x) \in \mathbb{N} - A$. The rst xed point theorem furnishes an integer n such that

$\phi_n^1 = \phi_{f(n)}^1$; this implies that $n \in A$ if and only if $f(n) \in A$, which is a contradiction.

(g) To show that Y is reducible to X, consider the partial function

$$\psi(x, y) = \phi^1(\beta_2^1(x), \beta_2^2(x))$$

and, for every integer n, set $\psi_n = \lambda y.\psi(n, y)$. If $n \in Y$, this function is total (and constant); otherwise it is not dened. Now if e is an index for ψ, then $s_1^1(e, n)$ is an index for ψ_n. Thus, if $n \in Y$, then $\phi^1(s_1^1(e, n), s_1^1(e, n))$ is dened and $s_1^1(e, n) \in X$. Conversely, if $n \notin Y$, then $\phi^1(s_1^1(e, n), s_1^1(e, n))$ is not dened and $s_1^1(e, n) \notin X$; this shows that $Y \leq X$.

24. That ψ is a partial recursive function derives from the fact that it was dened by cases, in a way which is sanctioned by Theorem 5.44. We also see that $g(x) = 0$ if $\phi^1(x, 0)$ is dened and that $g(x) = 1$ otherwise. In other words, g is the characteristic function of the set

$$\{x : \phi^1(x, 0) \text{ is not dened }\},$$

which is not recursive according to Rice's theorem; thus g is not recursive.

25. (a) To begin with, A is the domain of the partial recursive function $\lambda x.\phi^1(x, 0)$; so A is recursively enumerable. Consider the set

$$\mathcal{A} = \{f : f \in \mathcal{F}_1^*, f \text{ is recursive, and } f(0) \text{ is dened }\};$$

it is clear that $\mathcal{A}$ is neither empty nor equal to the set of all partial recursive functions of one variable; also, Rice's theorem allows us to conclude that A is not recursive. Since we already know that A is recursively enumerable, it follows from Theorem 5.38 that the complement of A is not recursively enumerable.

(b) Consider the partial function $H = \lambda xy.sg(1 + \phi^1(x, 0))$; it is recursive, so it has an index a. We have

$$H = \phi_a^2.$$

Next, consider, for each integer n, the function $H_n = \lambda y.H(n, y)$. The *smn* theorem tells us that $s_1^1(a, n)$ is an index of H_n; we can also easily see that if $n \in A$, then H_n is the constant function equal to 1, while if $n \notin A$, then H_n is the function whose domain is empty. Therefore,

$$n \in A \quad \text{if and only if} \quad s_1^1(a, n) \in B.$$

So for α, we may take the primitive recursive function $\lambda x.s_1^1(a, x)$.

The fact that B is not the complement of a recursively enumerable set is a consequence of the small lemma that follows and which will be used many times subsequently.

Lemma *Let $C \subseteq \mathbb{N}$, let $f \in \mathcal{F}_1$ be a total recursive function, and suppose that, for every integer n,*

$$n \in A \quad \textit{if and only if} \quad f(n) \in C;$$

then $\mathbb{N} - C$ is not recursively enumerable.

Proof Assume the contrary and let h be a partial recursive function whose domain is $\mathbb{N} - C$. Then $n \notin A$ if and only if $h(f(n))$ is dened; this implies that $\mathbb{N} - A$ is recursively enumerable (it is the domain of $h \circ f$). But this is false. ■

(c) Observe that if $n \notin A$, then $B^1(e, z, n)$ is not satised for any value of z and that, consequently, the function $\lambda y.F(n, y)$ is the constant function equal to 1. On the other hand, if $n \in A$, then $B^1(e, z, n)$ is satised for all values of z greater than or equal to the time required for the computation by the machine whose index is e with initial input n on its rst band. So in this case, the domain of the function $\lambda y.F(n, y)$ is nite. The function F is clearly recursive; suppose that b is an index for F. Then, by the *smn* theorem, $s_1^1(b, n)$ is an index for $\lambda y.F(n, y)$. So we see that

$$n \in A \quad \text{if and only if} \quad s_1^1(e, n) \in \mathbb{N} - B;$$

it then follows from the preceding lemma that B is not recursively enumerable.

(d) Set $B' = \{x : \phi_x^1 = f\}$. To show that neither B' nor its complement is recursively enumerable, we will construct two primitive recursive functions, γ and δ, of one variable such that, for every integer n,

$$n \in A \quad \text{if and only if} \quad \gamma(n) \in B';$$
$$n \in A \quad \text{if and only if} \quad \delta(n) \in \mathbb{N} - B'.$$

Consider the functions H' and F' dened by

$$H'(x) = f(y) \cdot H(x, y),$$
$$F'(x) = f(y) \cdot F(x, y),$$

where H and F are the functions dened in parts (b) and (c); let c and d be indices of these functions, respectively. As above, we see that, for every integer n, $s_1^1(c, n)$ and $s_1^1(d, n)$ are indices for the functions

$$H'_n = \lambda y.H'(n, y) \quad \text{and} \quad F'_n = \lambda y.F'(n, y).$$

Using what we know about the functions H and F, we see that if $n \in A$, then H'_n is equal to f and F'_n is a function whose domain is nite (and hence different from f); on the other hand, if $n \notin A$, then F'_n is equal to f and H'_n

is the function whose domain is empty. All of this proves that

$$n \in A \quad \text{if and only if} \quad s_1^1(c, n) \in B';$$
$$n \in A \quad \text{if and only if} \quad s_1^1(d, n) \in \mathbb{N} - B'.$$

The statement that was to be proved now follows from the lemma.

26. (a) Consider the function $\lambda nx.n$. It is recursive; hence there exists an integer i such that, for all n and x,

$$\phi^2(i, n, x) = n;$$

by setting $\delta = \lambda n.s_1^1(i, n)$, we see that the function $\phi^1_{\delta(n)}$ is surely the constant function equal to n.

(b) The third version of the xed point theorem tells us that there exists a primitive recursive function $h(n, t)$ such that, for all n and t,

$$\phi^1_{h(n,t)} = \phi^1_{\gamma(n,t,h(n,t))}.$$

If $h(n, t) \leq t$, then $\gamma(n, t, h(n, t)) = \delta(n)$; if not, then $\gamma(n, t, h(n, t)) = t$ and we obtain what was desired.

(c) The map from A_t into the set $\{0, 1, \ldots, t\}$ which assigns $h(n, t)$ to n is injective. To see this, suppose that n and m are in A_t and that $n \neq m$; then

$$\phi^1_{h(n,t)} = \phi^1_{\delta(n)} \neq \phi^1_{h(m,t)} = \phi^1_{\delta(m)}.$$

This shows that $h(n, t) \neq h(m, t)$ and that A_t has no more than $t + 1$ elements. So we are justied in dening

$$\alpha(t) = \mu n \leq t + 1\,(h(n, t) > t) \quad \text{and} \quad \beta(t) = h(\alpha(t), t).$$

Then $\beta(t) > t$ and $\phi^1_{\beta(t)} = \phi^1_{h(\alpha(t),t)} = \phi^1_t$.

27. (a) We will rst show that (i) implies (ii). Since the function ϕ^1 is partial recursive, there exists an integer i such that $\phi^1 = \psi^2_i$; hence, for all x and y,

$$\phi^1(x, y) = \psi^2(i, x, y) = \psi^1(\sigma_1^1(i, x), y) = \theta(\sigma_1^1(i, x), y).$$

So it sufces to choose $\beta = \lambda x.\sigma_1^1(i, x)$.

To show that (ii) implies (i), set

$$\psi^p(i, x_1, x_2, \ldots, x_p) = \theta(i, \alpha_p(x_1, x_2, \ldots, x_p)).$$

The property (enu) is easy to verify; suppose that f is a partial recursive function of p variables. Then the function

$$g = \lambda x.f(\beta_p^1(x), \beta_p^2(x), \ldots, \beta_p^p(x))$$

is also partial recursive and there exists an integer i such that $g = \theta_i$, so we see that $f = \psi_i^p$. Let us move on to the property (smn). We know there exists an integer e such that $\theta = \phi_e^2$ and, by setting $\alpha(i) = s_1^1(e, i)$, we see that $\theta_i = \phi^1_{\alpha(i)}$. We have

$$\begin{aligned}&\psi^{n+m}(i, x_1, x_2, \ldots, x_n, y_1, y_2, \ldots, y_m)\\ &\quad = \theta(i, \alpha_{n+m}(x_1, x_2, \ldots, x_n, y_1, y_2, \ldots, y_m))\\ &\quad = \phi^1(\alpha(i), \alpha_{n+m}(x_1, x_2, \ldots, x_n, y_1, y_2, \ldots, y_m)).\end{aligned}$$

Now consider the partial function

$$\lambda i x_1 x_2 \ldots x_n z. \phi^1(\alpha(i), \alpha_{n+m}(x_1, x_2, \ldots, x_n, \beta^1_m(z), \beta^2_m(z), \ldots, \beta^m_m(z))).$$

It has an index e', so, for all $i, x_1, x_2, \ldots, x_n$ and z,

$$\begin{aligned}&\psi^{n+m}(i, x_1, x_2, \ldots, x_n, \beta^1_m(z), \beta^2_m(z), \ldots, \beta^m_m(z))\\ &\quad = \phi^1(\alpha(i), \alpha_{n+m}(x_1, x_2, \ldots, x_n, \beta^1_m(z), \beta^2_m(z), \ldots, \beta^m_m(z)))\\ &\quad = \phi^{n+2}(e', i, x_1, x_2, \ldots, x_n, z)\\ &\quad = \phi^1(s^1_{n+2}(e', i, x_1, x_2, \ldots, x_n), z)\\ &\quad = \theta(\beta(s^1_{n+2}(e', i, x_1, x_2, \ldots, x_n)), z).\end{aligned}$$

When we replace z by $\alpha_m(y_1, y_2, \ldots, y_m)$ and set

$$\sigma^m_n(i, x_1, x_2, \ldots, x_n) = \beta(s^1_{n+2}(e', i, x_1, x_2, \ldots, x_n)),$$

we obtain

$$\begin{aligned}&\psi^{n+m}(i, x_1, x_2, \ldots, x_n, y_1, y_2, \ldots, y_m)\\ &\quad = \theta(\sigma^m_n(i, x_1, x_2, \ldots, x_n), \alpha_m(y_1, y_2, \ldots, y_m))\\ &\quad = \psi^m(\sigma^m_n(i, x_1, x_2, \ldots, x_n), y_1, y_2, \ldots, y_m).\end{aligned}$$

(b) The proof in Chapter 5 of the xed point theorems uses only the enumeration theorem and the *smn* theorem; here, we may use exactly the same proof.

(c) The function α has already been constructed and the function β is given by the hypotheses. It remains only to show that we may assume they are injective. We know how to do this for the function α; it sufces to use a function $\delta(n, x)$ that is strictly increasing in the rst variable and is such that, for all n and x,

$$\phi^1_{\delta(n,x)} = \phi^1_x.$$

We use the same line of argument for the function β. So we must show that there exists a function $\gamma(n, x)$ such that, for all n and x,

$$\theta_{\gamma(n,x)} = \theta_x.$$

To do this, we employ the proof given for Exercise 26 which uses only the xed point theorems and can therefore be applied to the family ψ.

(d) We will use the functions δ and γ mentioned in part (c). We are going to construct two sequences of functions f_n and g_n for $n \geq -1$ that will be approximations for the functions ε and ε^{-1}, respectively, that we are trying to construct. More precisely, we will notice, once the construction is completed, that, for $n \in \mathbb{N}$,

$$f_n(p) = \varepsilon(p) \quad \text{and} \quad g_n(p) = \varepsilon^{-1}(p) \qquad \text{if } p \leq n,$$
$$f_n(p) = g_n(p) = 0 \quad \text{if } p > n.$$

We will arrange things so that, in addition, for all p less than or equal to n, $\phi_p = \theta^1_{f_n(p)}$ and $\theta_p = \phi^1_{g_n(p)}$. These functions f_n and g_n are dened simultaneously by induction on n. For f_{-1} and g_{-1}, we set both equal to the function that is constantly equal to 0. Let us examine the case $n + 1$:

- $f_{n+1}(p) = f_n(p)$ except if $p = n + 1$;
- if there exists $a \leq n$ such that $g_n(a) = n + 1$, then $f_{n+1}(n + 1) = a$;
- otherwise, $f_{n+1}(n + 1)$ is the least integer m that does not belong to the (nite) set $\{1, 2, \ldots, n\} \cup \{f_n(0), f_n(1), \ldots, f_n(n)\}$ (this condition is to be ignored if $n = -1$) and is such that m equals $\gamma(k, \beta(n + 1))$ for some element k.

The denition of g is analogous.

- $g_{n+1}(p) = g_n(p)$ except if $p = n + 1$;
- if there exists $a \leq n + 1$ such that $f_{n+1}(a) = n + 1$, then $g_{n+1}(n + 1) = a$;
- otherwise, $g_{n+1}(n+1)$ is the least non-zero integer m that does not belong to the set $\{1, 2, \ldots, n + 1\} \cup \{g_n(0), g_n(1), \ldots, g_n(n)\}$ and is such that m equals $\delta(k, \alpha(n + 1))$ for some element k.

We leave it to the reader to verify that the functions $\lambda nx.f_n(x)$ and $\lambda nx.g_n(x)$ are recursive, as is the function $\varepsilon = \lambda x.f_x(x)$; the function $\lambda x.g_x(x)$ is the inverse of ε, which is therefore bijective and has the desired properties.

Solutions to the exercises for Chapter 6

1. (a) It sufces to verify axioms A_1, A_2, ... and A_7; this does not present any special difculty. We will treat A_7 for the sake of example. Let a and b belong to M and let us show that

$$a \times Sb = (a \times b) + a. \qquad (*)$$

We must distinguish several cases:

(i) a and b are both in $\mathbb{N}$; then $(*)$ is obvious since $\mathcal{M}$ is an extension of $\mathbb{N}$.

(ii) $a \in X \times \mathbb{Z}$, say $a = (x, n)$, and $b \in \mathbb{N}$; then $Sb = b + 1$, $a \times Sb = (x, n \times (b + 1))$.

If $b = 0$, $a \times Sb = (x, n) = a$ and $a \times b = 0$, so we do have $(a \times b) + a = a \times Sb$.

If $b \neq 0$, $a \times b = (x, n \times b)$ and $(a \times b) + a = (x, (n \times b) + n) = a \times Sb$.

(iii) $a \in \mathbb{N}$ and $b \in X \times \mathbb{Z}$, say $b = (y, m)$; then $Sb = (y, m + 1)$ and $a \times Sb = (y, a \times (m + 1))$. Also, $a \times b = (y, a \times m)$ and $(a \times b) + a = (y, (a + m) + a)$.

(iv) $a \in X \times \mathbb{Z}$ and $b \in X \times \mathbb{Z}$, say $a = (x, n)$ and $b = (y, m)$; then $Sb = (y, m + 1)$, $a \times Sb = (f(x, y), n \times (m + 1))$; on the other hand, $a \times b = (f(x, y), n \times m)$ and $(a \times b) + a = (f(x, y), (n + m) + n)$.

(b) We will make use of (a) to construct a model of $\mathcal{P}_0$ in which none of the given formulas is true. It is sufcient to take any X that has at least two elements, for example $X = \mathbb{N}$, and to take, for f, any non-associative function, for example $f(x, y) = x + 2y$. In the model $\mathcal{M}$ built from this data according to (a), we have, for example,

$$(1, 1) + (2, 0) = (1, 1) \quad \text{and} \quad (2, 0) + (1, 1) = (2, 1)$$

which shows that addition is not commutative, and

$$((1, 1) \times (2, 2)) \times (3, 3) = (5, 2) \times (3, 3) = (11, 6), \qquad \text{and}$$
$$(1, 1) \times ((2, 2) \times (3, 3)) = (1, 1) \times (8, 6) = (17, 6)$$

which shows that multiplication is not associative. For the third formula, we see, for example, that $(1, 0) \leq (1, 1)$ [because $(1, 1) + (1, 0) = (1, 1)$] and $(1, 1) \leq (1, 0)$ [because $(1, -1) + (1, 1) = (1, 0)$]. The fourth formula is not satised because, for example, $0 \times (1, 0) = (1, 0)$.

(c) In the models we have just constructed, addition is associative. We can use the same idea to show that the associativity of addition does not follow from $\mathcal{P}_0$. Here is a model of $\mathcal{P}_0$, among many others, in which addition is not associative. The base set is $\mathbb{N} \cup (\mathbb{N} \times \mathbb{Z})$ (so it is an extension of $\mathbb{N}$) and $\underline{S}$, $\underline{+}$, and $\underline{\times}$ are interpreted by

$$
\begin{aligned}
S(n, a) &= (n, a + 1);\\
(n, a) + m &= (n, a + m) = m + (n, a);\\
(n, a) + (m, b) &= (n + 2m, a + b) \quad \text{if } n \neq m;\\
(n, a) + (n, b) &= (n, a + b);\\
(n, a) \times m &= (n, am) = m \times (n, a) \quad \text{if } m \neq 0;\\
(n, a) \times 0 &= 0 \times (n, a) = 0;\\
(n, a) \times (m, b) &= (2nb, ab).
\end{aligned}
$$

Here, once more, it is not difcult to show that the seven axioms of $\mathcal{P}_0$ hold but that, for example,

$$((1,0)+(2,0))+(3,0)=(11,0);$$
$$(1,0)+((2,0)+(3,0))=(17,0).$$

2. (a) It is clear that the relation $\approx$ is symmetric; it is reexive because of axiom A_4. Let us prove it is transitive. If x, y, and z are elements of $\mathcal{M}$ and if there exist integers n, m, p, and q such that

$$\mathcal{M} \models x \,\underline{+}\, \underline{n} \simeq y \,\underline{+}\, \underline{m} \quad \text{and} \quad \mathcal{M} \models y \,\underline{+}\, \underline{p} \simeq z \,\underline{+}\, \underline{q},$$

then, because addition is associative and commutative in any model of $\mathcal{P}$,

$$\mathcal{M} \models x \,\underline{+}\, \underline{n+p} \simeq z \,\underline{+}\, \underline{m+q}.$$

(b) By hypothesis, we have integers n, m, p, and q such that

$$\mathcal{M} \models a \,\underline{+}\, \underline{n} \simeq a' \,\underline{+}\, \underline{m} \quad \text{and} \quad \mathcal{M} \models b \,\underline{+}\, \underline{p} \simeq b' \,\underline{+}\, \underline{q}$$

and, because addition is associative and commutative in any model of $\mathcal{P}$,

$$\mathcal{M} \models (a \,\underline{+}\, b) \,\underline{+}\, \underline{n+p} \;\simeq\; (a' \,\underline{+}\, b') \,\underline{+}\, \underline{m+q}.$$

(c) Reexivity is clear. Let us prove transitivity: so suppose x, y, and z are in E and that xRy and yRz. Thus there exist a in x, b and b' in y, and c in z such that $\mathcal{M} \models a \leq b \wedge b' \leq c$; also, there exists n in $\mathbb{N}$ such that

$$\mathcal{M} \models b \leq b' + n.$$

It follows that

$$\mathcal{M} \models a \leq c + n,$$

and hence that xRz because $c+n$ is also in z.

Let us now prove that R is antisymmetric: we assume there are points a and a' in $x \in E$ and b and b' in $y \in E$ such that

$$\mathcal{M} \models a \leq b \quad \text{and} \quad \mathcal{M} \models b' \leq a'.$$

We must show that $x = y$. The hypotheses translate as follows: there exist u and v in $\mathcal{M}$ and integers n, m, p, and q such that

$$\mathcal{M} \models a \,\underline{+}\, u \simeq b; \qquad \mathcal{M} \models b' \,\underline{+}\, v \simeq a';$$
$$\mathcal{M} \models a \,\underline{+}\, \underline{n} \simeq a' \,\underline{+}\, \underline{m}; \qquad \mathcal{M} \models b \,\underline{+}\, \underline{p} \simeq b' \,\underline{+}\, \underline{q}.$$

All this, together with the associativity and commutativity of addition, yields

$$\mathcal{M} \models a \,\underline{+}\, u \,\underline{+}\, v \,\underline{+}\, \underline{p+m} \simeq a \,\underline{+}\, \underline{n+q}.$$

It now follows from property (19) of the subsection on the ordering of the integers that $\mathcal{M} \models u \leq \underline{n+q}$ and, because $\mathbb{N}$ is an initial segment of $\mathcal{M}$, $u \in \mathbb{N}$ and $a \approx b$. Therefore, $x = y$.

The ordering R is indeed total: if x and y are elements of E, if $a \in x$ and $b \in y$, then $\mathcal{M} \models a \leq b$ or $\mathcal{M} \models b \leq a$ because the order $\leq$ is a total ordering in $\mathcal{M}$; so we do have xRy or yRx.

The standard elements are all equivalent and the equivalence class they form is less than all the others. By contrast, if a is a non-standard element, a and $a+a$ are not equivalent and the equivalence class of $a+a$ is strictly greater than that of a.

To show that R is a dense ordering of E, we rst show that

$$\mathcal{P} \vdash \forall v_0 \exists v_1 (v_1 \underline{+} v_1 \simeq v_0 \vee v_1 \underline{+} v_1 \simeq v_0 \underline{+} \underline{1}),$$

which is easily proved by induction on v_0.

If a and b are elements of $\mathcal{M}$ such that, say, $a \leq b$, and if c is the element such that $c + c = a + b$ or $c + c = a + b + 1$, then we easily see that $c \approx a$ if and only if $c \approx b$. It follows that if $a \approx b$ is false, then the equivalence class of c lies strictly between the equivalence class of a and that of b.

3. We prove by induction on n that if $(b_0, b_1, \ldots, b_n)$ is a sequence of integers that are pairwise relatively prime and if $(\alpha_0, \alpha_1, \ldots, \alpha_n)$ is another sequence of the same length, then there exists $a \in \mathbb{N}$ such that, for all i from 0 to n inclusive,

$$a \text{ is congruent to } \alpha_i \text{ modulo } b_i.$$

For $n = 0$, it sufces to take $a = \alpha_0$. Next, consider the case $n = 1$. Since b_0 and b_1 are relatively prime, the theorem of Bezout guarantees the existence of elements γ_0 and γ_1 in $\mathbb{Z}$ such that

$$\gamma_0 b_0 + \gamma_1 b_1 = 1,$$

which, when multiplied by $(\alpha_1 - \alpha_0)$, yields

$$(\alpha_1 - \alpha_0)\gamma_0 b_0 + \alpha_0 = (\alpha_0 - \alpha_1)\gamma_1 b_1 + \alpha_1,$$

so we obtain an element m in $\mathbb{Z}$, namely $(\alpha_1 - \alpha_0)\gamma_0 b_0 + \alpha_0$, that is congruent to α_0 modulo b_0 and to α_1 modulo b_1. To get an element of $\mathbb{N}$ that has this same property, it sufces to add to m some multiple kb_0b_1 where k is sufciently large.

Now consider the case $n + 1$. By the induction hypothesis, there exists an integer c such that, for all i from 0 to n inclusive,

$$c \text{ is congruent to } \alpha_i \text{ modulo } b_i.$$

But b_{n+1} and $b_0 b_1 \cdots b_n$ are also relatively prime. As we have just seen, this implies the existence of an $a \in \mathbb{N}$ such that a is congruent to c modulo $b_0 b_1 \cdots b_n$

and to α_{n+1} modulo b_{n+1}. This certainly implies that, for all i from 0 to $n+1$ inclusive,

$$a \text{ is congruent to } \alpha_i \text{ modulo } b_i.$$

4. Suppose that the formula $F[v_0, v_1, \ldots, v_p]$ represents a total function f from $\mathbb{N}^p$ into $\mathbb{N}$. Then by the denition of Drv_0,

$$x = f(n_1, n_2, \ldots, n_p) \quad \text{if and only if} \quad \text{there exists } y \text{ such that } (\#F[x, n_1, n_2, \ldots, n_p], y) \in \mathsf{Drv}_0.$$

So consider the function g:

$$g(n_1, n_2, \ldots, n_p) = \mu y\,((\#F[x, n_1, n_2, \ldots, n_p], y) \in \mathsf{Drv}_0).$$

This function is total recursive and $f(n_1, n_2, \ldots, n_p) = \beta_2^1(g(n_1, n_2, ..., n_p))$.

5. This involves a technique known as the method of pleonasm. We enumerate $T = \{F_n : n \in \mathbb{N}\}$ in such a way that the function $\lambda n.\#F_n$ is a total recursive function (see Section 5.4.1). For every $n \in \mathbb{N}$, let G_n denote the formula $F_1 \wedge F_2 \wedge \cdots \wedge F_n$. Set $T' = \{G_n : \in \mathbb{N}\}$. It is clear that the theories T and T' are equivalent and that the function $\lambda n.\#G_n$ is total recursive and strictly increasing; this implies that T' is a recursive theory (see Exercise 12 of Chapter 5).

6. We must rst convince ourselves that this question has a meaning, specically, that Fermat's last theorem is expressible by a formula of $\mathcal{L}_0$. This is not obvious, *a priori*, because of the exponentials x^t and so on. So we begin by eliminating these exponentials using the formulas that represent them. Thus, let $F[v_0, v_1, v_2]$ be a Σ formula such that, for all integers n, m, and p,

$$\mathcal{P}_0 \vdash F[\underline{n}, \underline{m}, \underline{p}] \quad \text{if and only if} \quad n = m^p.$$

We may then observe that the negation of Fermat's last theorem can be expressed in the language $\mathcal{L}_0$ by the following closed Σ formula:

$$G = \exists v_0 \exists v_1 \exists v_2 \exists v_3 \exists v_4 \exists v_5 \exists v_6 (v_2 \geq 1 \wedge v_4 \geq 1 \wedge v_6 \geq 1 \wedge v_0 \geq 3 \\ \wedge F[v_1, v_2, v_0] \wedge F[v_3, v_4, v_0] \wedge F[v_5, v_6, v_0] \wedge v_1 \,\underline{+}\, v_3 \simeq v_5).$$

If G is true in $\mathbb{N}$, then it is derivable in $\mathcal{P}_0$ (Proposition 6.34) so Fermat's last theorem is refutable in $\mathcal{P}_0$.

7. (a) It is clear that if $\mathbb{N} \models \exists v_1 Drv[\underline{\#F}, v_1]$, then there exists an integer n such that $\mathbb{N} \models Drv[\underline{\#F}, \underline{n}]$ and hence that $(\#F, n) \in \mathsf{Drv}$. It follows that the formula F is derivable in $\mathcal{P}$ and is therefore true in $\mathbb{N}$.

(b) The proof of the second incompleteness theorem supplies a model $\mathcal{M}$ of $\mathcal{P}$ and a closed formula F such that

$$\mathcal{M} \models \exists v_1 Drv[\underline{\#F}, v_1] \wedge \exists v_2 Drv[\underline{\#\neg F}, v_2].$$

This shows that (b) cannot be simultaneously satised by F and by $\neg F$.

(c) We will suppose that (c) is true for every closed formula F and obtain a contradiction. Since, in $\mathbb{N}$, either F is true or $\neg F$ is true, then either F is derivable in $\mathcal{P}$ or $\neg F$ is derivable in $\mathcal{P}$, in other words, $\mathcal{P}$ is complete; but we know this is false.

(d) This is obviously false since (d) implies (c).

8. For the formula $F[v_0]$ we may take the formula $Drv[\underline{\#\underline{0} \simeq \underline{1}},v_0]$, and for H we may take the formula

$$G[v_0, v_1, \ldots, v_n] \not\Leftrightarrow Drv[\underline{\#\underline{0} \simeq \underline{1}}, v_0].$$

9. Suppose that $\mathcal{P} \vdash \exists v_0 Drv[\underline{\#F}, v_0] \Rightarrow F$; then by taking the contrapositive,

$$\mathcal{P} \vdash \neg F \Rightarrow \neg\exists v_0 Drv[\underline{\#F}, v_0];$$

equivalently,

$$\mathcal{P} \cup \{\neg F\} \vdash \neg\exists v_0 Drv[\underline{\#F}, v_0].$$

But $\neg\exists v_0 Drv[\underline{\#F}, v_0]$ means that F is not derivable in $\mathcal{P}$, in other words, that $\mathcal{P} \cup \{\neg F\}$ is a consistent theory. So we have

$$\mathcal{P} \cup \{\neg F\} \vdash Con(\mathcal{P} \cup \{\neg F\}),$$

which implies, according to Godel's second incompleteness theorem, that $\mathcal{P} \cup \{\neg F\}$ is not a consistent theory; hence $\mathcal{P} \vdash F$.

10. (a) We will prove, by induction on the height of the formula $G[v_1, v_2, \ldots, v_p]$, that, for all elements $a_1, a_2, \ldots, a_n$ of N, we have

$$\mathcal{M} \models G[a_1, a_2, \ldots, a_p] \quad \text{if and only if} \quad \mathcal{N} \models G[a_1, a_2, \ldots, a_p].$$

If G is an atomic formula, this is true because $\mathcal{N}$ is a substructure of $\mathcal{M}$. The propositional connectives do not present any problem. For the sake of example, let us take the case of $\wedge$. Assume that $G[v_1, v_2, ..., v_p] = G_1 \wedge G_2$ and that $a_1, a_2, ..., a_n$ are points in N. Then,

$$\mathcal{M} \models G[a_1, a_2, \ldots, a_p]$$

if and only if

$$\mathcal{M} \models G_1[a_1, a_2, \ldots, a_p] \wedge \mathcal{M} \models G_2[a_1, a_2, \ldots, a_p].$$

Now by the induction hypothesis,

$$\mathcal{M} \models G_1[a_1, a_2, \ldots, a_p] \quad \text{if and only if} \quad \mathcal{N} \models G_1[a_1, a_2, \ldots, a_p], \quad \text{and}$$
$$\mathcal{M} \models G_2[a_1, a_2, \ldots, a_p] \quad \text{if and only if} \quad \mathcal{N} \models G_2[a_1, a_2, \ldots, a_p];$$

this clearly implies that

$$\mathcal{M} \models G[a_1, a_2, \ldots, a_p] \quad \text{if and only if} \quad \mathcal{N} \models G[a_1, a_2, \ldots, a_p].$$

Now let us deal with the existential quantier. We assume that

$$G[v_1, v_2, \ldots, v_p] = \exists v_0 F[v_0, v_1, \ldots, v_p]$$

and that $a_1, a_2, \ldots, a_p$ are points in N. If we suppose that $\mathcal{N} \models G[a_1, a_2, \ldots, a_p]$, then there exists a point a_0 of N such that $\mathcal{N} \models F[a_0, a_1, \ldots, a_p]$; so by the induction hypothesis, $\mathcal{M} \models F[a_0, a_1, \ldots, a_p]$ and thus $\mathcal{N} \models G[a_1, a_2, \ldots, a_p]$.

Conversely, suppose that the points $a_1, a_2, \ldots, a_p$ are in N and that there exists a point a_0 in M such that $\mathcal{M} \models F[a_0, a_1, \ldots, a_p]$. Consider the formula

$$\begin{aligned} H[v_0, v_1, ..., v_p] = (\neg\exists v_{p+1} F[v_{p+1}, v_1, ..., v_p] \Rightarrow v_0 \simeq \underline{0}) \\ \wedge(\exists v_{p+1} F[v_{p+1}, v_1, ..., v_p] \Rightarrow (F[v_0, v_1, ..., v_p] \\ \wedge \forall v_{p+2} < v_0 \neg F[v_{p+2}, v_1, ..., v_p])). \end{aligned}$$

Thus, the formula H denes the following function f from M^p into M:

- if there exists at least one element $x \in M$ such that

$$\mathcal{M} \models F[x, a_1, a_2, \ldots, a_p],$$

  then $f(a_1, a_2, \ldots, a_p)$ is the least element that satises this formula (which exists by the induction scheme);
- otherwise, $f(a_1, a_2, \ldots, a_p) = 0$.

By hypothesis (because $\mathcal{N}$ is closed under denable functions), $f(a_1, a_2, \ldots, a_p)$ is an element of N. From the denition of H, it follows that

$$\mathcal{M} \models F[f(a_1, a_2, \ldots, a_p), a_1, a_2, \ldots, a_p]$$

and, by the induction hypothesis,

$$\mathcal{N} \models F[f(a_1, a_2, \ldots, a_p), a_1, a_2, \ldots, a_p],$$

and hence $\mathcal{N} \models G[a_1, a_2, \ldots, a_p]$.

(b) If X_1 and X_2 are subsets of M that are denable by the formulas $F_1[v_0]$ and $F_2[v_0]$, respectively, then $X_1 \cap X_2$ is denable by the formula $F_1 \wedge F_2$. The analogous facts for union and complementation are also true; this shows that the collection of denable subsets of M form a Boolean subalgebra of the algebra of all subsets of M.

If f and g are functions that are denable by $G_1[v_0, v_1]$ and $G_2[v_0, v_1]$, respectively, then

$$\begin{aligned} &\{a \in M : \ f(a) = g(a)\} \\ &\quad = \{a \in M : \ \mathcal{M} \models \exists v_0 (G_1[v_0, a] \wedge G_1[v_0, a])\}; \end{aligned}$$

this is certainly a denable subset.

(c) Suppose, once again, that f and g are functions denable by $G_1[v_0, v_1]$ and $G_2[v_0, v_1]$, respectively. Now consider $f + g$, for example; it is denable by the formula

$$\exists v_2 \exists v_3 (v_0 \simeq v_2 + v_3 \wedge G_1[v_2, v_1] \wedge G_2[v_3, v_1]).$$

Similar conclusions for $f \times g$ and Sf are just as easy.

(d) Suppose that f, g, and h are in $\mathcal{F}$. Then,

$$\begin{aligned} &\{a \in M : f(a) = g(a)\} \cap \{a \in M : g(a) = h(a)\} \\ &\quad \subseteq \{a \in M : f(a) = h(a)\}, \end{aligned}$$

which shows that if the rst two of these sets belong to $\mathcal{U}$, then so does the third; this proves that the relation $\approx$ is transitive. It is obvious that $\approx$ is symmetric and reexive. Also,

$$\begin{aligned} &\{a \in M : f(a) = f'(a)\} \cap \{a \in M : g(a) = g'(a)\} \\ &\quad \subseteq \{a \in M : (f + g)(a) = (f' + g')(a)\}, \end{aligned}$$

so if $f \approx f'$ and $g \approx g'$, then $f + g \approx f' + g'$. Similar arguments apply to the successor and product operations.

(e) We must simply verify that if a and b are elements of M, then

$$\bar{a} + \bar{b} = \overline{a + b}, \qquad \bar{a} \times \bar{b} = \overline{a \times b}, \quad \text{and} \quad S\bar{a} = \overline{Sa};$$

these are all obvious.

(f) The argument is by induction on the complexity of F. As examples, we will treat the cases of $\neg$ and $\exists$.

- For $\neg$: we assume that $F[v_1, v_2, \ldots, v_p] = \neg G[v_1, v_2, \ldots, v_p]$ and that, for all $f_1, f_2, \ldots, f_p$ in $\mathcal{F}$,

  $$\mathcal{F}/\mathcal{U} \models G[f_1/\mathcal{U}, f_2/\mathcal{U}, \ldots, f_p/\mathcal{U}]$$

  if and only if

  $$\{a \in M : \mathcal{M} \models G[f_1(a), f_2(a), \ldots, f_p(a)]\} \in \mathcal{U}.$$

  But we see that

  $$\mathcal{F}/\mathcal{U} \models G[f_1/\mathcal{U}, f_2/\mathcal{U}, \ldots, f_p/\mathcal{U}]$$

  if and only if

  $$\mathcal{F}/\mathcal{U} \not\models F[f_1/\mathcal{U}, f_2/\mathcal{U}, \ldots, f_p/\mathcal{U}]$$

  and, since $\mathcal{U}$ is an ultralter,

  $$\{a \in M : \mathcal{M} \models G[f_1(a), f_2(a), \ldots, f_p(a)]\} \in \mathcal{U}$$

if and only if

$$\{a \in M : \mathcal{M} \models F[f_1(a), f_2(a), \ldots, f_p(a)]\} \notin \mathcal{U}.$$

- For $\exists$: this time, we assume that $F[v_1, v_2, \ldots, v_p] = \exists v_0 G[v_0, v_1, \ldots, v_p]$ and that G satises the induction hypothesis. Suppose rst that

$$\mathcal{F}/\mathcal{U} \models F[f_1/\mathcal{U}, f_2/\mathcal{U}, \ldots, f_p/\mathcal{U}].$$

So there exists a function $f_0 \in \mathcal{F}$ such that

$$\mathcal{F}/\mathcal{U} \models G[f_0/\mathcal{U}, f_1/\mathcal{U}, \ldots, f_p/\mathcal{U}]$$

and, by the induction hypothesis,

$$\{a \in M : \mathcal{M} \models G[f_0(a), f_1(a), \ldots, f_p(a)]\} \in \mathcal{U};$$

it follows from this that

$$\{a \in M : \mathcal{M} \models F[f_1(a), f_2(a), \ldots, f_p(a)]\} \in \mathcal{U}.$$

Conversely, suppose that

$$A = \{a \in M : \mathcal{M} \models F[f_1(a), f_2(a), \ldots, f_p(a)]\} \in \mathcal{U}.$$

Once again [see (a)], consider the formula

$$\begin{aligned} H[v_0, v_1, ..., v_p] = (\neg\exists v_{p+1} G[v_{p+1}, v_1, ..., v_p] \Rightarrow v_0 \simeq \underline{0}) \\ \wedge(\exists v_{p+1} G[v_{p+1}, v_1, ..., v_p] \Rightarrow (G[v_0, v_1, ..., v_p] \\ \wedge \forall v_{p+2} < v_0 \neg G[v_{p+2}, v_1, ..., v_p])). \end{aligned}$$

We can then dene a mapping f_0 from M into M as follows: for all $a \in M$, $f_0(a)$ is the unique element of M such that

$$\mathcal{M} \models H[f_0(a), f_1(a), \ldots, f_p(a)];$$

we then see that, for all $a \in M$,

$$\mathcal{M} \models G[f_0(a), f_1(a), \ldots, f_p(a)],$$

and hence, by the induction hypothesis,

$$\mathcal{F}/\mathcal{U} \models G[f_0/\mathcal{U}, f_1/\mathcal{U}, \ldots, f_p/\mathcal{U}]$$

and

$$\mathcal{F}/\mathcal{U} \models F[f_1/\mathcal{U}, f_2/\mathcal{U}, \ldots, f_p/\mathcal{U}].$$

Now let $d_1, d_2, \ldots, d_p$ be points of M and let $\overline{d_1}, \overline{d_2}, \ldots, \overline{d_p}$ be the corresponding constant functions. Then

$$\{a \in M : \mathcal{M} \models F[\overline{d_1}(a), \overline{d_2}(a), \ldots, \overline{d_p}(a)]\}$$

is equal to the whole of M if $\mathcal{M} \models F[d_1, d_2, \ldots, d_p]$ and is empty otherwise. Hence,

$$\{a \in M : \mathcal{M} \models F[\overline{d_1}(a), \overline{d_2}(a), \ldots, \overline{d_p}(a)]\} \in \mathcal{U}$$

if and only if $\mathcal{M} \models F[d_1, d_2, \ldots, d_p]$; this means precisely that the map from $\mathcal{M}$ into $\mathcal{F}/\mathcal{U}$ that sends a into $\bar{a}$ is elementary.

(g) Let $F[v_0, v_1, w_0, w_1, \ldots, w_p]$ be a formula of $\mathcal{L}_0$. Let

$$Func_F[w_0, w_1, \ldots, w_p]$$

denote the formula

$$\forall v_0 \exists ! v_1 F[v_0, v_1, w_0, w_1, \ldots, w_p];$$

this is the formula which expresses that F denes a function, once parameters are substituted for the variables w_i. We have to show that

$$\begin{aligned} \mathcal{M} \models \forall w_0 \forall w_1 \ldots \forall w_p (&Func_F[w_0, w_1, \ldots, w_p] \\ &\Rightarrow \forall v_2 \exists v_3 \forall v_0 \forall v_1 ((v_0 < v_2 \wedge F[v_0, v_1, w_0, w_1, \ldots, w_p]) \\ &\Rightarrow v_1 < v_3)). \end{aligned}$$

Since $\mathcal{M}$ is an elementary extension of $\mathbb{N}$, it is sufcient to show that this same formula is true in $\mathbb{N}$. So, if for integers $m_0, m_1, \ldots, m_p$, the formula $F[v_0, v_1, m_0, m_1, \ldots, m_p]$ denes a function, say f, with domain $\mathbb{N}$ and if n_2 is an integer, then there certainly exists an integer n_3, namely $\sup\{f(x) + 1 : x < n_2\}$, such that

$$\mathbb{N} \models \forall v_0 \forall v_1 ((v_0 < v_2 \wedge F[v_0, v_1, m_0, m_1, \ldots, m_p]) \Rightarrow v_1 < n_3).$$

(h) Let $\mathcal{B}$ denote the Boolean algebra of subsets of M that are denable with parameters in M and consider the following subset of $\mathcal{B}$:

$$\{[a, b] : a \in \mathbb{N}, \ b \in M - \mathbb{N}\},$$

where $[a, b]$ denotes the set of points of M that are between a and b inclusive. This set is closed under nite intersections and does not contain the empty set. It follows that there is an ultralter $\mathcal{U}$ of $\mathcal{B}$ that includes it. Let $\mathcal{N}$ denote the structure $\mathcal{F}/\mathcal{U}$ as constructed above, considered as an elementary extension of $\mathcal{M}$. We will show that it has the desired property.

Let $f/\mathcal{U} \in \mathcal{N}$. Choose an arbitrary non-standard element c of $\mathcal{M}$. We know, from the preceding question, that there exists $d \in M$ such that if $x \in M$ and $x < c$, then $f(x) < d$. Let $\bar{d}$ be the constant function equal to d and recall that we have identied $\bar{d}/\mathcal{U} \in \mathcal{N}$ and d. Then,

$$[0, c] \subseteq \{a \in M : f(a) < \bar{d}(a)\},$$

and, because $[0, c] \in \mathcal{U}$, we do, according to (f), have $f/\mathcal{U} < d$.

11. (a) Let H denote the conjunction of the seven axioms of $\mathcal{P}_0$ and let T be a theory that is satised by $\mathbb{N}$. If T were decidable, then the set

$$\{\#F : \#(H \Rightarrow F) \in \mathsf{Th}(T)\}$$

would be recursive; but this set is precisely $\mathsf{Th}(T \cup \mathcal{P}_0)$ which is a consistent theory (it has $\mathbb{N}$ as a model) that includes $\mathcal{P}_0$; this contradicts Godel's rst theorem.

(b) We will show how to construct the formula F^* from F. The procedure we will describe is effective and it is not a problem to prove the existence of a primitive recursive function α that, with the Godel number of F, associates the Godel number of F^*.

This construction is by induction on the height of F. We must begin with the atomic formulas which, for the case of $\mathcal{L}_0$, are of the form $t \simeq s$, where s and t are terms of $\mathcal{L}_0$. We dispense rst with the situation in which s and t are simple terms, i.e. where F has one of the following forms:

$F = v_i \simeq \underline{0}$, then $F^* = G_1[v_i]$;

$F = v_i \simeq v_j$, then $F^* = G_0[v_i] \wedge G_0[v_j] \wedge v_i \simeq v_j$;

$F = v_i \simeq \underline{S} v_j$, then $F^* = G_0[v_i] \wedge G_0[v_j] \wedge G_2[v_i, v_j]$;

$F = v_i \simeq v_j \underline{+} v_k$, then

$$F^* = G_0[v_i] \wedge G_0[v_j] \wedge G_0[v_k] \wedge G_3[v_i, v_j, v_k];$$

$F = v_i \simeq v_j \underline{+} v_k$, then

$$F^* = G_0[v_i] \wedge G_0[v_j] \wedge G_0[v_k] \wedge G_4[v_i, v_j, v_k].$$

We then deal with formulas of the form $v_i \simeq t$, where t is a term. This is done by induction on t. As an example, let us treat the case where $F = v_i \simeq t_1 \underline{+} t_2$ under the assumption that we have already constructed the formulas $(v_i \simeq t_1)^*$ and $(v_i \simeq t_2)^*$. Choose variables w_0 and w_1 that do not occur in v_i, t_1, and t_2. Set

$$F^* = \exists w_0 \exists w_1 ((w_0 \simeq t_1)^* \wedge (w_1 \simeq t_2)^* \wedge (v_i \simeq w_0 \underline{+} w_1)^*).$$

We nish with the atomic formulas by setting

$$(t_1 \simeq t_2)^* = \exists w_0 ((w_0 \simeq t_1)^* \wedge (w_1 \simeq t_2)^*),$$

where, once again, w_0 is a variable that does not occur in t_1 or in t_2.

There is then no problem with the further induction on the height of F :

- $(\neg F)^* = \neg F^*$;
- $(F_1 \wedge F_2)^* = F_1^* \wedge F_2^*$;
- $(F_1 \vee F_2)^* = F_1^* \vee F_2^*$;
- $(F_1 \Rightarrow F_2)^* = F_1^* \Rightarrow F_2^*$;

- $(F_1 \Leftrightarrow F_2)^* = F_1^* \Leftrightarrow F_2^*$;
- $(\exists w F[w])^* = \exists w_0(G_0[w_0] \wedge F[w_0]^*)$, where w_0 is a variable that does not occur in G_0 or in F;
- $(\forall w F[w])^* = \forall w_0(G_0[w_0] \wedge F[w_0]^*)$, where w_0 is a variable that does not occur in G_0 or in F.

(c) It is obvious that (1) implies (2) and (3) implies (1). So it remains to show that (2) implies (3). We will use the completeness theorem and show, assuming $T^- \vdash G$, that $T \cup \{\neg G^*\}$ does not have a model. Assume the contrary and let $\mathcal{N}$ be the $\mathcal{L}_0$-structure denable in a model $\mathcal{M}$ of $T \cup \{\neg G^*\}$. Since $\mathcal{M}$ is a model of T, $\mathcal{N}$ is [according to (b)] a model of T^- and hence of G; but since $\mathcal{M}$ is a model of $\neg G^*$, $\mathcal{N}$ is a model of $\neg G$; this is impossible.

(d) We will rst show that every consistent theory in the language $\mathcal{L}$ that extends $T_0 \cup \{H^*\}$ is undecidable (recall that H is the conjunction of the axioms of $\mathcal{P}_0$). So let T be such a theory and, as previously, let us consider

$$T^- = \{F : \ F \text{ is a closed formula of } \mathcal{L}_0 \text{ and } T \vdash F^*\}.$$

This is a consistent theory that extends $\mathcal{P}_0$, so it is undecidable. Now if T were decidable, then T^- would also be decidable because $T^- \vdash F$ if and only if $T \vdash F^*$ (and the passage from F to F^* is effective).

Suppose now that $\mathbb{N}$ is denable in $\mathcal{M}$ and let T be a theory in $\mathcal{L}$ that has $\mathcal{M}$ as a model. Our goal is to show that T is undecidable. Let K be the conjunction of the formulas of $T_0 \cup \{H^*\}$ (this is a nite theory). We then see that $\mathcal{M}$ is a model of $T' = T \cup \{K\}$, which is, therefore, consistent and, from what we have seen, undecidable. But for every formula F of $\mathcal{L}$,

$$T' \vdash F \quad \text{if and only if} \quad T \vdash K \Rightarrow F,$$

which shows that T is also undecidable.

(e) It is not difcult to dene $\mathbb{N}$ in $\mathbb{Z}$; for example, by using the following formulas:

- $G_0[v_0]$ $= \exists v_1 \exists v_2 \exists v_3 \exists v_4 (v_0 \simeq ((v_1 \underline{\times}\, v_1) \underline{+} (v_2 \underline{\times}\, v_2) \underline{+} (v_3 \underline{\times}\, v_3) \underline{+} (v_4 \underline{\times}\, v_4)))$; (this is where we invoke Lagrange's theorem.)
- $G_1[v_0] = v_0 \simeq \underline{0}$;
- $G_2[v_0, v_1] = G_0[v_0] \wedge G_0[v_1] \wedge \exists v_2 \forall v_3 (v_2 \underline{\times}\, v_3 \simeq v_3 \wedge v_0 \simeq v_1 \underline{+} v_2)$;
- $G_3[v_0, v_1, v_2] = G_0[v_0] \wedge G_0[v_1] \wedge G_0[v_2] \wedge v_0 \simeq v_1 \underline{+} v_2$;
- $G_4[v_0, v_1] = G_0[v_0] \wedge G_0[v_1] \wedge G_0[v_2] \wedge v_0 \simeq v_1 \underline{\times}\, v_2$.

We conclude from this that $\mathbb{Z}$ is strongly undecidable and that every theory in $\mathcal{L}$ that has $\mathbb{Z}$ as a model is undecidable; for example, the theory of rings, the theory of commutative rings, etc.

(f) We notice, rst of all, that if x is an element of M that belongs to $\mathbb{N} \times \mathbb{N}$, say $x = (n, m)$, there are exactly two elements y of M, namely m and $(n + m, n \cdot m)$ such that $(x, y) \in R^{\mathcal{M}}$; on the other hand, if $x \in \mathbb{N}$, the set of elements y of M such that $(x, y) \in R^{\mathcal{M}}$ is innite. This permits us to dene $\mathbb{N}$ in $\mathcal{M}$ by the formula $G_0[v_0]$ that is equal to

$$\exists v_1 \exists v_2 \exists v_3 (R v_0 v_1 \wedge R v_0 v_2 \wedge R v_0 v_3 \wedge \neg(v_1 \simeq v_2) \\ \wedge \neg(v_2 \simeq v_3) \wedge \neg(v_3 \simeq v_1)).$$

Addition and multiplication are then easy to dene:

$$G_3[v_0, v_1, v_2] = G_0[v_0] \wedge G_0[v_1] \wedge G_0[v_2] \\ \wedge \exists v_3 \exists v_4 (R v_1 v_3 \wedge R v_3 v_2 \wedge R v_3 v_4 \wedge R v_0 v_4);$$
$$G_4[v_0, v_1, v_2] = G_0[v_0] \wedge G_0[v_1] \wedge G_0[v_2] \\ \wedge \exists v_3 \exists v_4 (R v_1 v_3 \wedge R v_3 v_2 \wedge R v_3 v_4 \wedge R v_4 v_0).$$

Then zero and one can be dened as the identity elements for addition and multiplication, respectively, and the successor function can be dened with the help of addition.

It follows that $\mathcal{M}$ is strongly undecidable and that the empty theory in the language that consists of a single binary predicate symbol is undecidable.

(g) Since we have addition at our disposal, we are able to dene the ordering of the integers, the elements 0 and 1 and the successor function. It is then sufcient to show that multiplication is denable in $\mathcal{M}$. We begin by dening the least common multiple, (lcm) of two integers by the formula

$$G_5[v_0, v_1, v_2] = D v_1 v_0 \wedge D v_2 v_0 \\ \wedge \forall v_3 ((D v_1 v_3 \wedge D v_2 v_3) \Rightarrow D v_0 v_3).$$

Since the lcm of y and $y + 1$ is always $y \cdot (y + 1)$, we may dene the relation $x = y \cdot (y + 1)$ by the formula

$$G_6[v_0, v_1] = G_5[v_0, v_1, v_1 \underline{+} \underline{1}].$$

Observe that, for every x and y in $\mathbb{N}$, we have

$$(x + y) \cdot (x + y + 1) = x \cdot (x + 1) + y \cdot (y + 1) + 2xy;$$

therefore we may set $\psi_4[v_0, v_1, v_2]$ equal to

$$\exists v_3 \exists v_4 \exists v_5 \exists v_6 (v_3 \simeq v_0 \underline{+} v_0 \wedge G_6[v_4, v_1 \underline{+} v_2] \\ \wedge G_6[v_5, v_1] \wedge G_6[v_6, v_2] \wedge v_4 \simeq (v_5 \underline{+} v_6) \underline{+} v_3).$$

12. (a) The proof is a routine induction on the height of the formula F. It involves showing that the class of recursively enumerable sets is closed under conjunction, disjunction, existential quantication, and bounded universal quantication.

(b) Let f be a recursive function from $\mathbb{N}$ into $\mathbb{N}$; according to the second representation theorem (Theorem 6.33), there is a Σ formula $F[v_0, v_1]$ that represents f. Conversely, if

$$\begin{aligned} Graph(f) &= \{(n, f(n)) : \in \mathbb{N}\} \\ &= \{(n, m) : \mathbb{N} \models F[m, n]\} \\ &= \{(n, m) : \mathcal{P}_0 \vdash F[\underline{m}, \underline{n}]\}, \end{aligned}$$

where F is a Σ formula, then, from part (a), $Graph(f)$ is also recursively enumerable. Since f is a total function, we also have that

$$(n, m) \notin Graph(f)$$

if and only if

$$\text{there exists } m' \neq m \text{ such that } (n, m') \in Graph(f),$$

which means that the complement of $Graph(f)$ is also recursively enumerable; hence $Graph(f)$ is recursive (Theorem 5.38). Thus, f is recursive (Chapter 5, Exercise 10).

(c) If $F[v_0, v_1]$ is a Σ formula, then for all integers n and m,

$\mathbb{N} \models F[m, n]$ if and only if
there exists a derivation of F $[\underline{m}, \underline{n}]$ in $\mathcal{P}_0$

(Proposition 6.34); *a fortiori,*

$\mathbb{N} \models F[m, n]$ if and only if
there exists a derivation of F $[\underline{m}, \underline{n}]$ in $\mathcal{P}$.

Let α be the function of three variables dened as follows:

- If a is the Godel number of a Σ formula with two free variables, say $F[v_0, v_1]$, then $\alpha(a, m, n) = \#F[\underline{m}, \underline{n}]$.
- If not, then $\alpha(a, m, n) = 0$.

This function α is primitive recursive and, if a is the Godel number of the formula F, then for all integers n and m, we have

$\mathbb{N} \models F[m, n]$ if and only if
there exists $b \in \mathbb{N}$ such that $(\alpha(a, m, n), b) \in \mathsf{Drv}$.

We can then dene the partial function k:

$$k(a, n) = \mu y\,(\alpha(a, \beta_2^1(y), n), \beta_2^2(y)) \in \mathsf{Drv}$$

and set

$$h(a, n) = \beta_2^1(k(a, n)).$$

(d) It is not difcult to see that g is recursive; it is dened by cases and the relations

a is the Godel number of a Σ formula,

or

b is the Godel number of a derivation in $\mathcal{P}$ of the formula ... ,

and so on are all recursive.

Let us prove that g is total; let a, b, and n be integers and assume that a is the Godel number of a Σ formula, say $F[v_0, v_1]$, and that b is the Godel number of a derivation in $\mathcal{P}$ of the formula $\forall v_1 \exists v_0 F[v_0, v_1]$. The problem is to show the existence of an $m \in \mathbb{N}$ such that $\mathcal{P} \vdash F[\underline{m}, \underline{n}]$. But since $\mathbb{N}$ is a model of $\mathcal{P}$, we have

$$\mathbb{N} \models \forall v_1 \exists v_0 F[v_0, v_1],$$

so there exists an integer m such that $\mathbb{N} \models F[\underline{m}, \underline{n}]$; and since $F[\underline{m}, \underline{n}]$ is a Σ formula, we have

$$\mathcal{P} \vdash F[\underline{m}, \underline{n}]$$

by Proposition 6.34.

(e) It follows from all that has preceded that the set of functions

$$\mathcal{E} = \{\lambda n.g(a, b, n) : \ a \text{ and } b \text{ are in } \mathbb{N}\}$$

is exactly equal to the set of all recursive functions that are provably total. We then apply a diagonal argument to this set: the function

$$\lambda n.g(\beta_2^1(n), \beta_2^2(n), n) + 1$$

is total recursive but cannot belong to $\mathcal{E}$.

13. (a) If $\{F_1, F_2, \ldots, F_n\}$ is a nite set of closed formulas,

$$\mathcal{P} \cup \{F_1, F_2, \ldots, F_n\}$$

is a recursive theory. If it is consistent, it cannot be complete by Godel's rst incompleteness theorem (Theorem 6.30).

(b) The construction is by induction on the length of s: we assume that s is in $\{\mathbf{0}, \mathbf{1}\}^n$ and that the formulas $F_\emptyset$, $F_{(s(0))}$, $F_{(s(0),s(1))}$, $\ldots$, $F_{(s(0),s(1),\ldots,s(n-1))}$

have already been constructed in such a way that

$$\mathcal{P} \cup \{F_\emptyset, F_{(s(0))}, F_{(s(0),s(1))}, \ldots, F_{(s(0),s(1),\ldots,s(n-1))}\}$$

is a consistent theory, and we will construct the formulas $F_{(s(0),s(1),\ldots,s(n-1),0)}$ and $F_{(s(0),s(1),\ldots,s(n-1),1)}$. Since

$$\mathcal{P} \cup \{F_\emptyset, F_{(s(0))}, F_{(s(0),s(1))}, \ldots, F_{(s(0),s(1),\ldots,s(n-1))}\}$$

is not a complete theory [as we saw in part (a)], there exists a formula G that is neither derivable nor refutable in this theory; we set

$$F_{(s(0),s(1),\ldots,s(n-1),0)} = G \quad \text{and} \quad F_{(s(0),s(1),\ldots,s(n-1),1)} = \neg G.$$

(c) For each σ in $\{\mathbf{0}, \mathbf{1}\}^{\mathbb{N}}$, set

$$T_\sigma = \mathcal{P} \cup \{F_\emptyset, F_{(s(0))}, F_{(s(0),s(1))}, \ldots, F_{(s(0),s(1),\ldots,s(n-1))}, \ldots\}.$$

Every nite subset of T_σ is included in a set of the form

$$\mathcal{P} \cup \{F_\emptyset, F_{(s(0))}, F_{(s(0),s(1))}, \ldots, F_{(s(0),s(1),\ldots,s(n-1))}\},$$

where $n \in \mathbb{N}$ and $s \in \{\mathbf{0}, \mathbf{1}\}^n$, and is therefore consistent. Now let σ and τ be two distinct elements of $\{\mathbf{0}, \mathbf{1}\}^{\mathbb{N}}$ and let n be the least integer such that $\sigma(n) \neq \tau(n)$. Without loss of generality, suppose $\sigma(n) = 0$ and $\tau(n) = 1$. Then the formula $F_{(\tau(0),\tau(1),\ldots,\tau(n-1),\tau(n))}$ which belongs to T_τ and is equal to the formula $F_{(\tau(0),\tau(1),\ldots,\tau(n-1),1)}$ is the negation of the formula $F_{(\sigma(0),\sigma(1),\ldots,\sigma(n-1),0)}$ which belongs to T_σ; thus T_σ and T_τ are not equivalent.

We have thereby found $2^{\aleph_0}$ theories in $\mathcal{L}_0$ (as many as there are elements of $\{\mathbf{0}, \mathbf{1}\}^{\mathbb{N}}$), that all extend $\mathcal{P}$ and are pairwise inequivalent.

14. (a) If $\mathcal{M}$ is countable, then there are only countably many formulas with parameters from $\mathcal{M}$; so there cannot be more than this many subsets of $\mathbb{N}$ that are denable in $\mathcal{M}$.

(b) There is a formula $F[v_0, v_1]$ in $\mathcal{L}_0$ such that for all integers n and m,

$$\mathbb{N} \models F[n, m] \quad \text{if and only if} \quad \text{the } (n+1)\text{st prime number divides } m.$$

Let X be a subset of $\mathbb{N}$. Add a new constant symbol c to the language $\mathcal{L}_0$ and consider the following theory T_X in the language obtained thereby:

$$\begin{aligned} T_X = \{G[\underline{n_0}, \underline{n_1}, \ldots, \underline{n_p}] :\ & p \text{ is an integer, } G[v_0, v_1, \ldots, v_p] \text{ is a formula of } \mathcal{L}_0, \\ & n_0, n_1, \ldots, n_p \text{ are integers and } \mathbb{N} \models G[\underline{n_0}, \underline{n_1}, \ldots, \underline{n_p}]\} \\ & \cup \{F[n, c] : n \in X\} \cup \{\neg F[n, c] : n \notin X\}. \end{aligned}$$

This theory is consistent by the compactness theorem: observe that every nite subset of T_X is included in a set of the form

$$\begin{aligned} T_Y = &\{G[\underline{n_0}, \underline{n_1}, \ldots, \underline{n_p}] : \\ &p \text{ is an integer, } G[v_0, v_1, \ldots, v_p] \text{ is a formula of } \mathcal{L}_0, \\ &n_0, n_1, \ldots, n_p \text{ are integers and } \mathbb{N} \models G[\underline{n_0}, \underline{n_1}, \ldots, \underline{n_p}]\} \\ &\cup \{F[n, c] : n \in Y\} \cup \{\neg F[n, c] : n \notin Y\}, \end{aligned}$$

where Y is a nite subset of X. The structure $\mathbb{N}$ in which the constant c is interpreted by

$$\prod_{k \in Y} \pi(k),$$

where $\pi(k)$ is the $(k+1)$st prime number, is a model of T_Y. It follows (see Chapter 8, Lemma 8.13) that T_X has a countable model which we will call $\mathcal{M}$. We may even suppose (Chapter 8, Lemma 8.13) that this model is an elementary extension of $\mathbb{N}$. Let us abuse language by denoting the interpretation of c in $\mathcal{M}$ by c. Then

$$X = \{n \in \mathbb{N} : \mathcal{M} \models F[n, c],$$

which shows that X is denable in $\mathcal{M}$.

(c) For any countable elementary extension $\mathcal{M}$ of $\mathbb{N}$, let us consider

$$S(\mathcal{M}) = \{X : X \subseteq \mathbb{N} \text{ and } X \text{ is denable in } \mathcal{M}\}.$$

In part (a) we saw that $S(\mathcal{M})$ is a countable subset of $\mathcal{P}(\mathbb{N})$, and in part (b) that

$$\mathcal{P}(\mathbb{N}) = \bigcup \{S(\mathcal{M}) : \mathcal{M} \succ \mathbb{N} \text{ and } \mathcal{M} \text{ is countable}\}.$$

If λ is the cardinality of the set $\{S(\mathcal{M}) : \mathcal{M} \succ \mathbb{N}$ and $\mathcal{M}$ is countable$\}$, then $\lambda \times \aleph_0 = 2^{\aleph_0}$, from which it follows that $\lambda = 2^{\aleph_0}$. Now if $\mathcal{M}$ and $\mathcal{N}$ are two elementary extensions of $\mathbb{N}$ and if f is an isomorphism from $\mathcal{M}$ onto $\mathcal{N}$, then the image under f of a subset of $\mathbb{N}$ denable in $\mathcal{M}$ will also be denable in $\mathcal{N}$ (by means of the same formula). Thus if $S(\mathcal{M})$ and $S(\mathcal{N})$ are different, then $\mathcal{M}$ and $\mathcal{N}$ cannot be isomorphic. There are therefore $2^{\aleph_0}$ countable elementary extensions of $\mathbb{N}$ that are pairwise not isomorphic.

15. (a) Epimenides cannot be telling the truth because, being a Cretan, he must lie. But if he is lying, it is false that Cretans are liars, so he is telling the truth.

In fact, it is not difcult to counter this argument; rst of all, a liar may occasionally tell the truth; also, it is possible that Epimenides is lying, the truth being that certain Cretans, including himself, are liars.

(b) This barber is a woman; otherwise, one could not respond to the question `does this barber shave himself?' without leading to a contradiction.

Solutions to the exercises for Chapter 7

1. (a) In the solution to this exercise, we will write ε instead of ε_ϕ. Let us verify the axioms of ZF$^-$.

- *Extensionality*. Let x and y be two integers such that, for every integer z, $z \,\varepsilon\, x$ if and only if $z \,\varepsilon\, y$, i.e. such that $z \in \phi(x)$ if and only if $z \in \phi(y)$; it follows that $\phi(x) = \phi(y)$ and, since ϕ is bijective, that $x = y$.
- *Pairs*. Let x and y be two integers; we wish to nd an integer z such that the set (in the intuitive sense) of integers t that satisfy $t \,\varepsilon\, z$ [equivalently, $t \in \phi(z)$] is equal to the pair (in the intuitive sense) $\{x, y\}$; so we need to have $\phi(z) = \{x, y\}$. This denes a unique integer z since ϕ is a bijection.
- *Unions*. Let x be an integer; set $z = \bigcup_{t\in\phi(x)} \phi(t)$ and $y = \phi^{-1}(z)$. We see that, for every integer u, $u \in \phi(y)$ if and only if there exists an integer t such that $t \in \phi(x)$ and $u \in \phi(t)$. When this is re-expressed using the relation ε, it means that, for every u, $u \,\varepsilon\, y$ if and only if there exists t such that $t \,\varepsilon\, x$ and $u \,\varepsilon\, t$. Thus, in the universe $\langle \mathbb{N}, \varepsilon\rangle$, y is the union of the elements of x.
- *Subsets*. Let x be an integer; we seek an integer y such that, for every integer z, $z \in \phi(y)$ if and only if, for every t belonging to $\phi(z)$, t belongs to $\phi(x)$; in other words, for every z, $z \in \phi(y)$ if and only if $\phi(z) \in \wp(\phi(x))$. It is easy to see that $\wp(\phi(x))$ is a nite subset of W, the set of all nite subsets of $\mathbb{N}$. Its inverse image under the bijection ϕ is therefore a nite subset of $\mathbb{N}$, hence is an element of W, which itself has a unique preimage under ϕ. So the set we seek is $y = \phi^{-1}(\bar{\phi}^{-1}(\wp(\phi(x))))$.
- *Replacement*. Let x be an integer and let $F[v_0, v_1]$ be a formula of the language of set theory that is functional in v_0 (in the universe $\langle \mathbb{N}, \varepsilon\rangle$). To simplify the presentation, we have taken, here, a formula without parameters; the presence of parameters would not substantially change the proof that follows. We need to produce an integer y that is `the image of x under F', i.e. is such that, for every integer z, $z \,\varepsilon\, y$ if and only if there exists t such that $t \,\varepsilon\, x$ and $\langle \mathbb{N}, \varepsilon\rangle \models F[t, z]$. Let h denote the partial function from $\mathbb{N}$ into $\mathbb{N}$ dened as follows:

  for all integers n and m,
  $$h(n) = m \text{ if and only if } \langle \mathbb{N}, \varepsilon\rangle \models F[n, m].$$

  (This is a partial function since F is functional.) We then see that by setting $y = \phi^{-1}(\bar{h}(\phi(x)))$, we have the desired set. [We will have noted that the set $\bar{h}(\phi(x))$, the direct image of the nite set $\phi(x)$ under the partial function h, is indeed a nite subset of $\mathbb{N}$, so it does have a unique preimage under ϕ.]
- *Negation of the axiom of innity*. We argue by contradiction and assume that $\langle \mathbb{N}, \varepsilon\rangle$ does satisfy the axiom of innity. So there exist integers a

and f such that

$$\langle \mathbb{N}, \varepsilon \rangle \models \text{`}f \text{ is a mapping from } a \text{ into itself that is injective but not surjective'}.$$

We easily see that the set

$$\{(x, y) \in \phi(a)^2 : \langle \mathbb{N}, \varepsilon \rangle \models \text{`}y = f(x)\text{'}\}$$

is a mapping from $\phi(a)$ into itself that is injective but not surjective; this is impossible since $\phi(a)$ is a nite set.

Next, assume that, for all integers x and y, $x \in \phi(y)$ implies $x < y$. In particular, this implies that $\phi(\mathbf{0})$ is the empty set.

- *Foundation.* Let x be an integer greater than $\mathbf{0}$ (thus distinct from the empty set in $\langle \mathbb{N}, \varepsilon \rangle$). We seek an integer y such that, in $\langle \mathbb{N}, \varepsilon \rangle$, $y \mathrel{\varepsilon} x$ and $y \cap x = \emptyset$ (this last condition means that, for all t, if $t \mathrel{\varepsilon} y$, then $t \not\mathrel{\varepsilon} x$). It sufces to take y to be the least element (in the sense of the usual ordering $\leq$ on $\mathbb{N}$) of $\phi(x)$; note that $\phi(x) \neq \emptyset$ since $x \neq \mathbf{0}$. We do have $y \mathrel{\varepsilon} x$ since $y \in \phi(x)$ and, if $t \mathrel{\varepsilon} y$, then $t < y$; thus, in view of the way y was chosen, $t \notin \phi(x)$, i.e. $t \not\mathrel{\varepsilon} x$.

(b) The fact that ζ is a bijection from the set of nite subsets of $\mathbb{N}$ onto $\mathbb{N}$ is easily proved, as well as the fact that if x and y are two integers, then $x \in \zeta(y)$ implies $x < y$. It then follows from part (a) that $\mathcal{M}_\theta$ is a model of $\mathrm{ZF}^- + \mathrm{AF}$.

(c) It is sufcient to make a slight change in the denition of ζ. Consider the mapping ξ from W into $\mathbb{N}$ dened as follows:

- if x is a nite subset of $\mathbb{N}$ different from $\emptyset$ and from $\{\mathbf{0}\}$, then $\xi(x) = \zeta(x)$;
- $\xi(\emptyset) = \mathbf{1}$;
- $\xi(\{\mathbf{0}\}) = \mathbf{0}$.

The mapping ξ is again a bijection from W onto $\mathbb{N}$ and, from part (a), $\mathcal{M}_{\xi^{-1}}$ is a model of ZF^-; however, it does not satisfy AF since $\mathbf{0} \mathrel{\varepsilon_{\xi^{-1}}} \mathbf{0}$ (see Remark 7.76).

2. Let x be an ordinal and let y be a transitive subset of x that is distinct from x. According to Proposition 7.17 and Corollary 7.22, y, being a transitive set of ordinals, is itself an ordinal. Thus $y \subseteq x$ means that $y \leq x$, i.e. $y = x$ or $y \in x$. But $y = x$ is false by hypothesis, hence $y \in x$.

Conversely, let x be a set in the class On'. There are denitely some ordinals that are not included in x, otherwise, by applying the comprehension axiom to the set $\wp(x)$, the class of ordinals would be a set, which it is not. Let β denote the least ordinal that is not included in x. We can then choose an element $\alpha \in \beta$ (this element will obviously be an ordinal) such that $\alpha \notin x$. Since α is less than β, we have $\alpha \subseteq x$ by denition of β; moreover, α is a transitive set (it is an ordinal). If α were distinct from x, we would conclude, because x is in the class On', that $\alpha \in x$; but α was chosen precisely so this does not happen. Consequently, $\alpha = x$, which proves that x is an ordinal.

3. It sufces to revise the second proof of Theorem 7.69. Consider the class of well-orderings of subsets of x; this is a set, by the axiom of comprehension. The replacement scheme then guarantees that $\Gamma(x)$ is a set. This set is an ordinal (by Proposition 7.17 and Corollary 7.22) since it is clearly a transitive set of ordinals. It cannot be subpotent to x for this would imply that $\Gamma(x) \in \Gamma(x)$, which is absurd since $\Gamma(x)$ is an ordinal. Every ordinal strictly less than $\Gamma(x)$ belongs to $\Gamma(x)$, so is subpotent to x. It follows that $\Gamma(x)$ is the least ordinal that is not subpotent to x. Because it is not subpotent to x, $\Gamma(x)$ cannot be equipotent with any ordinal that is subpotent to x, and hence with any ordinal $\beta < \Gamma(x)$. This conrms that $\Gamma(x)$ is a cardinal.

Observe that in the proof of Theorem 7.69, the set that played the role that is played here by x was an ordinal; however, this fact did not intervene in any way in the proof of that theorem.

If the universe $\mathcal{U}$ satises the axiom of choice, then x (just as any other set) has a cardinality, say λ, that is the greatest cardinal subpotent to x. Since $\Gamma(x)$ is a cardinal and is the least ordinal that is not subpotent to x, we conclude immediately that $\Gamma(x) = \lambda^+$.

4. AC $\Rightarrow$ (a): Assume that a is a set and that I is the set of non-empty subsets of a. Let $(a_i)_{i\in I}$ be the family of sets such that, for every $i \in I$, $a_i = i$. By the denition of I, all the a_i are non-empty, so by AC, the product $\prod_{i\in I} a_i$ is non-empty; let x be an element of this product. Then x is a mapping from I into $\bigcup_{i\in I} a_i$ such that, for every $i \in I$, $x(i) \in a_i$. When we notice that $\bigcup_{i\in I} a_i = a$, we see that x is a mapping from I into a such that, for every non-empty subset of a, $x(i) \in i$, i.e. it is a choice function on a.

We should note that it is possible that a is the empty set and that, in this case, we need not invoke the axiom of choice to prove the existence of a choice function on a; the empty mapping will do perfectly well (the set of non-empty subsets of a is the empty set in this case).

(a) $\Rightarrow$ (b): Let x and y be two sets and let g be a surjection from x onto y. Let ϕ be a choice function on x and dene a map h from y into x in the following way:

$$\text{for every } t \in y, \quad h(t) = \phi(\bar{g}^{-1}(\{t\})).$$

This denition is legitimate since the fact that g is surjective guarantees that, for every element $t \in y$, the inverse image of $\{t\}$ under g is a non-empty subset of x. It is immediate that, for every $t \in y$, $g(h(t)) = t$; thus $g \circ h$ is the identity on y.

(b) $\Rightarrow$ (c): Let a be a set with the property that, for every pair of distinct elements x and y of a, $x \neq \emptyset$ and $y \neq \emptyset$ and $x \cap y = \emptyset$; set $w = \bigcup a$. By hypothesis, for every element $t \in w$, there exists a unique element $x \in a$ such that $t \in x$. So if we set $g(t) = x$, we dene in this way a mapping g from w into a; g is surjective since the empty set does not belong to a. Condition (b) then produces a mapping h from a into w such that $g \circ h$ is the identity on a. Let b be the image of this mapping h and observe that, for every $x \in a$, $h(x) \in x$ [since $x = g(h(x))$]. It follows that, for every $x \in a$, $h(x) \in b \cap x$ and, for every element y of a other than x (and so, by hypothesis, disjoint from x), $h(y) \notin x$. This proves that, for every element x of a, $h(x)$ is the unique element of $b \cap x$. So we have found a set whose intersection with each element of a is a singleton.

(c) $\Rightarrow$ AC: Let $(a_i)_{i \in I}$ be a family of non-empty sets. Set $b_i = \{i\} \times a_i$ for every $i \in I$ and set $a = \{b_i : i \in I\}$. The elements of a are non-empty and pairwise disjoint; so there exists a set b such that, for every $i \in I$, $b \cap b_i$ is a singleton. Set $c = \bigcup_{i \in I} a_i$ and set $b' = b \cap (I \times c)$; b' is then a subset of $I \times c$ and we see that, for every $i \in I$, there exists one and only one element of b' (namely, the unique element of $b \cap b_i$) whose rst projection is i. Thus b' is a mapping from I into c and, for every $i \in I$, $b(i)$, which is the second projection of the unique element of $b \cap b_i$, belongs to a_i. So b' belongs to $\prod_{i \in I} a_i$, which is consequently non-empty.

AC $\Rightarrow$ (d): Let a and b be two sets. We use Zermelo's theorem (see Theorem 7.41). We know that there exist ordinals α and β that are equipotent with a and b, respectively. Besides, Corollary 7.22 tells us that either α is included in β (in which case α is subpotent to β, which implies that a is subpotent to b) or else β is included in α (and in this case, β is subpotent to α and b is subpotent to a).

(d) $\Rightarrow$ AC: Once again, we will replace AC by its equivalent, Zermelo's theorem. We will show that an arbitrary set x can be well-ordered. We use the Hartog cardinal of x (the least ordinal that is not subpotent to x) dened in Exercise 3 and denoted by $\Gamma(x)$. Since (d) is satised by hypothesis, it must be that x is subpotent to $\Gamma(x)$. Let ϕ be an injective map from x into $\Gamma(x)$ and set

$$r = \{(u, v) \in x \times x : \phi(u) \subseteq \phi(v)\}.$$

It is routine to verify that r is a well-ordering of x; we have done nothing more than `import' the well-ordering of $\Gamma(x)$ using the injection ϕ.

5. It is obvious that in ZF, Zermelo's theorem (so also the axiom of choice) implies each of the statements (a), (b), and (c) since every set is then well-orderable. In ZF again, it is trivial that (a) implies (b) and it is not hard to check that (b) implies (a); to do this, let (x, R) be a well-ordered set and let α be the unique ordinal that is isomorphic to (x, R). The isomorphism between α and x induces a bijection between $\wp(\alpha)$ and $\wp(x)$ which allows us to `transfer' a well-ordering on $\wp(\alpha)$ [which exists, by (b)] into a well-ordering on $\wp(x)$. It is also true in ZF that (c) implies (b). For this, it sufces to prove that, for every ordinal α, the set $\wp(\alpha)$ is totally ordered. Set

$$r = \{(u, v) \in \wp(\alpha)^2 : u = v \text{ or the least element of the symmetric difference } u \Delta v \text{ belongs to } u\}.$$

It is elementary to show that r is a total order on $\wp(\alpha)$.

To conclude, we will prove that in ZF + AF, statement (a) implies Zermelo's theorem (and hence the axiom of choice). First, observe that in ZF + AF, Zermelo's theorem is equivalent to the following statement:

$$\text{for every ordinal } \alpha, \quad V_\alpha \text{ is well-orderable.} \tag{Ü}$$

Indeed, it is clear that this statement follows from Zermelo's theorem. Conversely, if it is satised, then for any set x, we can choose an ordinal α such that $x \in V_\alpha$ since the axiom of foundation is satised. But then $x \subseteq V_\alpha$ because V_α is transitive and, as we can well-order V_α by hypothesis, the restriction of such a well-ordering to x will be a well-ordering of x.

We will now argue by contradiction by supposing that (a) is satised simultaneously with the negation of (Ü). We may then consider the least ordinal α such that V_α is not well-orderable and we see that α must be a limit ordinal, for if $\alpha = \beta + \mathbf{1}$, then $V_\alpha = \wp(V_\beta)$; this would mean that V_β is well-orderable while $\wp(V_\beta)$ is not: this contradicts (a).

So we know that for every ordinal $\beta < \alpha$, there is a well-ordering of V_β. We will use (a) to show that there exists a family $(s(\beta) : \beta < \alpha)$, where, for every $\beta < \alpha$, $s(\beta)$ is a well-ordering of V_β. For each $\beta < \alpha$, set

$$X_\beta = \{\gamma : \gamma \text{ is the ordinal of a well-ordering of } V_\beta\}.$$

(This is a set; see the proof of Theorem 7.69.) Next, let $X = \bigcup_{\beta \in \alpha} X_\beta$ and let δ be the least upper bound of X. Let r be a well-ordering of $\wp(\delta)$ [by (a), such exists].

The family $(s(\beta) : \beta < \alpha)$ can now be dened by induction.

- If $\beta = \mathbf{0}$, there is no problem; $s(\beta) = \mathbf{0}$.
- If β is a limit ordinal, we know that $V_\beta = \bigcup_{\gamma \in \beta} V_\gamma$. It is then easy to verify that the relation $s(\beta)$ dened on V_β as follows is a well-ordering: for all elements x and y of V_β,

  $$x\, s(\beta)\, y \quad \text{if and only if} \begin{cases} rk(x) < rk(y), & \text{or} \\ rk(x) = rk(y) = \gamma & \text{and } x\, s(\gamma)\, y \end{cases}$$

  [note that we must have $\gamma < \beta$, so $s(\gamma)$ is already dened].
- If β is a successor ordinal, say $\beta = \gamma + \mathbf{1}$, let δ_γ denote the unique ordinal such that $(V_\gamma, s(\gamma))$ is isomorphic to $(\delta_\gamma, \in)$ and let $f\gamma$ denote the unique isomorphism from $(V_\gamma, s(\gamma))$ onto $(\delta_\gamma, \in)$. It is clear that $\delta_\gamma \leq \delta$ and that f_γ induces a bijection g from V_β [which is equal to $\wp(V_\gamma)$] onto $\wp(\delta_\gamma)$ [which

is included in $\wp(\delta)$]. We may then dene $s(\beta)$ by transferring the ordering $r \restriction \wp(\delta_\gamma)$ onto V_β by declaring, for elements x and y of V_β, that $x\ s(\beta)\ y$ if and only if $g(x)\ r\ g(y)$.

- Once the family $(s(\beta) : \beta < \alpha)$ is available, we can dene a well-ordering s on V_α as follows: for all elements x and y of V_α,

$$x\,s\,y \quad \text{if and only if} \begin{cases} rk(x) < rk(y), & \text{or} \\ rk(x) = rk(y) = \beta & \text{and } x\,s(\beta)\,y. \end{cases}$$

So we have arrived at a contradiction.

6. The initials CB will refer to the CantorñBernstein theorem (Theorem 7.43).

(1) $\Rightarrow$ (2): Suppose that x is a denumerable subset of a, that ϕ is a bijection from ω onto x, that c is the image under ϕ of the set of even integers, and that b is the image under ϕ of the set of odd integers. We dene a mapping f from a into a as follows: the restriction of f to $a - x$ is the identity and, for every $t \in x$, $f(t) = \phi(2\phi^{-1}(t))$. It is easy to verify that f is a bijection from a onto $a - b$ and that b is a denumerable subset of a.

(2) $\Rightarrow$ (3): Suppose that b is an arbitrary denumerable set, that x is a denumerable subset of a, that f is a bijection from a onto $a - x$, and that ϕ is a bijection from b onto x. The mapping g from $a \cup b$ into a which agrees with ϕ on b and with f on $a - (a \cap b)$ is obviously an injection. Also, the identity is obviously an injection from a into $a \cup b$. It follows from CB that a and $a \cup b$ are equipotent.

(3) $\Rightarrow$ (4): Suppose that x is a nite set and that b is a denumerable set that includes x (for example, $\omega \cup x$). We have a bijection from $a \cup b$ into a whose restriction to $a \cup x$ is an injection from $a \cup x$ into a. Since the identity is an injection from a into $a \cup x$, it follows from CB that a and $a \cup x$ are equipotent.

(4) $\Rightarrow$ (5): Suppose that x is a nite subset of a, that y is a set that is disjoint from a and equipotent with x, and that f is a bijection from $a \cup y$ onto a. Let z denote the direct image of y under f and set

$$t = x \cap z, \qquad u = x - t, \qquad v = z - t, \quad \text{and} \quad w = a - (x \cup z).$$

It is easy to verify that the sets t, u, v, and w constitute a partition of a, that x and z are equipotent, and that u and v are equipotent. Choose a bijection ϕ from u onto v and consider the map h from a into a that agrees with the identity on $w \cup t$, with ϕ on u, and with ϕ^{-1} on v. The map h is a bijection from a onto a that interchanges x and z; the composition $g = h \circ f$ is a bijection from $a \cup y$ onto y such that the image of y is x and the image of a is $a - x$. The restriction of g to a is a bijection from a onto $a - x$, so these two sets are equipotent.

(5) $\Rightarrow$ (6): This is obvious; any non-zero integer will do.

(6) $\Rightarrow$ (7): Since a is non-empty, it follows from (6) that, for every singleton subset y of a, a and $a - y$ are equipotent. Consider such a subset, for example $y = \{t\}$, where $t \in a$ and let f be a bijection from a onto $a - y$. Now let x be

an arbitrary set containing a single element, say $x = \{u\}$; if $u \in a$, it is rather obvious that a and $a \cup x$ are equipotent; if $u \notin a$, the mapping g from $a \cup x$ into a that agrees with f on a and sends u to t is a bijection. This shows that a and $a \cup x$ are equipotent and proves (7) [with $n = 1$, i.e. in fact (8)!].

(7) $\Rightarrow$ (8): Suppose that $n > \mathbf{0}$ is an integer provided by (7), that t is an arbitrary set, and that y is a set that does not contain t and whose cardinality is $n - 1$. Set $x = y \cup \{t\}$; we then have that $a \subseteq a \cup \{t\} \subseteq a \cup x$ and, according to (7), that a and $a \cup x$ are equipotent. This proves, invoking CB, that a and $a \cup \{t\}$ are equipotent.

(8) $\Rightarrow$ (9): Choose a set u such that $u \notin a$. Let f be a bijection from $a \cup \{u\}$ onto a and set $t = f\{u)$; we see that the restriction of f to a is a bijection from a onto $a - \{t\}$, so these two sets are equipotent.

(9) $\Rightarrow$ (10): Suppose that t is a set with $t \in a$ and that a is equipotent with $b = a - \{t\}$. Since a is non-empty and b is equipotent with a, b is non-empty. Also, $b \neq a$ since $t \in a$ but $t \notin b$. Thus b is a subset of a that is non-empty, distinct from a, and equipotent with a.

(10) $\Rightarrow$ (11): This is obvious since `equipotent' is stronger than `subpotent'.

(11) $\Rightarrow$ (1): Suppose the set b is such that $b \subseteq a$, $b \neq \emptyset$, $b \neq a$, and that a is subpotent to b. Let f be an injection from a into b. By induction on the integers, we dene a sequence $(x_n)_{n \in \omega}$ of element of a as follows: x_0 is an arbitrary element of $a - b$ (there are some) and, for every $n \in \omega$, $x_{n+1} = f(x_n)$. Let c be the image of this sequence, i.e.

$$c = \{t \in a : (\exists n \in \omega)\,(t = x_n)\};$$

we will show that c is a denumerable set. To do this, we must establish, for any distinct integers n and m, that $x_n \neq x_m$. Suppose this is false and let k denote the least element of the (consequently non-empty) set

$$Z = \{n \in \omega : (\exists m \in \omega)\,(m > n \wedge x_m = x_n)\};$$

let h be an integer such that $h > k$ and $x_h = x_k$. If $k \neq 0$, we have $x_h = f(x_{h-1}) = x_k = f(x_{k-1})$ and, since f is injective, $x_{h-1} = x_{k-1}$; this proves that $k - 1 \in Z$ and contradicts the denition of k. It follows that $k = 0$, but this also leads to a contradiction, for if $x_0 \notin b$ while x_h, which belongs to the image of f since $h > 0$, does belong to b, the equality $x_0 = x_k$ is violated.

The alert reader will have understood that the sets which satisfy one, hence all, of these eleven equivalent conditions are the innite sets. To be precise, these sets are innite **in the strong sense**. A set is innite **in the weak sense** if it is not equipotent with an integer. These two notions coincide in universes that satisfy the axiom of choice; but in the absence of AC, we cannot prove that a set which is not equipotent with an integer must then include a denumerable subset.

7. For this exercise and the next three, we will be content to provide the answers, occasionally accompanied by sketchy hints. The reader can (protably) supply complete proofs.

We will make use of the following facts:

- $\mathsf{card}(\wp(\mathbb{N})) = \mathsf{card}(\mathbb{N}^{\mathbb{N}}) = \mathsf{card}(\mathbb{Q}^{\mathbb{N}}) = \mathsf{card}(\mathbb{R}) = \mathsf{card}(\mathbb{R}^{\mathbb{N}}) = 2^{\aleph_0}$;
- the set $\wp_f(\mathbb{N})$ of nite subsets of $\mathbb{N}$ is denumerable;
- the set $\wp_\infty(\mathbb{N})$ of innite subsets of $\mathbb{N}$ has cardinality $2^{\aleph_0}$.

$\mathsf{card}(x_1) = 2^{\aleph_0}$: With each innite subset of $\mathbb{N}$, we associate the sequence of its elements taken in increasing order; this constitutes a bijection from $\wp_\infty(\mathbb{N})$ onto x_1, the set of strictly increasing sequences of integers.

$\mathsf{card}(x_2) = 2^{\aleph_0}$: x_2 is the set of bounded sequences of integers; it includes the set $2^{\mathbb{N}}$.

$\mathsf{card}(x_3) = 2^{\aleph_0}$: x_3 is the set of strictly increasing sequences of rationals; it includes x_1.

$\mathsf{card}(x_4) = 2^{\aleph_0}$: x_4 is the set of bounded sequences of rationals; it includes x_2.

$\mathsf{card}(x_5) = 2^{\aleph_0}$: With each element $f \in x_1$, we associate the map g from $\mathbb{N}$ into $\mathbb{Q}$ which, with each integer n, assigns the value $-1/(1 + f(n))$; g is a bounded, strictly increasing sequence of rationals, so we have an injection from x_1 into x_5.

$\mathsf{card}(x_6) = \aleph_0$: For every $n \in \mathbb{N}$, set

$$z_n = \{f \in \mathbb{Q}^{\mathbb{N}} : (\forall p \in \mathbb{N})(n \leq p \Rightarrow f(n) = f(p))\};$$

thus, $x_6 = \bigcup_{n\in\mathbb{N}} z_n$. Now each z_n is equipotent with $\mathbb{Q}^{n+1}$, so it follows that x_6 is equipotent with $\bigcup_{n\in\mathbb{N}} \mathbb{Q}^{n+1}$ which, by item (3) from Theorem 7.61, is a denumerable set.

$\mathsf{card}(x_7) = 2^{\aleph_0}$: x_7 is the set of unbounded sequences of reals; it includes x_1.

8. $\mathsf{card}(E_0) = \mathsf{card}(\mathbb{N}^{\mathbb{N}}) = 2^{\aleph_0}$.

$\mathsf{card}(E_1) = (2^{\aleph_0})^{\aleph_0} = 2^{\aleph_0^2} = 2^{\aleph_0}$.

$\mathsf{card}(E_2) = 2^{\aleph_0}$: Use the fact that the set of strictly increasing sequences of integers has cardinality $2^{\aleph_0}$ (see Exercise 7); with such a sequence u, we associate the sequence of rationals v dened by $v(n) = 1/(1 + u(n))$ for all n.

$\mathsf{card}(E_3) = 2^{\aleph_0}$: $E_2 \subseteq E_3 \subseteq E_0$.

$\mathsf{card}(E_4) = 2^{\aleph_0}$: $E_2 \subseteq E_4 \subseteq E_0$.

$\mathsf{card}(E_5) = 2^{\aleph_0}$: Every strictly increasing sequence of integers is an unbounded sequence of rationals and $E_5 \subseteq E_0$.

$\mathsf{card}(E_6) = \mathsf{card}(E_1) = 2^{\aleph_0}$.

$\mathsf{card}(E_7) = \mathsf{card}(E_6) = 2^{\aleph_0}$: If two continuous maps from $\mathbb{R}$ into $\mathbb{R}$ have the same restriction to $\mathbb{Q}$, then they are equal.

$\mathsf{card}(E_8) = 2^{\aleph_0}$: E_8 is obviously subpotent to $\mathbb{R} \times \mathbb{R}$.

$\mathsf{card}(E_9) = 2^{\aleph_0}$: We invoke the following classical result: every open subset of $\mathbb{R}$ is the union of a family of pairwise disjoint open intervals that is indexed by the integers; we conclude from this that there exists an injection from E_9 into $(E_8)^{\mathbb{N}}$.

9. A map from ω into ω will be called a sequence.

card$(a_1) = 1$: a_1 contains only the empty sequence.

card$(a_2) = 2^{\aleph_0}$: a_2 is equal to ω^ω [take $p = f(n)$].

card$(a_3) = 2^{\aleph_0}$: a_3 is the set of sequences that assume the value $\mathbf{0}$ at least once; it includes the set of sequences f such that $f(\mathbf{0}) = \mathbf{0}$, which is equipotent with $\omega^{\omega-\{\mathbf{0}\}}$.

card$(a_4) = 2^{\aleph_0}$: $a_2 \subseteq a_4 \subseteq \omega^\omega$ (in fact, $a_4 = \omega^\omega$).

card$(a_5) = 2^{\aleph_0}$: a_5 is the set of bounded sequences; it includes 2^ω.

card$(a_6) = 2^{\aleph_0}$: $a_3 \subseteq a_6 \subseteq \omega^\omega$ (in fact, $a_6 = a_3$).

card$(b_1) = \mathbf{0}$: Every sequence satises the negation of the property in question.

card$(b_2) = 2^{\aleph_0}$: b_2 is equal to ω^ω [take $p = f(n)$].

card$(b_3) = \mathbf{0}$: Every sequence satises the negation of the property in question.

card$(b_4) = 2^{\aleph_0}$: $b_2 \subseteq b_4 \subseteq \omega^\omega$ (in fact, $b_4 = \omega^\omega$).

card$(b_5) = 2^{\aleph_0}$: $b_5 = \omega^\omega$ (take $p = \mathbf{0}$).

card$(b_6) = 2^{\aleph_0}$: b_6 is the set of unbounded sequences; it includes the set of strictly increasing sequences (x_1 from Exercise 7).

10. card$(y_1) = \mu$: With each element $x \in b$, we associate the map from a into b whose value at every point is x; this denes a bijection from b onto y_1.

card$(y_2) = \mu$: y_2 is equal to y_1 since, for every $x \in \wp(a)$ and for every $f \in b^a$, we have card$(\bar{f}(x)) \leq$ card$(\bar{f}(a))$; this shows that $y_1 \subseteq y_2$. Conversely, if $f \in y_2$, we have card$(\bar{f}(a)) \leq 1$, hence card$(\bar{f}(a)) = 1$ because a is non-empty; this proves $y_2 \subseteq y_1$.

card$(y_3) = 2^\lambda$: For every map f from a into b, we have $\bar{f}^{-1}(b) = a$, thus card$(\bar{f}^{-1}(b)) = \lambda$; this shows that $y_3 = b^a$ and that card$(y_3) = \mu^\lambda = 2^\lambda$ (since $2 \leq \mu \leq \lambda$).

card$(y_4) = 2^\lambda$: Let x and y be distinct elements of b (b is innite). We have $\{x, y\}^a \subseteq y_4 \cup y_1 \subseteq b^a$, therefore $2^\lambda \leq \sup($card$(y_4), \mu) \leq \mu^\lambda = 2^\lambda$; also, $\mu < 2^\lambda$. This proves the result.

card$(y_5) = \lambda$: Since g is injective, the cardinality of $\bar{g}(b)$ is that of b, i.e. is equal to μ. Because card$(a) = \lambda > \mu$, the difference $a - \bar{g}(b)$ has cardinality λ (Proposition 7.74).

card$(y_6) = 2^\lambda$: When we associate with each element of y_6 its restriction to the set y_5, we obtain a bijection from y_6 onto b^{y_5}.

card$(y_7) = 2^\lambda$: Every element of y_6 is a surjective map from a onto b and is thus an element of y_7; it follows that $y_6 \subseteq y_7 \subseteq b^a$.

11. (a) Let n be a non-zero integer and set $b = a \times n$. We have card$(b) = \lambda$ (Corollary 7.72). So we can choose a bijection h from b onto a. For each $i \in \{1, 2, \ldots, n\}$, let a_i denote the image under h of the subset $a \times \{i - 1\}$ of b. The cardinality of each a_i is obviously λ and the a_i ($1 \leq i \leq n$) constitute a partition of a.

(b) For every $x \in \wp(a)$, we have $\mathsf{card}(a) = \sup(\mathsf{card}(x), \mathsf{card}(a - x))$ by Corollary 7.72. Thus, if $x \in \wp^*(a)$, $\mathsf{card}(x) = \lambda$.

(c) Consider a partition of a into three subsets a_1, a_2, a_3, each of cardinality λ [which is justied by part (a)]. The map $x \mapsto a_2 \cup x$ from $\wp(a_1)$ into $\wp(a)$ is injective and its image is a subset of $\wp^*(a)$, for if $x \subseteq a_1$, then the inclusions $a_2 \subseteq a_2 \cup x \subseteq a$ and $a_3 \subseteq a - (a_2 \cup x) \subseteq a$ prove that

$$\mathsf{card}(a_2 \cup x) = \mathsf{card}(a - (a_2 \cup x)) = \lambda.$$

Therefore, $\mathsf{card}(\wp(a_1)) \leq \mathsf{card}(\wp^*(a)) \leq \mathsf{card}(\wp(a))$. Conclusion:

$$\mathsf{card}(\wp^*(a)) = 2^\lambda.$$

(d) Let $a_1 \in \wp^*(a)$ and set $b = a - a_1$. Since $\mathsf{card}(b) = \lambda$, we can nd a partition of b into two sets b_1 and b_2 that each have cardinality λ [by part (a)]. The sets a_1, b_1, and b_2 constitute a partition of a. Choose a bijection ϕ from b_1 onto b_2 and dene a mapping h_{a_1} from a into a as follows:

- the restriction of h_{a_1} to a_1 is the identity;
- the restriction of h_{a_1} to b_1 is the map ϕ;
- the restriction of h_{a_1} to b_2 is the map ϕ^{-1}.

It is easy to verify that h_{a_1} is a bijection from a onto a whose set of xed points is a_1 [for $x \in b_1$, $h_{a_1}(x) \in b_2$, hence $h_{a_1}(x) \neq x$; the same argument applies if $x \in b_2$].

(e) The preceding question shows that the map $a_1 \mapsto h_{a_1}$ is an injection from $\wp^*(a)$ into the set $S(a)$ of bijections from a onto a. Since $S(a)$ is included in the set of all maps from a into a, it follows that

$$2^\lambda \leq \mathsf{card}(S(a)) \leq \lambda^\lambda = 2^\lambda.$$

So the cardinality of the set of bijections from a onto a is 2^λ.

(f) Let $b \in \wp^*(a)$ and let X be the set of bijections from a onto a whose restriction to b is the identity on b. We dene a bijection from X onto $S(a - b)$ (the set of bijections from $a - b$ onto $a - b$) by associating, with each element of X, its restriction to $a - b$. The cardinality of X is therefore that of $S(a - b)$, which is 2^λ since $\mathsf{card}(a - b) = \lambda$ [see part (e)].

(g) With each bijection f from a onto a, we can associate an injective mapping ϕ_f from a into $\wp(a)$ by setting $\phi_f(x) = \{f(x)\}$ for every $x \in a$. It is immediate that the map $f \mapsto \phi_f$ is an injection from $S(a)$ into the set of injective mappings from a into $\wp(a)$, which, in turn, is included in $\wp(a)^a$. But $\mathsf{card}(S(a)) = 2^\lambda$ and $\mathsf{card}(\wp(a)^a) = (2^\lambda)^\lambda = 2^{\lambda \times \lambda} = 2^\lambda$. The cardinal we seek is therefore 2^λ.

12. By induction on $\beta \in \alpha$, we will dene a family $(f_\beta)_{\beta \in \alpha}$ of injections from X_β into λ such that for all ordinals $\beta \in \alpha$ and $\gamma \in \alpha$, if $\beta < \gamma$, then $f_\beta \subseteq f_\gamma$.

- f_0 is an arbitrary injection from X_0 into λ [there are some, since $\mathsf{card}(X_0) < \lambda$].
- If $\beta \in \alpha$ and $\beta = \gamma + \mathbf{1}$, then f_β is dened as follows: the restriction of f_β to X_γ is f_γ and the restriction of f_β to $X_{\gamma+\mathbf{1}} - X_\gamma$ is an injection from $X_{\gamma+\mathbf{1}} - X_\gamma$ into λ whose image is disjoint from that of f_γ. Such an injection exists since the image of f_γ is a subset of λ whose cardinality is $\mathsf{card}(X_\gamma) < \lambda$, hence $\mathsf{card}(\lambda - \mathsf{Im}(f_\beta)) = \lambda > \mathsf{card}(X_{\gamma+\mathbf{1}} - X_\gamma)$; besides, the axiom of choice is satised.
- It is easy to verify that the family of mappings just dened has the desired properties. The map $f = \bigcup_{\beta\in\alpha} f_\beta$ is an injection from $\bigcup_{\beta\in\alpha} X_\beta$ into λ.

  Conclusion: $\mathsf{card}(\bigcup_{\beta\in\alpha} X_\beta) \leq \lambda$.

13. By denition, $\sum_{\alpha\in\kappa} \lambda_a = \mathsf{card}(\bigcup_{\alpha\in\kappa}(\lambda_\alpha \times \{\alpha\}))$. It follows from item (2) of Corollary 7.72 that $\sum_{\alpha\in\kappa} \lambda_a \leq \sup(\kappa, \sup_{\alpha\in\kappa} \lambda_\alpha)$. To obtain the reverse inequality, we need the following two remarks:

- The mapping $\alpha \mapsto (\mathbf{0}, \alpha)$ is an injection from κ into $\bigcup_{\alpha\in\kappa}(\lambda_\alpha \times \{\alpha\})$.
- The mapping from $\bigcup_{\alpha\in\kappa} \lambda_\alpha = \sup_{\alpha\in\kappa} \lambda_\alpha$ into $\bigcup_{\alpha\in\kappa}(\lambda_\alpha \times \{\alpha\})$ which, with each element x, associates the unique pair (x, α) such that $\alpha \in \kappa$, $x \in \lambda_\alpha$ and for all $\beta < \alpha$, $x \notin \lambda_\beta$, is also injective.

  It follows that $\kappa \leq \sum_{\alpha\in\kappa} \lambda_\alpha$ and $\sup_{\alpha\in\kappa} \lambda_\alpha \leq \sum_{\alpha\in\kappa} \lambda_\alpha$.
  Finally, $\sum_{\alpha\in\kappa} \lambda_\alpha = \sup(\kappa, \sup_{\alpha\in\kappa} \lambda_\alpha)$.

14. (a) We have $2 \leq \mu \leq \lambda$, hence $2^\lambda \leq \mu^\lambda \leq \lambda^\lambda = 2^\lambda$; this proves that $2^\lambda = \mu^\lambda = \lambda^\lambda$. Moreover, $\aleph_\mathbf{0} \leq \mu \leq \lambda$, hence $\lambda^{\aleph_\mathbf{0}} \leq \lambda^\lambda = 2^\lambda$. So it sufces to show that $2^\lambda \leq \lambda^{\aleph_\mathbf{0}}$. To do this, we verify that

$$2^\lambda = \prod_{n\in\omega} 2^{\lambda_n}, \tag{á}$$

and we notice that, for every $n \in \omega$, $2^{\lambda_n} \leq \lambda$.

To prove (á), consider a family $(X_n)_{n\in\omega}$ of pairwise disjoint sets such that, for every n, X_n has cardinality λ_n; set $X = \bigcup_{n\in\omega} X_n$. The cardinality of X is λ, hence that of $\wp(X)$ is 2^λ. With each $Y \subseteq X$, we associate the sequence $(Y_n)_{n\in\omega}$ such that, for all n, $Y_n = Y \cap X_n$. In this way, we have a mapping from $\wp(X)$ into $\prod_{n\in\omega} \wp(X_n)$ that is clearly bijective (since the X_n are pairwise disjoint); this proves (á).

(b) Let γ be a cardinal.

- If $\aleph_\mathbf{0} \leq \gamma \leq \lambda$, then

$$\lambda^{\aleph_\mathbf{0}} \leq \lambda^\gamma \leq \lambda^\lambda = 2^\lambda;$$

  but we have just seen that $2^\lambda \leq \lambda^{\aleph_\mathbf{0}}$, which proves the equalities $\lambda^{\aleph_\mathbf{0}} = \lambda^\gamma = \lambda^\lambda$.
- If γ is greater than or equal to λ, we have $2 \leq \lambda \leq \gamma$, hence

$$2^\gamma \leq \lambda^\gamma \leq \gamma^\gamma = 2^\gamma.$$

(c) It sufces to choose $\alpha = \delta = \lambda$, $\beta = 2^{\lambda}$, and $\gamma = \aleph_0$. Cantor's theorem (Theorem 7.49) tells us that $\alpha < \beta$; the fact that $\mu \geq \aleph_0$ and $\lambda \geq 2^{\mu} \geq 2^{\aleph_0} > \aleph_0$ implies $\gamma < \delta$. Now let us compute α^{γ} and β^{δ}:

- $\alpha^{\gamma} = \lambda^{\aleph_0} = \lambda^{\lambda} = 2^{\lambda}$ [from part (b)];
- $\beta^{\delta} = (2^{\lambda})^{\lambda} = 2^{\lambda \times \lambda} = 2^{\lambda}$.

The desired equality follows.

15. (a) For every ordinal α, α is conal with α; note that the identity is a strictly increasing, unbounded mapping from α into α (this is true even for $\alpha = \mathbf{0}$ for the empty mapping also has these properties). So the relation is reexive. To prove that it is transitive, consider three ordinals α, β, and γ such that α is conal with β and β is conal with γ; so we have mappings $f : \alpha \to \beta$ and $g : \beta \to \gamma$ that are strictly increasing and without strict upper bounds. Then $g \circ f$ is a strictly increasing map from α into γ. Let us prove that its image does not have a strict upper bound. Let δ be an element of γ; we can nd an element $\xi \in \beta$ such that $g(\xi) \geq \delta$, and then nd an element $\zeta \in \alpha$ such that $f(\zeta) \geq \xi$. Since g is increasing, we have

$$(g \circ f)(\zeta) = g(f(\zeta)) \geq g(\xi) \geq \delta.$$

The ordinal ω is conal with $\aleph_{\omega}$; the map $n \mapsto \aleph_n$ from ω into $\aleph_{\omega}$ is strictly increasing with no strict upper bound. But $\aleph_{\omega}$ is not conal with ω since there cannot exist an injective map (so certainly not a strictly increasing one) from $\aleph_{\omega}$ into ω. The relation 'is conal with' is therefore not symmetric.

The ordinals with which $\mathbf{1}$ is conal are the successor ordinals; if $\alpha = \beta + \mathbf{1}$, then the map f from $\mathbf{1}$ $(= \{\mathbf{0}\})$ into α dened by $f(\mathbf{0}) = \beta$ is strictly increasing and its image is not strictly bounded (for if $\gamma \in \alpha$, then $f(\mathbf{0}) = \beta \geq \gamma$); thus, $\mathbf{1}$ is conal with α. Conversely, if $\mathbf{1}$ is conal with the ordinal α and if f is a map from $\mathbf{1}$ into α whose image is not strictly bounded, then (because $\mathbf{0}$ is the only element of $\mathbf{1}$) it must be that $f(\mathbf{0}) \geq \delta$ for every $\delta \in \alpha$. This means that $f(\mathbf{0})$ must be the greatest element in α; this situation is possible only if α is a successor.

(b) If α and β are ordinals and β is conal with α, then the existence of a strictly increasing map from β into α proves that β is less than or equal to α. The class of ordinals β such that β is conal with α is therefore the set $\{\beta \in \alpha + \mathbf{1} : \beta \text{ is conal with } \alpha\}$ (axiom of comprehension). This set is not empty since it contains α; its least element, $cof(\alpha)$, is therefore less than or equal to α.

To prove that $cof(\alpha)$ is regular, we must show that $cof(cof(\alpha)) = cof(\alpha)$. From what we have just observed, $cof(cof(\alpha)) \leq cof(\alpha)$ so it remains to prove the reverse inequality. So suppose that γ is an ordinal that is conal with $cof(\alpha)$; then by transitivity, γ is conal with α. Because $cof(\alpha)$ is the least ordinal that is conal with α, we have $cof(\alpha) \leq \gamma$. This shows that $cof(\alpha) \leq cof(cof(\alpha))$.

(c) First suppose that $\beta \geq \mathit{cof}(\alpha)$. Let f be a strictly increasing map from $\mathit{cof}(\alpha)$ into α whose image is not strictly bounded. Then the map g from β into α that agrees with f on the subset $\mathit{cof}(\alpha)$ of β and that is equal to $\mathbf{0}$ elsewhere is not strictly bounded in α.

Conversely, suppose there exists a map f from β into α whose image is not strictly bounded. This means that the upper bound of the image of f is the ordinal α. Let δ be the least ordinal in the set

$$\{\gamma \leq \beta : \sup_{\xi \in \gamma} f(\xi) = \alpha\}.$$

(This set is not empty since, as we have just seen, it contains β.) We then dene a map g from δ into α as follows: for every $\gamma \in \delta$,

$$g(\gamma) = \sup_{\xi \in \gamma} f(\xi).$$

It is easy to check that g is increasing. The denition of δ guarantees, on the one hand, that the values assumed by g are in α and, on the other hand, that its image has no strict upper bound in α. Let X denote the image of g and let h be the mapping from X into δ whose value, for $x \in X$, is the least ordinal ξ such that $x = g(\xi)$. One is easily convinced that h is an isomorphism from X (with the order $\in$) onto its image, which is some subset Y of δ. Y is isomorphic to a unique ordinal $\sigma \leq \delta$ by Remark 7.23. Let ϕ denote the isomorphism from σ onto Y. We see that $h^{-1} \circ \phi$, an isomorphism from σ onto X, is a strictly increasing map from σ into α whose image is not strictly bounded. We conclude that σ is conal with α. As a result, β, which satises $\beta \geq \delta \geq \sigma$, is greater than or equal to the conality of α.

(d) It follows from part (a) that α is a successor ordinal if and only if $\mathit{cof}(\alpha) = \mathbf{1}$; thus $\mathbf{1}$ is the only successor ordinal that can be regular, and it is indeed regular since $\mathbf{0}$ is not conal with $\mathbf{1}$. So the unique regular successor ordinal is a cardinal. Since $\mathbf{0}$ is conal with $\mathbf{0}$, it is regular; it too is a cardinal. It remains to consider the regular limit ordinals. Let α be such an ordinal and let λ be its cardinality. Of course, $\lambda \leq \alpha$; we can also choose a bijection f from λ onto α. Since α is a limit ordinal, the image of f is not strictly bounded in α. Because of question (c), we may conclude that $\lambda \geq \mathit{cof}(\alpha) = \alpha$. The conclusion is that $\alpha = \lambda$ so α is a cardinal. Thus, every regular ordinal is a cardinal.

Let λ be a cardinal that is a regular ordinal and let X be a subset of λ such that $\mathrm{card}(X) < \lambda$. With the well-ordering $\in$, X is isomorphic to an ordinal $\delta \leq \lambda$ by Remark 7.23; but this inequality must be strict since $\mathrm{card}(\delta) = \mathrm{card}(X) < \lambda$ and λ is itself a cardinal. Let ϕ be an isomorphism from δ onto X; then ϕ is a strictly increasing map from $\delta < \lambda$ into λ, which is regular by hypothesis. It follows immediately that ϕ is strictly bounded in λ or, equivalently, that the least upper bound of X is strictly less than λ. Thus,

λ is a regular cardinal in the sense of Denition 7.87. Conversely, suppose that λ is a regular cardinal in the sense of Denition 7.87 and consider an ordinal $\alpha < \lambda$ and a strictly increasing map f from α into λ. The image of f is a subset Y of λ whose cardinality is strictly less than λ (since its cardinality is the same as that of α). It follows that the least upper bound of Y is an element of λ; this shows that the image of f is strictly bounded in λ, so α is not conal with λ. Thus λ is a regular ordinal.

(e) We now assume that the universe satises the axiom of choice. Let α be an ordinal and let λ be the conality of $\aleph_{\alpha+\mathbf{1}}$. There exists a strictly increasing map f from λ into $\aleph_{\alpha+\mathbf{1}}$ whose image is not strictly bounded. For every $\beta \in \lambda$, $f(\beta) \in \aleph_{\alpha+\mathbf{1}}$, i.e. $f(\beta) < \aleph_{\alpha+\mathbf{1}}$, thus $\mathsf{card}(f(\beta)) \leq \aleph_\alpha$. Moreover, since f is not strictly bounded in $\aleph_{\alpha+\mathbf{1}}$ which is a limit ordinal, we have

$$\aleph_{\alpha+\mathbf{1}} = \sup_{\beta\in\lambda} f(\beta) = \bigcup_{\beta\in\lambda} f(\beta).$$

We conclude that

$$\mathsf{card}\Big(\bigcup_{\beta\in\lambda} f(\beta)\Big) = \aleph_{\alpha+\mathbf{1}}.$$

But we also have (Corollary 7.72, with AC)

$$\mathsf{card}\Big(\bigcup_{\beta\in\lambda} f(\beta)\Big) \leq \sup\Big(\lambda,\ \sup_{\beta\in\lambda}\mathsf{card}(f(\beta))\Big).$$

Consequently, $\aleph_{\alpha+\mathbf{1}} \leq \sup(\lambda,\ \sup_{\beta\in\lambda}\mathsf{card}(f(\beta)))$. Also, for every $\beta \in \lambda$, $\mathsf{card}(f(\beta)) \leq \aleph_\alpha$, so $\sup_{\beta\in\lambda}\mathsf{card}(f(\beta)) \leq \aleph_\alpha < \aleph_{a+1}$, which implies that $\lambda \geq \aleph_{\alpha+\mathbf{1}}$. But $\mathit{cof}(\aleph_{\alpha+\mathbf{1}}) \leq \aleph_{\alpha+\mathbf{1}}$. Conclusion: $\aleph_{\alpha+\mathbf{1}}$ is regular.

Suppose that α is a limit ordinal and let δ be its conality. Let f be a strictly increasing map from δ into α whose image is not strictly bounded. Then the map g from δ into $\aleph_\alpha$ that sends β ($\in \delta$) to $\aleph_\beta$ is also strictly increasing, with no strict upper bound; this shows that δ is conal with $\aleph_\alpha$ and that $\mathit{cof}(\aleph_\alpha) \leq \delta$. To show that $\mathit{cof}(\aleph_\alpha) = \delta$, we will use question (c) and show that if β is an ordinal strictly less than δ and if h is a map from β onto $\aleph_\alpha$, then h is strictly bounded in $\aleph_\alpha$. So let k be a map whose domain is β and whose value, for $\gamma \in \beta$, is the unique ordinal ε such that $\aleph_\varepsilon$ is the cardinality of $h(\gamma)$; k is in fact a map from β into α and, from part (c), since β is strictly less than the conality of α, k is strictly bounded by some ordinal $\zeta \in \alpha$. It follows that, for every $\gamma \in \beta$, $h(\gamma) < \aleph_{\zeta+\mathbf{1}}$. Because α is a limit ordinal, $\zeta + \mathbf{1} < \alpha$ and $\aleph_{\zeta+\mathbf{1}} < \aleph_\alpha$; so h is strictly bounded in $\aleph_\alpha$.

(f) The map $n \mapsto \omega + n$ from ω into $\omega + \omega$ (the ordinal sum) is clearly strictly increasing and not strictly bounded. Thus ω is conal with $\omega + \omega$. It is also clear that ω is not conal with any of the ordinals $\omega+n$ ($\mathbf{0} < n < \omega$) because any strictly increasing map from ω into $\omega+n$ must be strictly bounded by ω. The least ordinal strictly greater than ω with which ω is conal is therefore $\omega+\omega$. When it comes to the least cardinal strictly greater than ω with which ω is conal, we must assume that the universe satises the axiom of choice to assert that it is $\aleph_\omega$; this follows from part (e).

16. (a) Let δ be the conality of λ. If $\delta = \lambda$ (in other words, if λ is regular), then by Cantor's theorem (Theorem 7.49), $\lambda^\delta > \lambda$. If not, there exists a limit ordinal α such that $\lambda = \aleph_\alpha$ and $cof(\alpha) = cof(\lambda)$ and there exists a strictly increasing map f from δ into α that is not strictly bounded [part (e) of Exercise 15]. For every $\beta \in \delta$, set $\lambda_\beta = \aleph_{f(\beta)}$ and $\mu_\beta = \lambda$. So for every $\beta \in \delta$, we have $\lambda_\beta < \mu_\beta$ and, according to Konig's theorem (Theorem 7.75),

$$\text{card}\left(\bigcup_{\beta\in\delta} \lambda_\beta\right) = \lambda < \text{card}\left(\prod_{\beta\in\delta} \mu_\beta\right) = \lambda^\delta.$$

(b) Because $\aleph_\mathbf{0} = \aleph_\mathbf{0} \times \aleph_\mathbf{0}$, we have $2^{\aleph_0} = 2^{\aleph_0 \times \aleph_0} = (2^{\aleph_0})^{\aleph_0}$. So it follows from the preceding question that the conality of $2^{\aleph_0}$ is strictly greater than ω.

(c) • Suppose $\mu < \delta$ ($\leq \lambda$). Then every element $f \in \lambda^\mu$ is strictly bounded in λ [part (c) of Exercise 15]. This means that λ^μ is included in $\bigcup_{\kappa\in\lambda} \kappa^\mu$.

(i) If $\lambda = \mu^+ = 2^\mu$ (so λ is regular), then $\lambda^\mu = (2^\mu)^\mu = 2^{\mu\times\mu} = 2^\mu = \lambda$.

(ii) If not, then for every cardinal κ strictly between μ and λ, we have

$$\kappa^\mu \leq (2^\kappa)^\mu = 2^{\kappa\mu} = 2^\kappa = \kappa^+ \leq \lambda;$$

so of course, if $\kappa \leq \mu$, we will have $\kappa^\mu \leq \lambda$.

As a result,

$$\lambda^\mu \leq \text{card}\left(\bigcup_{\kappa\in\lambda} \kappa^\mu\right) = \sup_{\kappa\in\lambda} \kappa^\mu \leq \lambda.$$

The inequality $\lambda \leq \lambda^\mu$ is obvious ($\mu \neq \mathbf{0}$), so we are able to conclude $\lambda^\mu = \lambda$.

• Suppose $\delta \leq \mu \leq \lambda$. From part (a) together with the GCH, we conclude $2^\lambda = \lambda^+ \leq \lambda^\delta \leq \lambda^\mu \leq \lambda^\lambda = 2^\lambda$. Thus, $\lambda^\mu = 2^\lambda$.

• Suppose $\lambda < \mu$. Then $2^\mu \leq \lambda^\mu \leq \mu^\mu = 2^\mu$; so $\lambda^\mu = 2^\mu$.

17. (a) We argue by contradiction. Let Φ be a (denable) strictly increasing function from On into On that does not have the property in question. Let δ denote the least ordinal for which $\Phi(\delta) < \delta$. Set $\beta = \Phi(\delta)$; thus $\beta < \delta$ and, by the choice of δ, $\Phi(\beta) \geq \beta$. This means that $\beta < \delta$ and $\Phi(\beta) \geq \Phi(\delta)$; thus Φ is not strictly increasing. Contradiction.

(b) Let α be an ordinal. By induction on the integers, we dene a family of ordinals $(\alpha_n : n \in \omega)$:

- α_0 is any ordinal strictly greater than α;
- for every $n \in \omega$, $\alpha_{n+1} = \Phi(\alpha_n)$.

We then distinguish two cases: either there exists an integer n such that $\alpha_{n+1} = \alpha_n$ (in which case α_n is a xed point of Φ that is strictly greater than α) or else the mapping $n \mapsto \alpha_n$ is a strictly increasing unbounded map from ω into the limit ordinal $\beta = \sup_{n\in\omega} \alpha_n$. Thus ω is conal with β, so Φ is continuous at β, which means that

$$\begin{aligned}\Phi(\beta) &= \sup_{\gamma\in\beta} \Phi(\gamma) = \sup_{n\in\omega} \Phi(\alpha_n) \text{ (because } \Phi \text{ is increasing)}\\ &= \sup_{n\in\omega} \alpha_{n+1} = \beta;\end{aligned}$$

once more, we have a xed point of Φ, namely β, that is strictly greater than α.

(c) The properties of being strictly increasing and of being continuous at limit ordinals are clearly preserved under composition. The construction in part (b) can be redone with the function $\Psi \circ \Phi$; starting with an arbitrary ordinal α, we choose an ordinal $\alpha_0 > \alpha$, then for every $n \in \omega$, we set $\alpha_{n+1} = (\Psi \circ \Phi)(\alpha_n)$. From question (a), we know that, for every $n \in \omega$, $\alpha_n \leq \Phi(\alpha_n) \leq (\Psi \circ \Phi)(\alpha_n) = \alpha_{n+1}$. It follows that if for some n, $\alpha_n = \alpha_{n+1}$, then α_n is a common xed point of Φ and Ψ that is strictly greater than α. In the opposite case, the ordinal $\beta = \sup_{n\in\omega} \alpha_n$ is a xed point of $\Psi \circ \Phi$. Because Φ and Ψ are strictly increasing, we have, from part (a), $\beta = (\Psi \circ \Phi)(\beta) \geq \Phi(\beta) \geq \beta$ and $\beta = \Phi(\beta) = \Psi(\beta)$. So β is a common xed point of Φ and Ψ that is greater than α.

(d) Let Φ and Ψ be the maps that, with each ordinal α, associate $\aleph_\alpha$ and V_α, respectively. It is immediate from their denitions that these functions are continuous at limit ordinals. It is obvious that Φ is strictly increasing. Also, Ψ is strictly increasing, for if $\alpha < \beta$, then $\alpha + \mathbf{1} \leq \beta$, so $\wp(V_\alpha) = V_{\alpha+\mathbf{1}} \subseteq V_\beta$, which shows that

$$\mathsf{card}(V_\alpha) < 2^{\mathsf{card}(V_\alpha)} = \mathsf{card}(V_{\alpha+\mathbf{1}}) \leq \mathsf{card}(V_\beta).$$

So we may apply part (c) to conclude that, for every ordinal α, there exists an ordinal $\beta > \alpha$ that is a common xed point of Φ and Ψ, i.e. such that

$$\aleph_\alpha = \mathsf{card}(V_\alpha) = \alpha.$$

18. (a) We use the results from Exercise 16. We have $\aleph_\omega = \sup_{n\in\omega} \aleph_n$; but

$$\omega^{\aleph_\omega} = 2^{\aleph_\omega} > \aleph_\omega = \sup_{n\in\omega} \aleph_{n+1} = \sup_{n\in\omega} 2^{\aleph_n} = \sup_{n\in\omega} \omega^{\aleph_n}.$$

Thus the rst function is not continuous at ω. Nor is the second one, for that matter:

$$(\aleph_\omega)^\omega = 2^{\aleph_\omega} > \aleph_\omega = \sup_{n\in\omega} \aleph_{n+1} = \sup_{n\in\omega}(\aleph_{n+1})^\omega.$$

(b) For the rst function, the answer is `yes' (refer to the denition of ordinal sum). The second function is not continuous at ω:

$$\left(\sup_{n\approx\omega} n\right) + \omega = \omega + \omega \quad \text{but} \ \sup_{n\in\omega}(n + \omega) = \omega.$$

The third function is continuous at every limit ordinal (refer to the denition of ordinal product). The fourth function is not continuous at ω:

$$2 \cdot \left(\sup_{n\in\omega} n\right) = 2 \cdot \omega = \omega + \omega \quad \text{but} \ \sup_{n\in\omega}(2 \cdot n) = \omega.$$

19. Assume that the universe $\mathcal{U}$ satises AF (so it coincides with the class $\mathcal{V}$). If a formula F with one free variable denes a non-empty class, in other words, if $\mathcal{U} \models \exists v_0 F[v_0]$, then we can choose from this class an element a of minimal rank; this means that there exists an ordinal α such that $a \in V_\alpha$ and that, for every ordinal $\beta < \alpha$, no element of V_β satises F. It is then clear that, for every set b such that $b \in a$, $\mathcal{U} \models \neg F[b]$ since the rank of b is strictly less than the rank of a. This proves that the proposed axiom scheme is satised in $\mathcal{U}$.

Conversely, suppose that the scheme is satised and consider the formula $F[v_0] = \forall v_1(On[v_1] \Rightarrow \neg v_0 \in V_{v_1})$ which denes the class $\mathcal{X}$ of sets that do not belong to $\mathcal{V}$. If the class $\mathcal{X}$ is not empty, we can nd, thanks to the axiom scheme, a set a in this class with the property that if $b \in a$, then b does not belong to $\mathcal{X}$. In other words, all the elements of a must belong to $\mathcal{V}$, and hence to V_α for some appropriate α; a would then be a subset of V_α, hence an element of $V_{\alpha+1}$, and hence an element of $\mathcal{V}$. This contradiction proves that the class $\mathcal{X}$ is empty or, equivalently, that the universe coincides with $\mathcal{V}$; this shows that it satises AF (see Theorem 7.80).

20. (a) It is clear that a closed conal set is not empty. To show that the set of closed conal subsets forms a lterbase, it will sufce to apply the result that will be proved next in part (b) to the special case of nite families of closed conal sets.

(b) Set $X = \bigcap_{i\in I} X_i$. It is clear that X is closed, for if $Y \subseteq X$ and $\mathrm{card}(Y) < \lambda$, then for all $i \in I$, $Y \subset X_i$ and $\sup Y \in X_i$; thus $\sup Y \in X$.

Let us prove that X is also conal. Let $a \in \lambda$. We dene a sequence $(f_n)_{n\in\omega}$ of mappings from I into λ by induction as follows: for all $i \in I$, $f_0(i)$ is the least element of X_i that is greater than a (there exists one because X_i is conal); for all $n \in \omega$ and for all $i \in I$, $f_{n+1}(i)$ is the least ordinal in X_i that is greater than $\sup\{f_n(j) : j \in I\}$. [Note that the cardinality of the set $\{f_n(j) : j \in I\}$ is strictly less than λ, so that $\sup\{f_n(j) : j \in I\}$ is also strictly less than λ.]

For every $i \in I$, $\{f_n(i) : n \in \omega\}$ is a denumerable subset of X_i; hence $\alpha_i = \sup\{f_n(i) : i \in I\}$ is an element of X_i. But for all i and j belonging to I, $\alpha_i = \alpha_j$; this is because by denition of the mappings f_n, we have $f_{n+1}(i) > f_n(j)$ for all $n \in \omega$, which shows that $\alpha_i \geq \alpha_j$. An analogous argument shows that $\alpha_j \geq \alpha_i$, so $\alpha_i = \alpha_j$. Consequently, the common value of all these α_i belongs to X.

(c) The implication (1) $\Rightarrow$ (2) is more or less obvious; if Y_1 and Y_2 are two disjoint stationary sets, then at least one of them cannot include a closed conal set, otherwise their intersection would not be empty.

The other implications are not any more difcult once we have observed the following fact: let $\mathcal{F}$ be the lter generated by the closed conal sets; then a subset Y of λ is stationary if and only if its complement does not belong to $\mathcal{F}$.

(d) We rst show that $\Delta(X)$ is closed. Let Y be a subset of $\Delta(X)$ whose cardinality is strictly less than λ. Y is well-ordered by the membership relation and is therefore isomorphic to an ordinal α that must be less than λ. There exists a strictly increasing bijection f from α onto Y. By denition of $\Delta(X)$, for every $\beta \in \alpha$, $f(\beta) \in X_{f(\beta)}$.

Let $\gamma = \sup Y$. We must show that $\gamma \in X_\gamma$. This is clear if $\gamma \in Y$. If not, then α is a limit ordinal and, for every $\beta \in \alpha$, $\gamma = \sup(f(\delta) : \beta \leq \delta < \alpha)$. But if $\beta \leq \delta < \alpha$, then $X_{f(\delta)} \subseteq X_{f(\beta)}$ [by property (2)], so $f(\delta) \in X_{f(\beta)}$. Since $X_{f(\beta)}$ is closed, it follows that $\gamma \in X_{f(\beta)}$; since this is true for all $\beta \in \alpha$, we see that $\gamma \in \bigcap_{\beta \in \alpha} X_{f(\beta)}$, which is equal to X_γ [by property (3)].

Now, let us show that $\Delta(X)$ is conal. Let $\alpha \in \lambda$. Dene a sequence $(\alpha_n : n \in \omega)$ by induction, setting $\alpha_0 = \alpha$ and α_{n+1} equal to the least ordinal that belongs to X_{α_n} and is strictly greater than α_n. Set $\beta = \sup(\alpha_n : n \in \omega)$. Once again using the same type of argument, we see that $\beta \in X_\beta$ and hence that $\beta \in \Delta(X)$.

(e) We will prove the contrapositive of Fodor's theorem. Let f be a map from λ into λ and assume that, for every $\alpha \in \lambda$, $\bar{f}^{-1}(\alpha)$ is not stationary. We will construct a closed conal set X such that, for every $\alpha \in X$, $f(\alpha) \geq \alpha$.

From the hypotheses together with the axiom of choice, we can nd a family $(T_\alpha : \alpha \in \lambda)$ of closed conal subsets of λ such that, for every $\alpha \in \lambda$ and for every $\beta \in T_\alpha$, $f(\beta) \neq \alpha$. We then dene another family $(X_\alpha : \alpha \in \lambda)$ of closed conal subsets of λ by induction as follows:

- $X_0 = \lambda$;
- if α is a limit ordinal, then $X_\alpha = \bigcap_{\beta \in \alpha} X_\beta$;
- if $\alpha = \beta + \mathbf{1}$, then $X_\alpha = X_\beta \cap T_\beta$.

The family $(X_\alpha : \alpha \in \lambda)$ satises conditions (1), (2), and (3) of the previous question and, moreover, for every $\alpha \in \lambda$ and for every $\beta \in X_\alpha$, $f(\beta) \geq \alpha$ (this is easy to check).

Let Y be the diagonal intersection of this family. This is a closed conal set and, if $\alpha \in Y$, then $\alpha \in X_\alpha$; hence $f(\alpha) \geq \alpha$.

(f) We will show that every closed conal subset X of λ contains an ordinal whose conality is $\aleph_0$. Dene a strictly increasing sequence of ordinals $(\alpha_n : n \in \omega)$ by induction, setting α_0 equal to the least ordinal belonging to X, and α_{n+1} equal to the least element of X strictly greater than α_n. It is then clear that the conality of $\beta = \sup(\alpha_n : n \in \omega)$ is $\aleph_0$ and that β belongs to X since X is closed. So it follows that the set of elements of λ that have conality $\aleph_0$ intersects every closed conal subset of λ; it is therefore stationary.

To show that the set of elements of λ that have conality $\aleph_1$ is also stationary, the argument is similar. If X is a closed conal set, we again dene by induction a strictly increasing sequence $(\alpha_i : i \in \aleph_1)$ of elements of X and we set $\beta = \sup(\alpha_i : i \in \aleph_1)$; then $\beta \in X$ and the conality of β is clearly less than or equal to $\aleph_1$. We nally prove by contradiction that it cannot equal $\aleph_0$; for if so, then there exists a strictly increasing sequence of ordinals $(\gamma_n : n \in \omega)$ such that $\sup(\gamma_n : n \in \omega) = \beta = \sup(\alpha_i : i \in \aleph_1)$. For each integer n, set

$$A_n = (i \in \aleph_1 : \alpha_i \leq \gamma_n\}.$$

Since the sequence $(\alpha_i : i \in \aleph_1)$ is strictly increasing and is not bounded by γ_n, A_n is a proper initial segment of $\aleph_1$ and is therefore denumerable. Besides, if $i \in \aleph_1$, there exists an integer n such that $\alpha_i \leq \gamma_n$, which proves that $\bigcup_{n\in\omega} A_n = \aleph_1$; but this is not possible by Corollary 7.72.

It is obvious that the conality of an ordinal cannot simultaneously equal $\aleph_0$ and $\aleph_1$; so we have found a pair of disjoint stationary sets.

(g) For every $\alpha \in \aleph_1$, $h_n(\alpha) \in \alpha$, so we may apply Fodor's theorem; there exists $\beta_n \in \aleph_1$ such that $h_n^{-1}(\beta_n)$ is a stationary set. Let Y_n denote this set (from which **0** has been removed if it was present). Then, for every $\gamma \in Y_n$, $f_\gamma(n) = h_n(\gamma) = \beta_n$.

Let $\gamma \in \bigcap_{n\in\omega} Y_n$. For every $n \in \omega$, $f_\gamma(n) = h_n(\gamma) = \beta_n$. Since f_γ is a surjective map from ω onto γ, we see that $\gamma = \{\beta_n : n \in \omega\}$. So we conclude that $\bigcap_{n\in\omega} Y_n$ is not a conal set, so it does not include any conal subset. Together with part (a), this shows that at least one of the sets Y_n does not belong to $\mathcal{F}$.

21. (a) The comments in Remark 7.82 show that if α is a limit ordinal, then $\langle V_\alpha, \in\rangle$ satises the axioms of extensionality, pairs, and subsets. Moreover, since α is strictly greater than ω, it also satises the axiom of innity, i.e. in $\langle V_\alpha, \in\rangle$, ω is an ordinal that is neither zero nor a successor.

It remains to prove the comprehension axioms. Let $F[v_0, v_1, \ldots, v_n]$ be a formula of L and let $b, a_1, a_2, \ldots, a_n$ be sets in V_α. We know that there exists a set c such that

$$\mathcal{U} \models \forall w(w \in c \Leftrightarrow (w \in b \wedge F^{V_\alpha}[w, a_1, \ldots, a_n])).$$

It is clear that $c \subseteq b$. Since there exists $\beta < \alpha$ such that $b \in V_\beta$, we have $c \in V_{\beta+1}$. Since α is a limit ordinal, $\beta + \mathbf{1} < \alpha$, so $c \in V_\alpha$. By construction of the formula F^{V_α}, we have

$$\mathcal{U} \models \forall w(w \in c \Leftrightarrow (w \in b \wedge F^{V_\alpha}[w, a_1, \ldots, a_n]))$$

if and only if

$$\langle V_\alpha, \in \rangle \models \forall w(w \in c \Leftrightarrow (w \in b \wedge F[w, a_1, \ldots, a_n])).$$

So the comprehension axioms are true in $\langle V_\alpha, \in \rangle$.

(b) If ZF is not consistent, then it is surely not a consequence of Z, which is consistent by hypothesis. So suppose that ZF is consistent and let $\mathcal{U}$ be a model of ZF. Consider the set $V_{\omega+\omega}$ dened inside this model $\mathcal{U}$. (Here, $\omega + \omega$ denotes the ordinal sum.) We have just seen that $\langle V_{\omega+\omega}, \in \rangle$ is a model of Z. But it is not a model of ZF, for otherwise we would be able to dene by induction a sequence of ordinals $(\alpha_n : n \in \omega)$ by setting

$$\alpha_0 = \omega;$$
$$\alpha_{n+1} = \alpha_n + 1.$$

We then see that, in $\langle V_{2\omega}, \in \rangle$, the set $\bigcup_{n\in\omega} \alpha_n$ is equal to the class of all ordinals; this violates Proposition 7.25.

22. To begin with, $\langle \mathcal{W}, \in \rangle$ satises the axiom of extensionality since this axiom is a universal statement and $\langle \mathcal{W}, \in \rangle$ is a substructure of $\mathcal{U}$. In addition, for all sets x and y, $\mathsf{cl}(\{x, y\}) = \mathsf{cl}(x) \cup \mathsf{cl}(y) \cup \{x, y\}$; thus $\mathsf{cl}(\{x, y\})$ is denumerable if $\mathsf{cl}(x)$ and $\mathsf{cl}(y)$ are. It follows that $\langle \mathcal{W}, \in \rangle$ satises the axiom of pairs. By the very denition of transitive closure, the inclusion $\mathsf{cl}(\bigcup_{t\in x} t) \subseteq \mathsf{cl}(x)$ is true for all x; this shows that $\langle \mathcal{W}, \in \rangle$ satises the axiom of unions.

We leave it to the reader to verify, by referring to the denition of the ordinals, that for every x in $\mathcal{W}$,

$$\langle \mathcal{W}, \in \rangle \models On(x) \quad \text{if and only if} \quad \mathcal{U} \models On(x).$$

Now $\mathsf{cl}(\omega) = \omega$, thus ω is in $\mathcal{W}$ and we see that

$$\langle \mathcal{W}, \in \rangle \models \text{`}\omega \text{ is an ordinal that is neither zero nor a successor'}.$$

So $\langle \mathcal{W}, \in \rangle$ satises the axiom of innity.

We are now left with the most difcult, the axioms of replacement. Let $W[v_0]$ abbreviate the formula `$\mathsf{cl}(v_0)$ is denumerable'.

Let $F[w_0, w_1, v_0, v_1, \ldots, v_n]$ be a formula of L and let $b, a_0, a_1, \ldots, a_n$ be sets in $\mathcal{W}$; suppose also that the formula $F[w_0, w_1, v_0, v_1, \ldots, v_n]$ is functional in v_0, in $\langle \mathcal{W}, \in \rangle$, in other words, that

$$\langle \mathcal{W}, \in \rangle \models \forall w_0 \forall w_1 \forall w_2 ((F[w_0, w_1, a_0, \ldots, a_n] \\ \wedge F[w_0, w_2, a_0, \ldots, a_n]) \Rightarrow w_1 \simeq w_2).$$

This implies that

$$\mathcal{U} \models \forall w_0 ((W[w_0] \wedge W[w_1] \wedge W[w_2] \\ \wedge F^{\mathcal{W}}[w_0, w_1, a_0, \ldots, a_n] \wedge F^{\mathcal{W}}[w_0, w_2, a_0, \ldots, a_n]) \\ \Rightarrow w_1 \simeq w_2);$$

so the formula

$$W[w_0] \wedge W[w_1] \wedge F^{\mathcal{W}}[w_0, w_1, a_0, a_1, \ldots, a_n]$$

is functional in w_0 in $\mathcal{U}$. By the axiom of replacement in $\mathcal{U}$, we conclude that there exists a set c such that

$$\mathcal{U} \models \forall v_0 (v_0 \in c \Leftrightarrow \exists w_0 \in b\, (W[w_0] \wedge W[v_0] \\ \wedge F^{\mathcal{W}}[w_0, v_0, a_0, a_1, \ldots, a_n])). \qquad \text{(Ü)}$$

The set c is denumerable because the formula $F^{\mathcal{W}}[w_0, w_1, a_0, a_1, \ldots, a_n]$ denes a surjective map from a subset of b, which is denumerable since it is in $\mathcal{W}$, onto c (see Proposition 7.62). Besides, all the elements of c are in $\mathcal{W}$ by denition. Since $\mathsf{cl}(c) = c \cup \bigcup_{t \in c} \mathsf{cl}(t)$, we see that c is in $\mathcal{W}$. From (Ü), we conclude that

$$\langle \mathcal{W}, \in \rangle \models \forall v_0 (v_0 \in c \Leftrightarrow \exists w_0 \in b\, F[w_0, v_0, a_0, a_1, \ldots, a_n]).$$

So the replacement axioms are satised in $\langle \mathcal{W}, \in \rangle$.

It is clear that all the subsets of ω are in $\mathcal{W}$; but $\wp(\omega)$, which is not denumerable, is not in $\mathcal{W}$; this shows that $\langle \mathcal{W}, \in \rangle$ does not satisfy the axiom of subsets.

23. Every real number r in the interval (0,1] has a decimal expansion. This means that there is a sequence $(a_i : i \in \omega)$ of integers, all between 0 and 9 inclusive, such that

$$r = \sum_{i \geq 1} \frac{a_i}{10^i}.$$

This expansion is not unique. However, if we set

$$S^* = \{s \in \{0, 1, 2, 3, 4, 5, 6, 7, 8, 9\}^{\omega} : \\ \text{for every } n \in \omega,\ \text{there exists } p \geq n \text{ such that } s(p) \neq 0\},$$

then, for every $r \in (0, 1]$, there exists one and only one element $(a_i : i \in \omega) \in S^*$ such that $r = \sum_{i \geq 1} a_i/10^i$.

To show that the interval $(0, 1]$ is not denumerable, it sufces to show that S^* is not denumerable; this amounts to showing that if $(s^i : i \in \omega)$ is a sequence of elements of S^*, then there exists an element s in S^* that is not equal to any of the s^i. We may dene such an s by

$$s(i) = 1 \quad \text{if } s^i(i) \neq 1;$$
$$s(i) = 2 \quad \text{if } s^i(i) = 1.$$

Solutions to the exercises for Chapter 8

1. (a) Begin by noting that the set $\mathbb{N}^* = \mathbb{N} - \{0\}$ is an example of a set that is disjoint from its power set (0 belongs to every element of $\mathbb{N}^*$ but to none of its subsets).

$\mathcal{M}_X \models H_0$ because X and $\wp(X)$ constitute a partition of $\mathcal{M}_X$; $\mathcal{M}_X \models H_1$ because, for every pair $(x, y) \in \bar{A}$, we have $x \in X$ and $y \in \wp(X)$; $\mathcal{M}_X \models H_2$ by extensionality; $\mathcal{M}_X \models H_3$ because the empty set is an element of $\wp(X)$; $\mathcal{M}_X \models H_4$ because every subset $x \in \wp(X)$ has a complement in X; $\mathcal{M}_X \models H_5$ because

$$\text{for all subsets } x \in \wp(X) \text{ and } y \in \wp(X), \quad x \cup y \in \wp(X);$$

nally, for every $n \geq 1$, $\mathcal{M}_X \models F_n$ because

$$\text{for all elements } x_1, x_2, \ldots, x_n \text{ of } X, \quad \{x_1, x_2, ..., x_n\} \in \wp(X).$$

(b) Yes. The language is denumerable, the theory T has innite models [for example, the model $\mathcal{M}_X$ from part (a) when X is innite]; thus by the downward LowenheimñSkolem theorem, T has a denumerable model. In fact, we can easily describe one by taking X to be some denumerable set in part (a); it is the substructure of $\mathcal{M}_X$ whose base set is

$$X \cup \wp_f(X) \cup \wp_{cof}(X),$$

where $\wp_f(X)$ is the set of nite subsets of X and $\wp_{cof}(X)$ is the set of conite subsets of X. It is easy to check that this is a denumerable model of T.

(c) They are all integers of the form $k + 2^k$, where $k \in \mathbb{N}$.

(d) We leave to the reader the task of proving, using H_5, that, for every integer $n \geq 1$, we have $\{H_0, H_1, H_2, H_3, H_4, H_5, F_1\} \vdash F_n$.

(e) Let $\mathcal{M}_0$ denote the model described in part (b) whose base set is

$$X \cup \wp_f(X) \cup \wp_{cof}(X);$$

to be precise, we will take $X = \mathbb{N}^*$. We are going to describe a denumerable model $\mathcal{M}_1$ of T that is not isomorphic to $\mathcal{M}_0$. Let C be an innite subset of X whose complement is also innite (for example, the set of even integers). Consider the Boolean subalgebra B of $\wp(X)$ generated by

$$\wp_f(X) \cup \wp_{cof}(X) \cup \{C\};$$

B is still denumerable. For $\mathcal{M}_1$, take the substructure of $\mathcal{M}_X$ whose base set is $X \cup B$; it is easy to verify that this is a model of T.

We will now prove that $\mathcal{M}_0$ is not isomorphic to $\mathcal{M}_1$. For every integer $p > 0$, consider the formula

$$F_p[v_0] = \exists v_1 \exists v_2 \dots \exists v_p \Biggl(\forall w_0 \Biggl(A w_0 v_0 \Rightarrow \Biggl(\bigvee_{1 \le i \le p} w_0 \simeq v_i \Biggr) \Biggr) \\ \vee \forall w_0 \Biggl(\neg A w_0 v_0 \Rightarrow \Biggl(\bigvee_{1 \le i \le p} w_0 \simeq v_i \Biggr) \Biggr) \Biggr).$$

Then $\mathcal{M}_0$ has the following property:

for every element x of $\mathcal{M}_0$, there exists an integer p such that $\mathcal{M}_0 \models F_p[x]$.

Obviously, this property must also be satised by any structure that is isomorphic to $\mathcal{M}_0$; but $\mathcal{M}_1$ does not have this property since C does not satisfy any of the formulas F_p in $\mathcal{M}_1$.

2. In what follows, the word `preliminaries' refers to results found in the solution to Exercise 15 from Chapter 3.

- (a) It is clear that the relation $\approx$ is reexive (take $m = n = p = q = 0$ in the denition) and symmetric. Suppose that $a, b, c \in M$, that $a \approx b$ and $b \approx c$. So there exist natural numbers m, n, p, q, r, s, t, and u such that

  $$\bar{d}^m(\bar{g}^n(a)) = \bar{d}^p(\bar{g}^q(b)) \quad \text{and} \quad \bar{d}^r(\bar{g}^s(b)) = \bar{d}^t(\bar{g}^u(c)).$$

  The fact that $\bar{d}$ and $\bar{g}$ commute implies (see the preliminaries)

  $$\bar{d}^{m+r}(\bar{g}^{n+s}(a)) = \bar{d}^{p+r}(\bar{g}^{q+s}(b)) = \bar{d}^{t+p}(\bar{g}^{u+q}(c)),$$

  which proves that $a \approx c$.
- Since for every element a of M, $a \approx \bar{d}(a)$ and $a \approx \bar{g}(a)$, the grill of a is closed under the mappings $\bar{d}$ and $\bar{g}$.
  Note that this property is stronger than mere compatibility of the relation $\approx$ with the mappings $\bar{d}$ and $\bar{g}$ [which would say that if $a \approx b$, then $\bar{d}(a) \approx \bar{d}(b)$ and $\bar{g}(a) \approx \bar{g}(b)$].
- Let G be a grill of $\mathcal{M}$. The fact that G is closed under $\bar{d}$ and $\bar{g}$ is sufcient to guarantee that $\langle G, \bar{d} \restriction G, \bar{g} \restriction G \rangle$ is a substructure $\mathcal{G}$ of $\mathcal{M}$. All

closed universal formulas of L that are satised in $\mathcal{M}$ are also satised in $\mathcal{G}$ (Corollary 8.38). So to show that $\mathcal{G}$ is a model of T, we need only check that $\mathcal{G}$ satises the formulas of T that are not universal, i.e. the formulas $\forall x \exists u$ $du \simeq x$ and $\forall x \exists v\ gv \simeq x$. So let a be an element of G. Because these formulas are true in $\mathcal{M}$, we can nd two elements b and c in M such that $a = \bar{d}(b) = \bar{g}(c)$. But we have just proved that $b \approx \bar{d}(b)$ and $c \approx \bar{g}(c)$. As a result, b and c are in the same grill as a, i.e. in G; this proves that the formulas in question are satised in $\mathcal{G}$.

- The standard model (see the preliminaries) has a unique grill; to see this, consider two pairs (i, j) and (k, l) in $\mathbb{Z}^2$ and set $m = \sup(j - l, 0)$, $n = \sup(i-k, 0)$, $p = \sup(l-j, 0)$, and $q = \sup(k-i, 0)$; it is then easy to see that

$$s_{d^m}(s_{g^n}(i, j)) = s_{d^p}(s_{g^q}(k, l)),$$

which shows that $(i, j) \approx (k, l)$.

We will now prove a property which will be of use in some later questions. In a model $\mathcal{M} = \langle M, \bar{d}, \bar{g}\rangle$ of T, every grill G determines an L-structure $\mathcal{G} = \langle G, \bar{d} \restriction G, \bar{g} \restriction G\rangle$ that is isomorphic to the standard structure $\mathcal{M}_0$.

Let a be an element of G. Consider the mapping ϕ from $\mathbb{Z} \times \mathbb{Z}$ into M which, with the pair (i, j), associates the element $\bar{d}^j(\bar{g}^i(a))$ of M. We leave it to the reader to verify that ϕ is injective, that its image is G, and that ϕ is an isomorphism from $\mathcal{M}_0$ onto $\mathcal{G}$.

(b) Let T' denote the theory in the language L' obtained by adding to T all the formulas G_{mnpq} (for $m, n, p, q \in \mathbb{N}$). Given an L'-structure $\mathcal{M}' = \langle M, \bar{d}, \bar{g}, \bar{\lambda}, \bar{\mu}\rangle$ that is a model of T', the underlying L-structure $\mathcal{M} = \langle M, \bar{d}, \bar{g}\rangle$ is obviously a model of T; and the elements $\bar{\lambda}$ and $\bar{\mu}$ are not $\approx$-equivalent, for if they were, one of the formulas G_{mnpq} would not be satised in $\mathcal{M}'$. So the model $\mathcal{M}$ contains at least two grills (those of $\bar{\lambda}$ and of $\bar{\mu}$) and is therefore not isomorphic to the standard model. Thus the existence of a model for the theory T' implies the existence of a non-standard model for the theory T.

So it remains to prove the existence of a model of T'; in view of the compactness theorem, this amounts to proving the existence of a model for an arbitrary nite subset of T. Let T_0 be a nite subset of T'. There exists a natural number N such that

$$T_0 \subseteq T_N = T \cup \{Gmnpq :\ \sup(m, n, p, q) \leq N\}.$$

Let $\mathcal{M}_N$ be the L'-structure obtained by enriching the standard L-structure $\mathcal{M}_0$ with the following interpretations of $\bar{\lambda}$ and $\bar{\mu}$:

$$\bar{\lambda} = (0, 0), \qquad \bar{\mu} = (N+1, N+1).$$

If m, n, p, and q are natural numbers less than or equal to N, we have

$$s_{d^m}(s_{g^n}(\bar{\lambda})) = (n, m) \quad \text{while} \quad s_{d^p}(s_{g^q}(\bar{\mu})) = (N+1, N+1),$$

which shows that the formula G_{mnpq} is satised in the structure $\mathcal{M}_N$; so $\mathcal{M}_N$ is a model of T_N and, *a fortiori*, of T_0.

(c) Consider the structure $\mathcal{M}_A = \langle M_A, d_A, g_A \rangle$, where

- $M_A = A \times \mathbb{Z} \times \mathbb{Z}$;
- for all $a \in A$ and for all $i, j \in \mathbb{N}$, $d_A((a, i, j)) = (a, i, j + 1)$ and $g_A((a, i, j)) = (a, i + 1, j)$.

It is easy to verify that $\mathcal{M}_A$ is a model of T. Also, if (a, i, j) and (a', i', j') are elements of $\mathcal{M}_A$, then $(a, i, j) \approx (a', i', j')$ if and only if $a = a'$; this is because the map that, with $a \in A$, associates the set

$$\{(a, i, j) : (i, j) \in \mathbb{Z} \times \mathbb{Z}\}$$

is a bijection from A onto the set of grills of $\mathcal{M}_A$.

It is natural to feel that this question renders the previous one superuous since we have done more here than prove the existence of a non-standard model of T : we have explicitly described one.

(d) Let $\mathcal{M} = \langle M, d_1, g_1 \rangle$ and $\mathcal{N} = \langle N, d_2, g_2 \rangle$ be two models of T and let σ be a bijection from the set $M/\approx$ of grills of $\mathcal{M}$ onto the set $N/\approx$ of grills of $\mathcal{N}$. Because, as we have seen, every grill is isomorphic to the standard model, we may conclude that we can always nd an isomorphism between any two grills (not necessarily extracted from the same model). For every grill $G \in M/\approx$, choose an isomorphism ϕ_G from G onto $\sigma(G)$ (notice that we are identifying a grill with its associated L-structure and that, besides, we are using the axiom of choice).

The union ϕ of all the maps ϕ_G for $G \in M/\approx$ is an isomorphism from $\mathcal{M}$ onto $\mathcal{N}$.

(e) If A is a set with one element and B is a set with two elements, we see that models $\mathcal{M}_A$ and $\mathcal{M}_B$, which are denumerable, are not isomorphic ($\mathcal{M}_A$ has one grill while $\mathcal{M}_B$ has two). Thus T is not $\aleph_0$-categorical.

Moreover, if $\mathcal{M}$ is a model of T and if we let C denote the set of its grills, we note, from part (d), that $\mathcal{M}$ is isomorphic to $\mathcal{M}_C$ and hence that

$$\mathsf{card}(M) = \mathsf{card}(C \times \mathbb{Z} \times \mathbb{Z}) = \sup(\aleph_0, \mathsf{card}(C)).$$

It follows that the cardinality of the set of grills of a denumerable model of T is either a positive integer or $\aleph_0$; it is therefore a non-zero element of the denumerable set $\omega + 1 = \omega \cup \{\omega\}$.

The mapping c from $\mathcal{X}$ into $\omega + 1$ that, with each element of $\mathcal{X}$, associates the cardinality of its set of grills is injective; to see this, note that if $\mathcal{M} \in \mathcal{X}$, $\mathcal{N} \in \mathcal{X}$, and $c(\mathcal{M}) = c(\mathcal{N})$, then from part (d), $\mathcal{M}$ and $\mathcal{N}$ are isomorphic, so according to the properties of $\mathcal{X}$, $\mathcal{M} = \mathcal{N}$. This proves that the cardinality of $\mathcal{X}$ is at most $\aleph_0$.

But we have also seen that, for every non-zero element x of $\omega + 1$, $\mathcal{M}_x$ is a model of T whose set of grills has the same cardinality as x. The properties of

$\mathcal{X}$ imply that we can nd a model that is isomorphic to $\mathcal{M}_x$. So the map c is surjective onto $(\omega + 1) - \{0\}$.

The cardinality of $\mathcal{X}$ is therefore exactly $\aleph_0$.

(f) Let κ be an uncountable cardinal. We have seen that if $\mathcal{M}$ is a model of T and if C is the set of its grills, then $\mathsf{card}(M) = \sup(\aleph_0, \mathsf{card}(C))$. This implies that if the cardinality of M is equal to κ, then the cardinality of C is also equal to κ. Thus, all models of cardinality κ are isomorphic and T is categorical in every uncountable cardinal.

3. (a) We argue by induction on t. If t is a variable x, then

$$T \vdash \forall x\, t \simeq f_e x.$$

If $t = f_\beta u$ and if we assume (induction hypothesis) that $T \vdash \forall x\, u \simeq f_\alpha x$, then we have $T \vdash \forall x\, t \simeq f_\beta f_\alpha x$, so, according to the second set of formulas in T, if $\gamma = \beta\alpha$, then

$$T \vdash \forall x\, t \simeq f_\gamma x.$$

This same argument also shows that, for the variable x, we can take the one that appears in the term t (there is only one since the language has only unary function symbols and no constant symbols).

(b) Since there are no relation symbols other than the symbol for equality, every atomic formula is of the form $t \simeq u$, where t and u are terms. Since each term involves at most one variable, each atomic formula can involve at most two variables.

Consequently, given an atomic formula $F[v_0, v_1]$, there are two terms t and u such that $F = t \simeq u$. We may assume that v_0 is the variable that occurs in t and, if we let x denote the variable that occurs in u, then we have $x = v_0$ or $x = v_1$. As seen in part (a), there exist elements β and γ of G such that

$$T \vdash \forall v_0\, t \simeq f_\beta v_0 \quad \text{and} \quad T \vdash \forall x\, u \simeq f_\gamma x.$$

Set $\alpha = \beta^{-1} \circ \gamma$.

- If $x = v_0$, we have

$$T \vdash \forall v_0 (F \Leftrightarrow f_\beta v_0 \simeq f_\gamma v_0);$$

and therefore,

$$T \vdash \forall v_0 (F \Leftrightarrow v_0 \simeq f_\alpha v_0);$$

if $\alpha = e$, we obtain $T \vdash \forall v_0 \forall v_1 F$ and if $\alpha \neq e$, we obtain

$$T \vdash \forall v_0 \forall v_1 \neg F.$$

- If $x = v_1$, we have

$$T \vdash \forall v_0 \forall v_1 (F \Leftrightarrow f_\beta v_0 \simeq f_\gamma v_1);$$

and therefore,

$$T \vdash \forall v_0 \forall v_1 (F \Leftrightarrow v_0 \simeq f_\alpha v_0).$$

(c) This presents no problem and is left to the reader.

(d) To show that $O(a)$ is a substructure of $\mathcal{M}$, we use the fact that, for all α and β in G, $\phi_\beta(\phi_\alpha(a)) = \phi_{\beta \cdot \alpha}(a)$. It is straightforward to check that the mapping whose value at α is $\phi_\alpha(a)$ is a monomorphism from $\mathcal{G}$ into $\mathcal{M}$ and that its image is $O(a)$.

Let us prove that X_M is a partition of M. Begin by noting that if $b \in O(a)$, then $a \in O(b)$ [for if $b = \phi_\alpha(a)$, then $a = \phi_{\alpha^{-1}}(b)$] and $O(a) = O(b)$. Suppose that $O(a)$ and $O(b)$ are not disjoint; then if c belongs to their intersection, $O(a) = O(c) = O(b)$. Two distinct elements of X_M are therefore disjoint and, in addition, it is clear that the union of all the elements of X_M is equal to the whole of M.

We now assume that the partitions X_M and X_N associated with two models $\mathcal{M}$ and $\mathcal{N}$ are equipotent. Thanks to the axiom of choice, we can nd two families $(a_x : x \in X_M)$ and $(b_y : y \in X_N)$ such that, for all $x \in X_M$, $a_x \in x$ and, for all $y \in X_N$, $b_y \in y$. There exists a bijection σ from X_M onto X_N, and the map τ from $(a_x : x \in X_M)$ into $(b_y : y \in X_N)$ dened for all $x \in X_M$ by

$$\tau(a_x) = b_{\sigma(x)}$$

is also a bijection. The reader can then verify that the mapping π dened for all $\alpha \in G$ and $x \in X_M$ by

$$\pi(\phi_\alpha(a_x)) = \phi_\alpha(\tau(a_x))$$

is an isomorphism from $\mathcal{M}$ onto $\mathcal{N}$.

(e) The proof is analogous to that of part (f) of Exercise 2. We show that if κ is an innite cardinal that is strictly greater than the cardinality of G, then T is κ-categorical. Every model of T contains a copy of G and is therefore innite. T is complete by Vaught's theorem.

(f) If the cardinality κ of G is nite, the preceding argument remains valid for all models whose cardinality, λ, is innite; the theory is λ-categorical. However, we may no longer conclude that the theory is complete since, in this case, it has nite models, the model $\mathcal{G}$ in particular, so Vaught's theorem is no longer applicable. The formula

$$\exists x_1 \exists x_2 \ldots \exists x_\kappa \left(\bigwedge_{1 \le i < j \le \kappa} \neg x_i \simeq x_j \wedge \forall x \bigvee_{1 \le i \le \kappa} x \simeq x_i \right)$$

is satised in the model $\mathcal{G}$ but is not satised in any innite model of T (such as the one that we would obtain by taking the union of denumerably many pairwise disjoint copies of $\mathcal{G}$). So in this case, the theory is not complete. Nonetheless, we do have a complete theory if we add formulas to T expressing that the base set is innite (see Section 3.5 of Chapter 3); we are then once again in a situation where Vaught's theorem applies.

4. (a) Let

$$EQ = \{\forall v_0 R v_0 v_0\} \cup \{\forall v_0 \forall v_1 (R v_0 v_1 \Rightarrow R v_1 v_0)\}$$
$$\cup \{\forall v_0 \forall v_1 \forall v_2 ((R v_0 v_1 \wedge R v_1 v_2) \Rightarrow R v_0 v_2)\}.$$

Also, for $k \in \mathbb{N} - \{0, 1\}$, let F_k and G_k, respectively, denote the following formulas:

$$\exists v_1 \exists v_2 \ldots \exists v_k \bigwedge_{1 \le i < j \le k} \neg R v_i v_j \quad \text{and}$$

$$\forall v_0 \exists v_1 \exists v_2 \ldots \exists v_k \left(\bigwedge_{1 \le i < j \le k} \neg v_i \simeq v_j \wedge \bigwedge_{1 \le i \le k} R v_0 v_i \right).$$

As our theory T we take

$$T = EQ \cup \{F_k : k \in \mathbb{N} - \{0, 1\}\} \cup \{G_k : k \in \mathbb{N} - \{0, 1\}\}.$$

(The rst of these sets expresses that the interpretation of R is an equivalence relation, the second that there are innitely many equivalence classes, and the third that each equivalence class is innite.)

Let A and B be two non-empty sets. Consider the L-structure

$$\mathcal{M}_{A,B} = \langle M_{A,B}, R_{A,B} \rangle,$$

where $M_{A,B} = A \times B$ and where for (a, b) and (a', b') in $M_{A,B}$, $R_{A,B}((a, b), (a', b'))$ if and only if $a = a'$. This is a model of EQ and, if A and B are both innite, is a model of T.

(b) Suppose that T is equivalent to a nite set of formulas, A. By compactness, each formula of A is a consequence of some nite subset of T and, hence, the whole of A itself is a consequence of some nite subset S of T. As T is a consequence of A, T is also a consequence of S; hence S and T are equivalent. We can nd an integer N such that S is included in

$$T_N = EQ \cup \{F_k : 1 < k < N\} \cup \{G_k : 1 < k < N\}.$$

We arrive at a contradiction when we consider two sets A and B of cardinality N; $\mathcal{M}_{A,B}$ is a model of T_N but is not a model of T.

(c) Let λ be an uncountable cardinal. Then the structures $\mathcal{M}_{\lambda,\omega}$ and $\mathcal{M}_{\omega,\lambda}$ are two models of T of cardinality λ that are not isomorphic; each equivalence class of the rst is denumerable while those of the second have cardinality λ. So the theory T is not categorical in any uncountable cardinal. We can also see that if $\mathcal{M}'$ is a model of T and if there exists an injection of $\mathcal{M}_{\lambda,\omega}$ into $\mathcal{M}'$, then $\mathcal{M}'$ has at least λ equivalence classes; so there cannot exist an elementary embedding of $\mathcal{M}_{\lambda,\omega}$ into $\mathcal{M}_{\omega,\lambda}$. Similarly, if there exists an injection from $\mathcal{M}_{\omega,\lambda}$ into $\mathcal{M}'$, then certain classes of $\mathcal{M}'$ (those that contain the image of some element of $\mathcal{M}_{\omega,\lambda}$) have cardinality at least λ; so there cannot exist an elementary embedding of $\mathcal{M}_{\omega,\lambda}$ into $\mathcal{M}_{\lambda,\omega}$.

Let $\langle M, R_M\rangle$ and $\langle N, R_N\rangle$ be two denumerable models of T. The sets M/R_M and N/R_N are innite, hence denumerable; so there exists a bijection

$$\phi : M/R_M \to N/R_N .$$

The equivalence classes for R_M and R_N are also denumerable; so for every $i \in M/R_M$, we can nd a bijection f_i from i onto $\phi(i)$.

The union of the f_i, i.e. the mapping $f : M \to N$ dened for all $a \in M$ by

$$f(a) = f_{cl(a)}(a)$$

[where $cl(a)$ is the equivalence class of a modulo R_M], is a bijection from M onto N and is an isomorphism from $\langle M, R_M\rangle$ onto $\langle N, R_N\rangle$.

So the theory T is $\aleph_0$-categorical and is not categorical in any innite cardinal other that $\aleph_0$.

(d) From all that has preceded, we can conclude using Vaught's theorem (after noting that T obviously has no nite models) that T is a complete theory.

(e) Every L-structure that is a model of T can be enriched to an L_∞-structure that is a model of T_+; to do this, it sufces to interpret the symbols c_n, for $n \in \mathbb{N}$, by elements from different equivalence classes (which is possible since there are innitely many such classes). For example, if we start with the model $\mathcal{M}_{A,B}$ constructed in part (a) with A and B innite, we may choose pairwise distinct points $a_n \in A$, for $n \in \mathbb{N}$, and a point $b \in B$ and interpret c_n by (a_n, b); in this way, we obtain a model of T_+.

To show that T_+ is not equivalent to a nite theory, show that every nite subset of T_+ has a nite model; this can be done using an argument that is analogous to the one used above in part (b).

(f) The argument employed above in part (c) to prove that the theory T is not categorical in any uncountable cardinal can be reused to arrive at the same conclusion for T_+; it is sufcient to enrich the two non-isomorphic models of cardinality λ from part (c) to models of T_+ by applying the method from part (e); the resulting L_∞-structures cannot be isomorphic.

But unlike T, the theory T_+ is not $\aleph_0$-categorical. Take, for example, the model $\mathcal{M}_{\mathbb{N},\mathbb{N}}$ from part (a) and enrich it to an L_∞-structure in two different ways. On the one hand, for every $n \in \mathbb{N}$, when we interpret the constant c_n by the pair $(n, 0)$, we obtain a denumerable model of T_+ which we will call $\mathcal{M}_0$. On the other hand, when, for every $n \in \mathbb{N}$, we interpret the constant c_n by the pair $(n+1, 0)$, we obtain another denumerable model which we will call $\mathcal{M}_1$. These models cannot be isomorphic for the following reason; every monomorphism h from $\mathcal{M}_0$ into $\mathcal{M}_1$ must send $(n, 0)$ to $(n+1, 0)$; so h must, for every $n \in \mathbb{N}$, send the set

$$\{x \in \mathbb{N} \times \mathbb{N} : \mathcal{M}_0 \vDash Rc_n x\} = \{(n, y) : y \in \mathbb{N}\}$$

into the set

$$\{x \in \mathbb{N} \times \mathbb{N} : \mathcal{M}_1 \vDash Rc_n x\} = \{(n+1, y) : y \in \mathbb{N}\};$$

thus, the point $(0, 0)$ does not belong to the image of h; so the monomorphism h from $\mathcal{M}_0$ into $\mathcal{M}_1$ is not surjective.

(g) Let R_1 and R_2 be the interpretations of R in $\mathcal{M}_1$ and $\mathcal{M}_2$, respectively, and, for every $p \in \mathbb{N}$, let a_p and b_p be the interpretations of c_p in $\mathcal{M}_1$ and $\mathcal{M}_2$, respectively. We begin by dening a bijection h from M_1/R_1 onto M_2/R_2 in such a way that

$$\text{for every integer } p, \text{ if } 0 \leq p \leq n, \quad \text{then } h(cl(a_p)) = cl(b_p).$$

It is clear that this can be done. [Here, $cl(x)$ denotes the equivalence class of x modulo R_1 or R_2 according as x is in M_1 or M_2.] Next, for every $\alpha \in M_1/R_1$, dene a bijection f_α from α onto $h(\alpha)$ in such a way that

$$\text{if } 0 \leq p \leq n \text{ and } \alpha = cl(a_p), \quad \text{then } f_\alpha(a_p) = b_p.$$

The union of the mappings f_α for $\alpha \in M_1/R_1$ is an isomorphism from $\mathcal{M}_1 \restriction L_n$ onto $\mathcal{M}_2 \restriction L_n$.

We are now in a position to prove by contradiction that T_+ is a complete theory in L_∞. If it is not complete, we can nd a closed formula F of L_∞, a denumerable model $\mathcal{M}_1$ of $T_+ \cup \{F\}$, and a denumerable model $\mathcal{M}_2$ of $T_+ \cup \{\neg F\}$. Since F involves only nitely many symbols, there exists an integer n such that $F \in L_n$. We have just seen that $\mathcal{M}_1 \restriction L_n$ and $\mathcal{M}_2 \restriction L_n$ are isomorphic, but this contradicts the fact that one satises F and the other satises $\neg F$.

5. (a) Let λ be an innite cardinal and let $\mathcal{M} = \langle M, \ldots \rangle$ be a λ-structure; so M is an innite set. The value of the formula $x \simeq x$ in $\mathcal{M}$ is M. This set is not nite; so it is a set of cardinality λ.

(b) In a structure $\mathcal{M} = \langle M, \ldots \rangle$ of cardinality $\aleph_0$, every subset of M is either nite or of cardinality $\aleph_0$. In particular, this is true for those subsets of M that are the value in $\mathcal{M}$ of some formula of $\mathcal{F}_1$; thus $\mathcal{M}$ is an $\aleph_0$-structure.

(c) Let λ be an innite cardinal. Adjoin to L a set of new constant symbols, C, of cardinality λ and, in the enriched language, consider the theory

$$T_F = T \cup \{F[c] : c \in C\} \cup \{\neg c \simeq d : c \neq d,\ c,\ d \in C\}.$$

It is clear that the reduct to the language L of any model of T_F is a model of T in which the value of the formula F has cardinality greater than or equal to λ. To show that T_F has at least one model, we can use a basic compactness argument; it sufces to show, for every nite subset C_0 of C, that the theory

$$T' = T \cup \{F[c] : c \in C_0\} \cup \{\neg c \simeq d : c \neq d, c, d \in C_0\}$$

has a model. By hypothesis, there exists a model $\mathcal{M}$ of T such that $Val(F, \mathcal{M})$ has cardinality greater than that of C_0. We enrich $\mathcal{M}$ to a model of T' by interpreting the symbols in C_0 by distinct points of $Val(F, \mathcal{M})$ (there are sufciently many); the remaining points of C can be interpreted arbitrarily.

According to the downward LowenheimñSkolem theorem, we can nd a model $\mathcal{N}$ of T_F whose cardinality is that of the enriched language which, in this case, is equal to λ since L is denumerable. The value of F in $\mathcal{N}$ must have cardinality λ. The reduct of $\mathcal{N}$ to the language L is therefore a model of T that answers the question.

(d) Let λ be an innite cardinal and let $\mathcal{M}_0$ be an innite model of T. Set

$$A = \{G \in \mathcal{F}_1 : \ Val(G, \mathcal{M}_0) \text{ is innite } \}.$$

For every formula F of A, let C_F be a set of constant symbols of cardinality λ chosen so that if F and G are distinct elements of A, then the sets C_F and C_G are disjoint. For $F \in A$, consider the following theory:

$$T_F = T \cup \{F[c] : c \in C\} \cup \{\neg c \simeq d : c \neq d, c, d \in C_F\}.$$

Let $T' = \mathsf{Th}(\mathcal{M}) \cup \bigcup_{F \in A} T_F$.

We begin by showing that T' is consistent; by compactness, it is sufcient to show that if A_0 is a nite subset of A and D_F is a nite subset of C_F, then

$$T'' = \mathsf{Th}(\mathcal{M}) \cup \{F[c] : F \in A_0 \text{ and } c \in D_F\} \cup \bigcup_{F \in A_0} \{\neg c \simeq d : c \neq d \text{ and } c, d \in D_F\}$$

has a model. We can enrich $\mathcal{M}$ to a model of T''; we simply need, for each $F \in A_0$, to interpret the symbols of D_F (there are only nitely many) by distinct points of $Val(F, \mathcal{M})$ (which is an innite set).

So we see that T' has a model and, as in part (c), that it has a model of cardinality λ. Let $\mathcal{N}'$ be such a model and let $\mathcal{N}$ be its reduct to L. We will now prove that $\mathcal{N}$ is a λ-model. Let $F \in \mathcal{F}_1$. If $F \in A$, then $Val(F, \mathcal{N})$ has cardinality λ by construction. If $F \notin A$, then $Val(F, \mathcal{N})$ is nite; let n be its cardinality. Thus the closed formula

$$H = \forall v_0 \forall v_1 \ldots \forall v_n \Bigg(\bigwedge_{0 \le i \le n} F[v_i] \Rightarrow \bigvee_{0 \le i < j \le n} v_i = v_j \Bigg)$$

is true in $\mathcal{M}$. Since $\mathcal{N}$ and $\mathcal{M}$ have the same theory, $Val(F, \mathcal{N})$ is nite; so $\mathcal{N}$ is a λ-model.

(e) We argue by contradiction. Suppose that S is not complete; then there is a closed formula F of L such that the theories $S \cup \{F\}$ and $S \cup \{\neg F\}$ are both consistent. Just as S, these theories only have innite models. According to part (d), each of these theories has a λ-model for every innite cardinal λ. If we choose λ so that all λ-models of S are isomorphic, we arrive at a contradiction since, among these isomorphic λ-models, one must satisfy F and another must satisfy $\neg F$, which is impossible. Thus, the theory S is complete.

6. (a) The structure $\langle \mathbb{Z}, n \mapsto n + 1 \rangle$ is obviously a model of T_1; we will call this the standard model. In every model $\mathcal{M} = \langle M, \phi \rangle$ of T_1, we can dene an equivalence relation $\sim$ by

$$a \sim b \quad \text{if and only if there exists } n \in \mathbb{N} \text{ such that } a = \phi^n(b) \text{ or } b = \phi^n(a).$$

The equivalence class of a for the relation $\sim$ will be called its orbit. Each orbit determines a substructure of $\mathcal{M}$ that is a model of T_1 isomorphic to the standard model. For two models of T_1 to be isomorphic, it is necessary and sufcient their sets of orbits be equipotent. If κ is an uncountable cardinal, the set of orbits of any model of T_1 of cardinality κ must also have cardinality κ since each orbit is denumerable. It follows that the theory T_1 is κ-categorical and, since all its models are innite, it is also complete (Vaught's theorem).

(b) A model $\mathcal{M} = \langle M, \phi, \Omega \rangle$ of T_2 is a model of T_1 enhanced by a `colouring' (with two colours) of its orbits; each orbit is either included in Ω (let us call these red) or is included in $M - \Omega$ (call these yellow). The formula $\forall x(Px \Leftrightarrow Pfx)$ excludes all other possibilities. Moreover, each color is effectively present ($\exists x\, Px$, $\exists x \neg Px$). For two models of T_2 to be isomorphic, it is necessary and sufcient that the sets of their red orbits are equipotent, as well as the sets of their yellow orbits. Let κ be an arbitrary innite cardinal. We obtain two non-isomorphic models of T_2 of cardinality κ by taking, on the one hand, a model that has one red orbit and κ yellow orbits and, on the other hand, a model with one yellow orbit and κ red orbits. To be precise, we could take $M = \mathbb{Z} \times \kappa$, $\phi = (n, \alpha) \mapsto (n + 1, \alpha)$, $\Omega_1 = \mathbb{Z} \times \{0\}$, and $\Omega_2 = \mathbb{Z} \times (\kappa - \{0\})$; then set

$$\mathcal{M}_1 = \langle M, \phi, \Omega_1 \rangle \quad \text{and} \quad \mathcal{M}_2 = \langle M, \phi, \Omega_2 \rangle.$$

This shows that T_2 is not κ-categorical.

(c) Let λ be an uncountable cardinal and let $\mathcal{M} = \langle M, \phi, \Omega \rangle$ be a λ-model of T_2 [see the previous exercise, especially part (d)]. The value of the formula Px

in $\mathcal{M}$ is not a nite set (there is at least one orbit included in Ω, a red orbit); it follows (because $\mathcal{M}$ is a λ-model) that $Val(Px, \mathcal{M})$ is a set of cardinality λ. A similar argument applies for the formula $\neg Px$. Consequently, the set of red orbits and the set of yellow orbits of $\mathcal{M}$ each have cardinality λ (since each orbit is denumerable while λ is not). As in part (b), we conclude that all λ-models of T_2 are isomorphic and, from part (e) of Exercise 5, that T_2 is complete (it is clear that T_2 has no nite models).

7. (a) The models of T_0 are the totally ordered sets in which every element has an immediate successor (i.e. a strict least upper bound) and predecessor (i.e. a strict greatest lower bound).

Set $\mathcal{M}_0 = \langle 2\mathbb{Z}, \leq \rangle$ and $\mathcal{M}_1 = \langle \mathbb{Z}, \leq \rangle$ ($2\mathbb{Z}$ is the set of even positive and negative integers). $\mathcal{M}_0$ and $\mathcal{M}_1$ are obviously models of T_0; in $\mathcal{M}_0$, the successor and predecessor of the element $2k$ are $2k + 2$ and $2k - 2$, respectively; in $\mathcal{M}_1$, the successor and predecessor of the element h are $h + 1$ and $h - 1$, respectively. $\mathcal{M}_0$ is a substructure of $\mathcal{M}_1$ but it is not an elementary substructure because, for example, the formula

$$\forall v_0 (R v_0 0 \vee R 2 v_0)$$

with parameters from $\mathcal{M}_0$ is satised in $\mathcal{M}_0$ but not in $\mathcal{M}_1$ ($\mathcal{M}_1$ does not satisfy $R10$ or $R21$).

(b) For every integer $i \in \mathbb{N}$ set

$$A_i = \{x \in \mathbb{Q} : 2^i x \in \mathbb{Z}\} \quad \text{and} \quad \{\mathcal{A}_i = \langle A_i, \leq \rangle\}.$$

The ordering in question is the usual order on $\mathbb{Q}$; A_i is the set of rational numbers of the form $a/2^i$, where $a \in \mathbb{Z}$; $\mathcal{A}_i$ is an L_0-structure. It is clear that, for $i \leq j$, $\mathcal{A}_i$ is a substructure of $\mathcal{A}_j$. In addition, each $\mathcal{A}_i$ is a model of T_0 and is in fact isomorphic to $\mathcal{M}_1$ (the map which, with each $a \in \mathbb{Z}$, associates $a/2^i$ is an isomorphism from $\mathcal{M}_1$ onto $\mathcal{A}_i$). So we are in the presence of a chain of models of T_0. The union of this chain is the set A of rational numbers whose denominator is a power of 2, together with the usual ordering. This is not a model of T_0 since, for any pair of elements a and b of A with $a < b$, b cannot be the immediate successor of a because the rational number $(a + b)/2$, which belongs to A, lies strictly between a and b.

So we may conclude from Theorem 8.43 that T_0 is not equivalent to any $\forall\exists$ theory in L_0.

8. (a) Let $\mathcal{M}$ be a prime model of a model-complete theory T. Given two arbitrary models $\mathcal{A}$ and $\mathcal{B}$ of T, there exist structures $\mathcal{A}'$ and $\mathcal{B}'$, isomorphic to $\mathcal{A}$ and $\mathcal{B}$, respectively (and hence models of T), such that $\mathcal{M} \subseteq \mathcal{A}'$ and $\mathcal{M} \subseteq \mathcal{B}'$. Because T is model-complete, we know that these inclusions are elementary, i.e. $\mathcal{M} \prec \mathcal{A}'$ and $\mathcal{M} \prec \mathcal{B}'$. In particular, $\mathcal{A}'$ and $\mathcal{B}'$ are elementarily equivalent to $\mathcal{M}$, hence $\mathcal{A}' \equiv \mathcal{B}'$. But since $\mathcal{A}$ is isomorphic to $\mathcal{A}'$ and $\mathcal{B}$ is isomorphic to $\mathcal{B}'$, we conclude that $\mathcal{A} \equiv \mathcal{B}$. We have proved that any two models of T are elementarily equivalent, so T is complete.

(b) (1) implies (2): We prefer to prove that the negation of (2) implies the negation of (1). Let $\mathcal{M}$ be a model of T and let F be a closed formula of L_M (the language obtained by adjoining to L a constant symbol for each element of M) such that F is not a consequence of $T \cup \Delta(\mathcal{M})$. So there exists a model $\mathcal{M}'$ of $T \cup \Delta(\mathcal{M}) \cup \{\neg F\}$ and we may even assume that $\mathcal{M} \subseteq \mathcal{M}'$ (see Lemma 8.13). However, $\mathcal{M}'$ is not an elementary extension of $\mathcal{M}$ since F is satised in $\mathcal{M}$ but not in $\mathcal{M}'$.

(2) implies (1): Suppose that $\mathcal{M} \subseteq \mathcal{M}'$ are two models of T. Then once $\mathcal{M}'$ has been enriched, in a natural way, to an L_M-structure, it is a model of $T \cup \Delta(\mathcal{M})$, and is thus a model of $D(\mathcal{M})$. It follows that $\mathcal{M} \prec \mathcal{M}'$.

(2) implies (3) is obvious.

(3) implies (2): We will prove that the negation of (2) implies the negation of (3). Let $\mathcal{M}$ be a model of T and let F be a formula of $D(\mathcal{M})$ that is not a consequence of $T \cup \Delta(\mathcal{M})$. So there exists a formula $G[v_0, v_1, \ldots, v_n]$ of L and elements $a_0, a_1, \ldots, a_n$ of M such that $F = G[a_0, a_1, \ldots, a_n]$ and $\mathcal{M} \models F$. By Theorem 8.9, there exists a denumerable elementary submodel $\mathcal{M}_0$ of $\mathcal{M}$ that contains the elements $a_0, a_1, \ldots, a_n$. Since F is not a consequence of $T \cup \Delta(\mathcal{M})$, neither is it a consequence of the subset $T \cup \Delta(\mathcal{M}_0) \subseteq T \cup \Delta(\mathcal{M})$. But $F \in D(\mathcal{M}_0)$; thus condition (3) is not satised.

(1) implies (4) is obvious.

(4) implies (3) is proved in the same way as (1) implies (2), taking care to only choose denumerable models (as the LowenheimñSkolem theorem allows us to do).

(c) It is sufcient to show that the class of models of T is closed under unions of chains (Theorem 8.43). Because T is model-complete, any chain of models of T satises the hypotheses of Theorem 8.21, so the union of such a chain is a model of T. The converse is clearly false; the empty theory is an $\forall\exists$ theory that is not model-complete! It is more difcult to nd a complete $\forall\exists$ theory that is not model-complete; this will be done in the last part of Exercise 9.

(d) The condition $(*)$ is veried for all existential formulas (see Theorem 8.39) and for all formulas that are equivalent modulo T to an existential formula. This proves the `if' direction.

For the opposite direction, we add, as suggested, new constant symbols $c_0, c_1, \ldots, c_n$ to the language and we consider the theory

$$\begin{aligned}\Psi = \{G[c_0, c_1, \ldots, c_n] :\ & G[v_0, v_1, \ldots, v_n] \text{ is a} \\ & \text{universal formula of } L \text{ and} \\ & T \vdash \neg F[c_0, c_1, \ldots, c_n] \Rightarrow G[c_0, c_1, \ldots, c_n]\}.\end{aligned}$$

Let $\mathcal{N}$ be a model of $T \cup \Psi$. We are going to prove that $\mathcal{N}$ has an extension $\mathcal{N}'$ that satises $\neg F[c_0, c_1, \ldots, c_n]$. To do this, we use the method of diagrams; it sufces to show that $T \cup \Delta(\mathcal{N}) \cup \{\neg F[c_0, c_1, \ldots, c_n]\}$ is consistent. For an argument by contradiction, we assume this is false; so there exists a formula H of $\Delta(\mathcal{N})$ such that

$$T \vdash H \Rightarrow F[c_0, c_1, \ldots, c_n].$$

Since H belongs to $\Delta(\mathcal{N})$, there exists a quantier-free formula

$$K[v_0, v_1, \ldots, v_{n+p}]$$

of L and points $a_1, a_2, \ldots, a_p$ in N such that

$$H = K[c_0, c_1, \ldots, c_n, a_1, a_2, \ldots, a_{n+p}].$$

Because the points $a_1, a_2, \ldots, a_p$ do not appear in T or in $F[c_0, c_1, \ldots, c_n]$, we may conclude that

$$\begin{aligned} T \vdash \forall v_1 \forall v_2 \ldots \forall v_p (&K[c_0, c_1, \ldots, c_n, v_1, v_2, \ldots, v_p] \\ &\Rightarrow F[c_0, c_1, \ldots, c_n]), \end{aligned}$$

or, equivalently,

$$\begin{aligned} T \vdash\ &\neg F[c_0, c_1, \ldots, c_n] \\ &\Rightarrow \forall v_1 \forall v_2 \ldots \forall v_p \neg K[c_0, c_1, \ldots, c_n, v_1, v_2, \ldots, v_p]. \end{aligned}$$

Thus, we see that the formula

$$\forall v_1 \forall v_2 \ldots \forall v_p \neg K[c_0, c_1, \ldots, c_n, v_1, v_2, \ldots, v_p]$$

belongs to Ψ, so is true in $\mathcal{N}$; but this contradicts the fact that

$$K[c_0, c_1, \ldots, c_n, a_1, a_2, \ldots, a_{n+p}]$$

belongs to $\Delta(\mathcal{N})$.

Next, suppose that the formula $F[v_0, v_1, \ldots, v_n]$ satises condition $(*)$. Since a model of $T \cup \{F[c_0, c_1, \ldots, c_n]\}$ cannot have an extension that satis-es $\neg F[c_0, c_1, \ldots, c_n]$, it follows from what we have just proved above that

$$T \cup \{F[c_0, c_1, \ldots, c_n]\} \cup \Psi$$

is contradictory. So by compactness, there exists a nite subset Ψ_0 of Ψ such that $\neg F[c_0, c_1, \ldots, c_n]$ is a consequence of $\Psi_0 \cup T$ and, since Ψ_0 is a consequence of $T \cup \{\neg F[c_0, c_1, \ldots, c_n]\}$, we have

$$T \vdash \neg F[c_0, c_1, \ldots, c_n] \Leftrightarrow \bigwedge_{G \in \Psi_0} G[c_0, c_1, \ldots, c_n].$$

Also, we know that a conjunction of universal formulas is equivalent to a universal formula and that the negation of a universal formula is equivalent to an existential formula. So there exists an existential formula $H[v_0, v_1, \ldots, v_n]$ of L such that

$$T \vdash F[c_0, c_1, \ldots, c_n] \Leftrightarrow H[c_0, c_1, \ldots, c_n],$$

and since the constants c_i do not appear in T,

$$T \vdash \forall v_0 \forall v_1 \ldots \forall v_n (F[v_0, v_1, \ldots, v_n] \Leftrightarrow H[v_0, v_1, \ldots, v_n]).$$

(e) To say that T is model-complete is to assert that condition $(*)$ is veried for every formula of L. According to part (d), it is sufcient to prove that

condition $(*)$ is satised for all formulas
if and only if
it is satised for all universal formulas .

Assume that $(*)$ is satised for every universal formula; we will prove that it is satised by every formula. We argue by induction. Let $F[v_0, v_1, \ldots, v_n]$ be a formula of L; we may assume that the only logical symbols appearing in F are $\neg$, $\wedge$, $\vee$, and $\forall$. The only case that poses a problem is when $F[v_0, v_1, \ldots, v_n] = \neg G[v_0, v_1, \ldots, v_n]$. By the induction hypothesis, we know that there exists an existential formula

$$H[v_0, v_1, \ldots, v_n]$$

such that

$$T \vdash \forall v_0 \forall v_1 \ldots \forall v_n (H[v_0, v_1, \ldots, v_n] \Leftrightarrow G[v_0, v_1, \ldots, v_n]);$$

therefore,

$$T \vdash \forall v_0 \forall v_1 \ldots \forall v_n (\neg H[v_0, v_1, \ldots, v_n] \Leftrightarrow \neg G[v_0, v_1, \ldots, v_n]).$$

But $\neg H[v_0, v_1, \ldots, v_n]$ is equivalent to a universal formula and, by hypothesis, there exists an existential formula $K[v_0, v_1, \ldots, v_n]$ such that

$$T \vdash \forall v_0 \forall v_1 \ldots \forall v_n (\neg H[v_0, v_1, \ldots, v_n] \Leftrightarrow K[v_0, v_1, \ldots, v_n]).$$

It follows from all this that

$$T \vdash \forall v_0 \forall v_1 \ldots \forall v_n (F[v_0, v_1, \ldots, v_n] \Leftrightarrow K[v_0, v_1, \ldots, v_n]).$$

9. (a) In part (d) of Exercise 8, we proved the equivalence of (2) with a condition that is *a priori* stronger than (1) (without cardinality restrictions). So we know that (2) implies (1). For the reverse direction, we adapt this proof taking care to choose only models of cardinality λ (which is possible thanks to the LowenheimñSkolem theorem).

(b) Begin with a model $\mathcal{M}$ of T of cardinality λ. We will rst construct an extension $\mathcal{M}^1$ of $\mathcal{M}$ that is a model of T and has the following property:

$$\begin{array}{l}\text{for all elements } a_0, a_1, \ldots, a_n \text{ of } M \text{ and}\\ \text{for every model } \mathcal{M}' \text{ of } T \text{ that is an extension of } \mathcal{M}^1,\\ \text{if } \mathcal{M}^1 \models F[a_0, a_1, \ldots, a_n], \text{ then } \mathcal{M}' \models F[a_0, a_1, \ldots, a_n]. \end{array} \quad (*)$$

To do this, we enumerate the set of sequences of elements of M of length n (this set has cardinality λ): $\{(a_0^i, a_1^i, \ldots, a_n^i) : i \in \lambda\}$. The model $\mathcal{M}^1$ will be the union of an increasing chain $(\mathcal{M}_i : i \in \lambda)$ of models of T which we will construct by induction on $i \in \lambda$ as follows:

- $\mathcal{M}_0 = \mathcal{M}$.
- If i is a limit ordinal, $\mathcal{M}_i = \bigcup_{j<i} \mathcal{M}_j$; we know that $\mathcal{M}_i$ is a model of T because T is an $\forall\exists$ theory (see Theorem 8.43).
- Suppose that $j = i + 1$. We distinguish two cases:

(i) (α) If, for every extension $\mathcal{N}$ of $\mathcal{M}_j$ that is a model of T, $\mathcal{N} \models F[a_0^i, a_1^i, \ldots, a_n^i]$, then we set $\mathcal{M}_i = \mathcal{M}_j$; we remark that this property will also be true for the model $\mathcal{M}^1$ that we will obtain at the end of this construction: for every extension $\mathcal{N}$ of $\mathcal{M}^1$ that is a model of T, $\mathcal{N} \models F[a_0^i, a_1^i, \ldots, a_n^i]$.

(ii) (β) If there exists an extension $\mathcal{N}$ of $\mathcal{M}_j$ that is a model of T, $\mathcal{N} \models \neg F[a_0^i, a_1^i, \ldots, a_n^i]$, then we set $\mathcal{M}_i$ equal to such an $\mathcal{N}$ of cardinality λ (again, the LowenheimñSkolem theorem guarantees that one exists); note that because F is universal, for every extension $\mathcal{N}$ of $\mathcal{M}_i$ (in particular for our target structure $\mathcal{M}^1$), $\mathcal{N} \models \neg F[a_0^i, a_1^i, \ldots, a_n^i]$.

Now, as we anticipated, by setting $\mathcal{M}^1 = \bigcup_{i\in\lambda} \mathcal{M}_i$, we obtain a model of T that has property $(*)$.

Next, start over! Using the same method, we construct a model $\mathcal{M}^2$ such that

$$\begin{array}{l}\text{for all elements } a_0, a_1, \ldots, a_n \text{ of } M_1 \text{ and}\\ \text{for every model } \mathcal{M}' \text{ of } \mathcal{T} \text{ that is an extension of } \mathcal{M}^2,\\ \text{if } \mathcal{M}^2 \models F[a_0, a_1, \ldots, a_n], \text{ then } \mathcal{M}' \models F[a_0, a_1, \ldots, a_n], \end{array}$$

then another model $\mathcal{M}^3$, and so on. If we set $\mathcal{M}' = \bigcup_{k\in\mathbb{N}} \mathcal{M}^k$, we see that $\mathcal{M}'$ is a model of T (again, because T is an $\forall\exists$ theory) of cardinality λ and that it satises the condition $(**)$ from the statement of the exercise (because every nite sequence of elements from the structure $\mathcal{M}'$ already appears in the base set of one of the structures $\mathcal{M}^k$).

(c) We have just seen that if T is an $\forall\exists$ theory, then it has a model $\mathcal{M}$ of cardinality λ that satises $(**)$. If, in addition, T is λ-categorical, then all models of T of cardinality λ are isomorphic to $\mathcal{M}$, so they also satisfy $(**)$. Thus, condition $(*)$ from part (b) is veried for any universal formula

$F[v_0, v_1, \ldots, v_n]$. It then follows from part (d) of Exercise 8 that every universal formula is equivalent modulo T to an existential formula, so by part (e) of that same exercise, T; is model-complete.

(d) The theory T_0 is universal, hence it is $\forall\exists$. Consider the following models of T_0 :

$$\mathcal{N}_0 = \langle \mathbb{N}, n \mapsto n+1 \rangle \quad \text{and} \quad \mathcal{N}_1 = \langle \mathbb{Z}, n \mapsto n+1 \rangle.$$

The closed formula $\forall v_0 \exists v_1 (v_0 \simeq f v_1)$ is true in $\mathcal{M}_1$ but not in $\mathcal{M}_0$, so these models are not elementarily equivalent; so T_0 is not complete. We have $\mathcal{M}_0 \subseteq \mathcal{M}_1$ but, clearly, $\mathcal{M}_0$ is not an elementary submodel of $\mathcal{M}_1$; so T_0 is not model-complete.

Add a denumerably innite set $c_0, c_1, \ldots, c_n, \ldots$ of new constant symbols to the language of T_0 and consider the theory

$$T = T_0 \cup \{\forall v_0 \neg c_n \simeq f v_0 : n \in \mathbb{N}\} \\ \cup \{c_n \neq c_m : n, m \in \mathbb{N} \text{ and } n \neq m\}.$$

The theory T is $\forall\exists$ since it is universal. We will show that it is not model-complete. Let $\mathcal{N}_0$ and $\mathcal{N}_1$ be the structures whose respective base sets are $\mathbb{N} \times \mathbb{N}$ and $\mathbb{Z} \times \mathbb{N}$, in which the interpretation of f is the mapping $(m, n) \mapsto (m+1, n+1)$, and in which the constant symbol c_n is interpreted by the pair $(n, 0)$; these are models of T that satisfy $\mathcal{N}_0 \subseteq \mathcal{N}_1$ but not $\mathcal{N}_0 \prec \mathcal{N}_1$ [for example, the formula $\exists v_0 f v_0 \simeq (0, 1)$, with parameters from $\mathcal{N}_0$, is satised in $\mathcal{N}_1$ but not in $\mathcal{N}_0$].

The most difcult part is to show that T is complete. We will be content to outline the main idea of the proof. We proceed with an analysis of models of T (as we have already done many times); a model $\mathcal{M} = \langle M, \bar{f}, (\bar{c}_n)_{n \in \mathbb{N}} \rangle$ of T decomposes into $\bar{f}$-orbits [these are the equivalence classes for the relation in which x and y are related if and only if there exist integers n and m such that $\bar{f}^n(x) = \bar{f}^m(y)$]; it is easy to see that for the initial language (without the c_n), each orbit is isomorphic either to $\langle \mathbb{Z}, n \mapsto n+1 \rangle$ (when the restriction of $\bar{f}$ is bijective) or to $\langle \mathbb{N}, n \mapsto n+1 \rangle$ (when it is not). For the enriched language, there are three types of orbits: of type $\mathbb{Z}$, of type $\mathbb{N}$ containing one of the elements $\bar{c}_n$, and nally of type $\mathbb{N}$ without constants. We prove, on the one hand, that any two denumerable models that have denumerably many orbits of each type are isomorphic and, on the other hand, that every denumerable model has an elementary extension that does have denumerably many orbits of each type.

10. Let L_0 be the language reduced to the single symbol R. We will make use of the following facts.

Every denumerable model of A is isomorphic to the ordered set of rationals (see Example 8.19).

If $\mathcal{M}$ and $\mathcal{N}$ are models of A and if $(m_0, m_1, \ldots, m_k)$ and $(n_0, n_1, \ldots, n_k)$ are two sequences of elements from $\mathcal{M}$ and $\mathcal{N}$, respectively, then the following two conditions are equivalent:

(1) For every formula $F[v_0, v_1, \ldots, v_k]$ of L_0,

$$\mathcal{M} \models F[m_0, m_1, \ldots, m_k] \quad \text{if and only if} \quad \mathcal{N} \models F[n_0, n_1, \ldots, n_k].$$

(2) For every quantier-free formula $F[v_0, v_1, \ldots, v_k]$ of L_0,

$$\mathcal{M} \models F[m_0, m_1, \ldots, m_k] \quad \text{if and only if} \quad \mathcal{N} \models F[n_0, n_1, \ldots, n_k]$$

(see Lemma 8.24).

In addition, here are a few remarks concerning the sequences α, β, and γ. These are three strictly increasing sequences of rationals: the rst is unbounded, the second is bounded and has the rational number 0 as its least upper bound, while the third is bounded but does not have a least upper bound in $\mathbb{Q}$ (its least upper bound in $\mathbb{R}$ is the number e which is not rational).

(a) Let $\mathcal{M}$ and $\mathcal{N}$ be two models of T and, for $k \in \mathbb{N}$, let m_k and n_k be the interpretations of c_k in $\mathcal{M}$ and $\mathcal{N}$, respectively. Let F be a closed formula of L. So for some integer k and some formula $G[v_0, v_1, \ldots, v_k]$ of L_0, $F = G[c_0, c_1, \ldots, c_k]$. Since the sequences

$$(m_0, m_1, \ldots, m_k) \quad \text{and} \quad (n_0, n_1, \ldots, n_k)$$

are strictly increasing, they satisfy condition (2) above, and hence condition (1) also. In other words,

$$\mathcal{A} \models G[c_0, c_1, \ldots, c_k] \quad \text{if and only if} \quad \mathcal{B} \models G[c_0, c_1, \ldots, c_k].$$

The structures $\mathcal{M}$ and $\mathcal{N}$ satisfy the same closed formulas, so they are elementarily equivalent and the theory T is complete.

(b) Let $\mathcal{M}$ be a denumerable model of T. Then $\mathcal{M} \restriction L_0$ is a denumerable model of A; so there exists an isomorphism ϕ from $\mathcal{M} \restriction L_0$ onto $\langle \mathbb{Q}, \leq \rangle$. We can enrich $\langle \mathbb{Q}, \leq \rangle$ to an L-structure $\mathcal{M}_1$ by declaring that, for all $n \in \mathbb{N}$, the interpretation of c_n is the image under ϕ of the interpretation of c_n in $\mathcal{M}$. We are guaranteed in this way that ϕ is an isomorphism from $\mathcal{M}$ onto $\mathcal{M}_1$.

It remains to prove that $\mathcal{M}_1$ is isomorphic to one of the three structures $\mathcal{A}$, $\mathcal{B}$, or $\mathcal{C}$. Let δ_n denote the interpretation of c_n in $\mathcal{M}_1$. Then

$$\Delta = (\delta_n : n \in \mathbb{N})$$

is a strictly increasing sequence of rationals for which there are three possibilities:

(1) The sequence Δ is unbounded. Here is an isomorphism ψ from $\mathcal{A}$ onto $\mathcal{M}_1$:

- If $x \leq 0$, then $\psi(x) = x + \delta_0$ (thus ψ is an increasing map and is a bijection from the interval $(-\infty, \alpha_0]$ onto the interval $(-\infty, \delta_0]$).

- If $n \leq x \leq n+1$, then $\psi(x) = (\delta_{n+1} - \delta_n)(x - n) + \delta_n$ (again, ψ is an increasing map from $[\alpha_n, \alpha_{n+1}]$ onto $[\delta_n \delta_{n+1}]$).

(2) The sequence Δ is bounded and its least upper bound is the rational number ε. This time, it is $\mathcal{B}$ that is isomorphic to $\mathcal{M}_1$. The same principle is used to construct the isomorphism ψ from $\mathcal{B}$ onto $\mathcal{M}_1$; we dene bijections between the intervals into which $\mathbb{Q}$ is decomposed by the sequences $(\beta_n : n \in \mathbb{N})$ and $(\delta_n : n \in \mathbb{N})$, respectively.

β_0 β_1 β_{n-1} β_n 0

- - - • óóóó • óó - - - óó • óóóó • óóñ - - - óóñ • óó - - -

$\downarrow \psi$ $\downarrow \psi$ $\downarrow \psi$ $\downarrow \psi$ $\downarrow \psi$

- - - • óóóó • óó - - - óó • óóóó • óóñ - - - óóñ • óó - - -

δ_0 δ_1 δ_{n-1} δ_n ε

ó If $x \leq \beta_0$, then $\psi(x) = x - \beta_0 + \delta_0$.

ó If $\beta_n \leq x \leq \beta_{n+1}$, then $\psi(x) = ((\delta_{n+1} - \delta_n)/(\beta_{n+1} - \beta_n))(x - \beta_n) + \delta_n$.

ó If $x \geq 0$, then $\psi(x) = x + \varepsilon$.

(3) The sequence Δ is bounded but its least upper bound is an irrational number ε. We dene a map ψ from $\mathcal{C}$ into $\mathcal{M}_1$ by the following:

ó If $x \leq \gamma_0$, then $\psi(x) = x - \gamma_0 + \delta_0$.

ó If $\gamma_n \leq x \leq \gamma_{n+1}$, then $\psi(x) = ((\delta_{n+1} - \delta_n)/(\beta_{n+1} - \beta_n))(x - \gamma_n) + \delta_n$.

ó It remains to dene ψ on $\mathbb{Q} \cap (e, +\infty)$; there exists an isomorphism θ from this interval onto $\mathbb{Q} \cap (\varepsilon, +\infty)$ (because these are two denumerable, dense linear orderings with no rst or last element). Set $\psi(x) = \theta(x)$.

It is more or less obvious that the structures $\mathcal{A}$, $\mathcal{B}$, and $\mathcal{C}$ are pairwise not isomorphic. The conclusion is that the only models of T, up to isomorphism, are $\mathcal{A}$, $\mathcal{B}$, and $\mathcal{C}$.

(c) This follows nearly immediately from Lemma 8.6.

(d) Since T is model-complete, we can replace `elementary extension' by `(simple) extension' in the question. Moreover, it is sufcient to prove that $\mathcal{A}$ has an extension that is isomorphic to $\mathcal{B}$, that $\mathcal{B}$ has an extension that is isomorphic to $\mathcal{C}$, and that $\mathcal{C}$ has an extension that is isomorphic to $\mathcal{B}$.

Instead of showing that $\mathcal{A}$ has an extension that is isomorphic to $\mathcal{B}$, we will show that $\mathcal{B}$ has a substructure that is isomorphic to $\mathcal{A}$ (using an argument analogous to the proof of Lemma 8.13). If $\mathcal{B}_0$ is the substructure of $\mathcal{B}$ whose base set is the interval $(-\infty, 0)$, we see that $\mathcal{B}_0$ is a model of T in which the sequence $(\beta_n : n \in \mathbb{N})$ is unbounded, so it is isomorphic to $\mathcal{A}$ according to part (b).

Similarly, there is a substructure $\mathcal{C}_0$ of $\mathcal{C}$ that is isomorphic to $\mathcal{B}$; it is the one whose base set is $(-\infty, e) \cup [3, +\infty)$. $\mathcal{C}_0$ is certainly a model of T and, in $\mathcal{C}_0$, the sequence $(\gamma_n : n \in \mathbb{N})$ is bounded and has a least upper bound, namely 3.

We can also nd a denumerable extension of $\mathcal{C}$ that is a model of T and in which the sequence $(\gamma_n : n \in \mathbb{N})$ is bounded and has a least upper bound: it is the substructure of $\langle \mathbb{R}, \leq \rangle$ whose base set is $\mathbb{Q} \cup \{e\}$.

11. (a) Let $\mathcal{M} = \langle M, \bar{f}, \bar{R} \rangle$ be a model of A. The rst formula expresses that $\leq$ is reexive, the second that it is total and antisymmetric [if x and y are distinct elements of M, then exactly one of $\bar{R}(x, y)$ or $\bar{R}(y, x)$ holds]; and the third that $\bar{R}$ is transitive. So we are dealing with a total ordering. Together with the fourth and fth formulas, we have that $\bar{f}$ is an isomorphism of the structure $\langle M, \bar{R} \rangle$ onto itself. The sixth and seventh formulas assert that $\bar{f}(x)$ is a strict bound for x and that it is the successor of x (i.e. the least of its strict upper bounds).

(b) It is not a problem to verify this.

(c) The fact that the relation $\leq$ is reexive and transitive follows easily from properties of $\bar{f}$ and $\bar{R}$. The fact that the relation $\approx$ is an equivalence relation is more or less obvious.

If $a \approx b$, then there exist integers n and p such that $\bar{f}^n(a) = \bar{f}^p(b)$. We then have

$$\mathcal{M} \models \neg R f^{n+1} a\, b \wedge \neg R f^{p+1} b\, a,$$

and it is false that $a \ll b$ and that $b \ll a$. Conversely, if $a \ll b$ and $b \ll a$ are both false, there exist integers n and p such that

$$\mathcal{M} \models \neg R f^n a\, b \wedge \neg R f^p b\, a.$$

Suppose, for example, that a is less than or equal to b (for $\bar{R}$) and that m is the least integer such that $\mathcal{M} \models \neg R f^m a\, b$; m is strictly positive and $\mathcal{M} \models R f^{m-1} a\, b$. Because $f^m a$ is the successor of $f^{m-1} a$, we see that $f^m a = b$ (if not, we would also have $\mathcal{M} \models R f^m a\, b$). This proves that $a \approx b$.

Since $\bar{f}$ is bijective, we can refer to $\bar{f}^{-1}$ and, indeed, to $\bar{f}^n$ for all $n \in \mathbb{Z}$. Let a be an element of M. It is easy to check that the map from $\mathbb{Z}$ into M whose value for $n \in \mathbb{Z}$ is $\bar{f}^n(a)$ is a monomorphism from $\mathbb{Z}$ into $\mathcal{M}$ and that its image is the equivalence class of a relative to $\approx$. It is also easy to see that this class is closed under $\bar{f}$, which shows that this is a substructure of $\mathcal{M}$.

We see that if a, b, and c are elements of M and if $a \approx b$, then $a \ll c$ if and only if $b \ll c$; indeed, suppose, for example, that $a = \bar{f}^n(b)$. It is then certainly clear that `for every $p \in \mathbb{N}$, $\bar{R}(\bar{f}^p(b), c)$' is equivalent to `for every $p \in \mathbb{N}$, $\bar{R}(\bar{f}^{n+p}(b), c)$', which is equivalent in turn to `for every

$p \in \mathbb{N}$, $\bar{R}(\bar{f}^p(a), c)$'. Thus, we can dene a binary relation $\lhd$ on $M/\approx$ as follows: for all a, b in M,

$$a/\approx \lhd b/\approx \quad \text{if and only if} \quad a \ll b.$$

The fact that $\lhd$ is irreexive and transitive follows immediately from the corresponding properties for $\ll$. To show that $\lhd$ is a total strict ordering, it is sufcient to show that if α and β are elements of $M/\approx$, then either $\alpha = \beta$ or $\alpha \lhd \beta$ or $\beta \lhd \alpha$; in other words, that if a and b are elements of M, then $a \approx b$ or $a \ll b$ or $b \ll a$; but that is what we did at the beginning of this question.

So let C be the set $M/\approx$ and $X = \langle C, \lhd \rangle$. For each $\alpha \in C$, choose a point c_α in the class α. It is then easy to verify that the map ϕ from $\mathcal{M}_X$ into $\mathcal{M}$ dened for all $\alpha \in C$ and for all $n \in \mathbb{Z}$ by

$$\phi((\alpha, n)) = \bar{f}^n(c_\alpha)$$

is an isomorphism.

(d) We leave it to the reader to check the next two facts.

- If ϕ is an isomorphism between two totally ordered sets X and Y, then the map ψ from $\mathcal{M}_X$ into $\mathcal{M}_Y$ dened for all $a \in M$ and for all $n \in \mathbb{Z}$ by

$$\psi((a, n)) = (\phi(a), n)$$

is an isomorphism.

- If ψ is an isomorphism from $\mathcal{M}_X$ onto $\mathcal{M}_Y$, then the set

$$\{(a, b) : \text{ there exist } n \text{ and } p \text{ in } \mathbb{Z} \text{ such that } \psi((a, n)) = (b, p)\}$$

is the graph of an isomorphism from X onto Y.

If λ is an innite cardinal, then we can nd two models of A of cardinality λ that are not isomorphic; for example, $X = \lambda$ and $Y = \lambda \cup \{\lambda\}$ are totally ordered sets of cardinality λ that are not isomorphic (the second has a greatest element but the rst does not). Consequently, $\mathcal{M}_X$ and $\mathcal{M}_Y$ are two models of cardinality λ that are not isomorphic. This shows that A is not categorical in any innite cardinal.

(e)(1) We use the method of diagrams. In the language L_M, consider the complete diagram $D(\mathcal{M})$ of $\mathcal{M}$; add a new constant symbol c and consider the theory

$$T = D(\mathcal{M}) \cup \{\neg Rcf^n\underline{a} : n \in \mathbb{N}\} \cup \{\neg R\underline{b}f^n c : n \in \mathbb{N}\}.$$

Using the compactness theorem we can show that this theory is consistent; if T_0 is a nite subset of T, there exists an integer k such that T_0 is included in

$$D(\mathcal{M}) \cup \{\neg Rcf^n\underline{a} : 0 \leq n \leq k\} \cup \{\neg R\underline{b}f^n c : 0 \leq n \leq k\}$$

and, to produce a model of T_0, it sufces to enrich $\mathcal{M}$ by taking $\bar{f}^{k+1}(a)$ as the interpretation of c.

We have seen that T has a model $\mathcal{M}_1$ whose reduct to L is an elementary extension of $\mathcal{M}$. If $\bar{c}$ is the interpretation of c in $\mathcal{M}$, we certainly have $a \ll \bar{c}$ and $\bar{c} \ll b$.

(2) The argument is analogous. It is sufcient to show that the theory

$$T_1 = D(\mathcal{M}) \cup \{\neg Rcf^n\underline{a} : n \in \mathbb{N}\} \cup \{\neg R\underline{a}f^n b : n \in \mathbb{N}\},$$

in the language enriched by two new constant symbols b and c, is consistent.

(3) This verication is left to the reader.

(4) If $c \approx a_i$ for some i between 1 and n inclusive, then there exists an element p in $\mathbb{Z}$ such that $c = \bar{f}^p(a_i)$. Then, using (3), we see that

$$P((\mathcal{M}, a_1, a_2, \ldots, a_n, c),\ (\mathcal{N}, b_1, b_2, \ldots, b_n, \bar{f}^p(b_i))).$$

If not, then we must distinguish several cases:

- For every i between 0 and n inclusive, $c \ll a_i$; in this case, we use (2) to produce an elementary extension $\mathcal{N}'$ of $\mathcal{N}$ and a point d in $\mathcal{N}'$ such that, for all i from 0 to n inclusive, $d \ll a_i$. Then, again using (3), we have

$$P((\mathcal{M}, a_1, a_2, \ldots, a_n, c),\ (\mathcal{N}', b_1, b_2, \ldots, b_n, d)).$$

- For every i between 0 and n inclusive, $a_i \ll c$; this time, we choose $\mathcal{N}' \succ \mathcal{N}$ and $d \in \mathcal{N}'$ such that, for all i from 0 to n inclusive, $b_i \ll d$. Again, we have

$$P((\mathcal{M}, a_1, a_2, \ldots, a_n, c),\ (\mathcal{N}', b_1, b_2, \ldots, b_n, d)).$$

- Finally, in the remaining case, suppose that in the set $\{a_1, a_2, \ldots, a_n\}$, a_i is the largest element that is less than c and a_j is the smallest element that is greater than c. So we have $a_i \ll c$ and $c \ll a_j$ and, because $P((\mathcal{M}, a_1, a_2, \ldots, a_n),\ (\mathcal{N}, b_1, b_2, \ldots, b_n))$, $b_i \ll b_j$. We can then use (1) to nd $\mathcal{N}' \succ \mathcal{N}$ and $d \in \mathcal{N}'$ such that $b_i \ll d$ and $d \ll b_j$. Once again, we have

$$P((\mathcal{M}, a_1, a_2, \ldots, a_n, c),\ (\mathcal{N}', b_1, b_2, \ldots, b_n, d)).$$

(5) We may always assume that the universal quantier does not occur in the formula $G[v_1, v_2, \ldots, v_n]$. In a proof by induction, the hypothesis provides the result for atomic formulas and the steps involving the propositional connectives are straightforward. So it remains to deal with the existential quantier.

So suppose that $P((\mathcal{M}, a_1, a_2, \ldots, a_n),\ (\mathcal{N}, b_1, b_2, \ldots, b_n))$ is satised, that $G[v_1, v_2, \ldots, v_n] = \exists v_0 F[v_0, v_1, \ldots, v_n]$, and that $\mathcal{M} \models G[a_1, a_2, \ldots, a_n]$. Consequently, there exists a point c in $\mathcal{M}$ such that

$\mathcal{M} \models F[c, a_1, \ldots, a_n]$. According to (4), there exists $\mathcal{N}' \succ \mathcal{N}$ and $d \in \mathcal{N}'$ such that

$$P((\mathcal{M}, a_1, a_2, \ldots, a_n, c),\ (\mathcal{N}', b_1, b_2, \ldots, b_n, d));$$

by the induction hypothesis, we have

$$\mathcal{N}' \models F[c, b_1, b_2, \ldots, b_n],$$

and hence

$$\mathcal{N}' \models G[b_1, b_2, \ldots, b_n].$$

Since $\mathcal{N}'$ is an elementary extension of $\mathcal{N}$, we also have

$$\mathcal{N} \models G[b_1, b_2, \ldots, b_n].$$

(6) Let $\mathcal{M}$ and $\mathcal{N}$ be two models of A. When we apply the preceding result to these two models and to the empty sequence, we see that $\mathcal{M}$ and $\mathcal{N}$ satisfy the same closed formulas. This shows that two arbitrary models of A are elementarily equivalent and that T is complete.

12. We will use Lemma 8.6. Add $n + 1$ new constant symbols $c_0, c_1, \ldots, c_n$ to the language L and consider the theory

$$\Phi = \{H[c_0, c_1, \ldots, c_n] : H[v_0, v_1, \ldots, v_n] \text{ is a quantier-free formula of } L \text{ and } T \vdash F[c_0, c_1, \ldots, c_n] \Rightarrow H[c_0, c_1, \ldots, c_n]\}.$$

Let $\mathcal{M}$ be a model of $\Phi \cup T$. Consider the theory

$$\Psi = \{K[c_0, c_1, \ldots, c_n] : K[v_0, v_1, \ldots, v_n] \text{ is a quantier-free formula of } L \text{ and } \mathcal{M} \models K[c_0, c_1, \ldots, c_n]\}.$$

We claim that $\Psi \cup T \cup \{F[c_0, c_1, \ldots, c_n]\}$ is consistent; if not, then for some nite subset Ψ_0 of Ψ, the theory $\Psi_0 \cup T \cup \{F[c_0, c_1, \ldots, c_n]\}$ is contradictory. There exists a quantier-free formula $J[v_0, v_1, \ldots, v_n]$ of L such that

$$J = \bigwedge_{K \in \Psi_0} K \quad \text{and} \quad T \vdash F[c_0, c_1, \ldots, c_n] \Rightarrow \neg J[c_0, c_1, \ldots, c_n].$$

It follows that $\neg J[c_0, c_1, \ldots, c_n]$ belongs to Φ; but this is absurd since $\mathcal{M}$ is a model of Φ and satises $J[c_0, c_1, \ldots, c_n]$.

So there exists a model $\mathcal{N}$ of $\Psi \cup T \cup \{F[c_0, c_1, \ldots, c_n]\}$. The interpretations of $c_0, c_1, \ldots, c_n$ in $\mathcal{M}$ and $\mathcal{N}$, respectively, satisfy the same atomic formulas; so by Lemma 8.6, they satisfy the same formulas and $\mathcal{M} \models F[c_0, c_1, \ldots, c_n]$. We have thereby proved that every model of $\Phi \cup T$ satises $F[c_0, c_1, \ldots, c_n]$.

Once more, we use the compactness theorem; there is some nite subset Φ_0 of Φ such that $F[c_0, c_1, \ldots, c_n]$ is a consequence of $\Phi_0 \cup T$. Let $H[v_0, v_1, \ldots, v_n]$

be a quantier-free formula of L such that

$$H[c_0, c_1, \ldots, c_n] = \bigwedge_{K \in \Phi_0} K.$$

Then $T \vdash F[c_0, c_1, \ldots, c_n] \Leftrightarrow H[c_0, c_1, \ldots, c_n]$ and, as a result,

$$T \vdash \forall v_0 \forall v_1 \ldots \forall v_n (F[v_0, v_1, \ldots, v_n] \Leftrightarrow H[v_0, v_1, \ldots, v_n]).$$

Remark When every formula is equivalent modulo T to a quantier-free formula, we say that T **admits elimination of quantiers**. The argument that we have just presented can be applied in a much more general context. In fact, it proves the following theorem.

Theorem *Assume that for all models $\mathcal{M}$ and $\mathcal{N}$ of T and for all sequences $(a_0, a_1, \ldots, a_n)$ and $(b_0, b_1, \ldots, b_n)$ from M and N, respectively, whenever $(a_0, a_1, \ldots, a_n)$ and $(b_0, b_1, \ldots, b_n)$ satisfy the same atomic formulas in $\mathcal{M}$ and $\mathcal{N}$, respectively, then they satisfy the same formulas. Then T admits elimination of quantiers.*

For example, the theory under consideration in Exercise 11 admits elimination of quantiers [thanks to the property proved in part (5) of (e)].

13. Let T be a theory in a language L that is equivalent to a nite theory. So T is equivalent to a single closed formula of L which we will call F. The class of L-structures that are not models of T is precisely the class of models of $\neg F$ and, by Äos' theorem (see Theorem 8.31), this class is closed under ultraproducts.

Conversely, suppose that T is not equivalent to a nite theory. Then for every nite subset X of T, there exists an L-structure $\mathcal{M}_X$ that is a model of X but not of T. Let P denote the set of nite subsets of T and, for every $X \in P$, set

$$O(X) = \{Y \in P : X \subseteq Y\}.$$

If $X_1, X_2, \ldots, X_n$ are elements of P, then $O(X_1) \cap O(X_2) \cap \cdots \cap O(X_n)$ includes $X_1 \cup X_2 \cup \cdots \cup X_n$ so it is not empty. According to Theorem 2.79, there exists an ultralter $\mathcal{U}$ that includes the set $\{O(X) : X \in P\}$. We claim that

$$\mathcal{M} = \prod_{X \in P} \mathcal{M}_X \Big/ \mathcal{U}$$

is a model of T. To see this, suppose F is a formula of T. So we have $O(\{F\}) \in \mathcal{U}$, $O(\{F\}) \subseteq \{X \in P : \mathcal{M}_X \models F\}$, and hence

$$\{X \in P : \mathcal{M}_X \models F\} \in \mathcal{U}.$$

Using Äos' theorem, this shows that $\mathcal{M}$ satises F. In this way, we have obtained structures that are not models of T but whose ultraproduct is a model of T.

14. (a) We noted in Example 8.50 that the theory of rings can be axiomatized by Horn formulas; therefore a reduced product of elds is a ring. If the lter is an ultralter, it follows from Äos' theorem that the ultraproduct will be a eld.

Let $\mathcal{F}$ be a lter on a set I and, for every $i \in I$, let K_i be a eld. Set

$$A = \prod_{i \in I} K_i \Big/ \mathcal{F}.$$

Suppose that $\mathcal{F}$ is not an ultralter. So there exists a nite subset I_0 of I such that neither I_0 nor $I - I_0$ belongs to $\mathcal{F}$. For each $i \in I$, let $\mathbf{0}_i$ and $\mathbf{1}_i$ denote the identity elements for addition and multiplication in the eld K_i. Consider the functions in $\prod_{i \in I} K_i$ dened in the following way:

- if $i \in I_0$, then $a_0(i) = \mathbf{0}_i$ and $a_1(i) = \mathbf{1}_i$;
- if $i \notin I_0$, then $a_0(i) = \mathbf{1}_i$ and $a_1(i) = \mathbf{0}_i$.

Let $\bar{a}_0$ and $\bar{a}_1$ be the corresponding elements of A. Then because

$$\{i \in I : K_i \models a_0(i) \simeq \mathbf{0}_i\} = I_0 \notin \mathcal{F},$$

we have $A \not\models \bar{a}_0 \simeq \mathbf{0}$ by the denition of reduced products; similarly, $A \not\models \bar{a}_1 \simeq \mathbf{0}$. By contrast,

$$\{i \in I : K_i \models a_0(i) \times a_1(i) \simeq \mathbf{0}_i\} = I,$$

so $A \models \bar{a}_0 \times \bar{a}_1 \simeq \mathbf{0}$. The ring A is therefore not an integral domain, so it is not a eld.

(b) This follows directly from the denitions.

15. (a) When we repeat the notations and the method from Exercise 18 of Chapter 3, we can easily see that the theory

$$\mathsf{Th}(\langle \alpha, \leq \rangle) \cup \{c_{n+1} < c_n : n \in \mathbb{N}\}$$

is consistent.

(b) By the LowenheimñSkolem theorem, we know that there exists a denumerable subset X_0 of $\aleph_1$ such that $\langle X_0, \leq \rangle \prec \langle \aleph_1, \leq \rangle$. Set $\alpha_0 = \sup X_0$. We can then nd a denumerable subset X_1 of $\aleph_1$ such that $\alpha_0 \subseteq X_1$ and

$$\langle X_0, \leq \rangle \prec \langle X_1, \leq \rangle \prec \langle \aleph_1, \leq \rangle.$$

Repeating this process, we dene by induction a sequence $(X_n : n \in \mathbb{N})$ of denumerable subsets of $\aleph_1$ such that if $\alpha_n = \sup X_n$,

$$\alpha_n \subseteq X_{n+1} \quad \text{and} \quad \langle X_n, \leq \rangle \prec \langle X_{n+1}, \leq \rangle \prec \langle \aleph_1, \leq \rangle.$$

Set $\alpha = \sup\{\alpha_n : n \in \mathbb{N}\} = \bigcup_{n \in N} X_n$. Since any union of elementary submodels is an elementary submodel, we conclude that $\langle \alpha, \leq \rangle \prec \langle \aleph_1, \leq \rangle$.

(c) Choose a denumerable ordinal α such that $\langle \alpha, \leq \rangle \prec \langle \aleph_1, \leq \rangle$ and repeat the construction from part (b) starting with a denumerable set X_0 such that

$\alpha \in X_0$. We obtain another denumerable ordinal β, strictly greater than α, such that

$$\langle \beta, \leq \rangle \prec \langle \aleph_1, \leq \rangle.$$

In fact, this argument shows that

$$\text{the set of ordinals } \alpha \text{ such that } \langle \alpha, \leq \rangle \prec \langle \aleph_1, \leq \rangle$$

is a closed, conal subset of $\aleph_1$ (see Exercise 20 of Chapter 7).

16. (a) Since every point of $\mathcal{N}$ is the interpretation of a constant symbol, T is in fact the complete diagram of $\mathcal{N}$ (or, more precisely, of the reduct of $\mathcal{N}$ to the language L minus its constant symbols). It follows from this, with the help of Lemma 8.13, that every model of T is isomorphic to an elementary extension of $\mathcal{N}$.

(b) Let A and B be two subsets of $\mathbb{N}$ such that $A \subseteq B$ and $A \in \mathcal{F}_a$. Then

$$\mathcal{N} \models \forall v_0(\underline{A}v_0 \Rightarrow \underline{B}_0),$$

therefore

$$\mathcal{M} \models \forall v_0(\underline{A}v_0 \Rightarrow \underline{B}_0),$$

and since $\mathcal{M} \models \underline{A}a$, we conclude $\mathcal{M} \models \underline{B}a$ and $B \in \mathcal{F}_a$.

Next, suppose that A and B both belong to $\mathcal{F}_a$. Set $C = A \cap B$. Then

$$\mathcal{N} \models \forall v_0((\underline{A}v_0 \wedge \underline{B}v_0) \Rightarrow \underline{C}v_0),$$

therefore

$$\mathcal{M} \models \forall v_0((\underline{A}v_0 \wedge \underline{B}v_0) \Rightarrow \underline{C}v_0),$$

and since $\mathcal{M} \models \underline{A}a \wedge \underline{B}a$, we see that $\mathcal{M} \models \underline{C}$ and $C \in \mathcal{F}_a$. In addition, the formula $\forall v_0(v_0 \notin \underline{\emptyset})$ is true in $\mathcal{N}$, hence also in $\mathcal{M}$, and it follows that $\emptyset \notin \mathcal{F}_a$.

This shows that $\mathcal{F}_a$ is a lter. To prove that it is an ultralter, suppose we are given a subset A of $\mathbb{N}$; denote its complement by B. Then

$$\mathcal{N} \models \forall v_0(\underline{A}v_0 \vee \underline{B}v_0),$$

and hence $\mathcal{M} \models \underline{A}a \vee \underline{B}a$; this shows that either A or B belongs to $\mathcal{F}_a$.

To prove that the ultralter $\mathcal{F}_a$ is non-trivial, we invoke Lemma 2.75; if $A = \{n\}$ for some $n \in \mathbb{N}$, then $\mathcal{N} \models \forall v_0(\underline{A}v_0 \Rightarrow v_0 \simeq \underline{n})$. Since $\mathcal{M} \not\models a \simeq \underline{n}$, we see that $A \notin \mathcal{F}$.

(c) If $f_r(n) = f_s(n)$ for innitely many integers n, then

$$p_r(n)/q_r(n) = p_s(n)/q_s(n)$$

for innitely many integers n; so the convergent sequences

$$(p_r(n)/q_r(n) : n \in \mathbb{N}) \quad \text{and} \quad (p_s(n)/q_s(n) : n \in \mathbb{N})$$

have a common subsequence and must consequently have the same limit and $r = s$.

(d) Let $\mathcal{M}$ be a model of T that is not isomorphic to $\mathcal{N}$. We wish to show that the cardinality of M is at least $2^{\aleph_0}\cdot$ According to part (a), we may assume that $\mathcal{M}$ is an elementary extension of $\mathcal{N}$. Let a be a point of $\mathcal{M}$ that does not belong to $\mathbb{N}$. We claim that if r and s are distinct positive reals, then

$$\mathcal{M} \models \neg \underline{f_r} a \simeq \underline{f_s} a;$$

this will show that the mapping from $\mathbb{R}_+$ into M whose value at r is $\underline{f_r}(a)$ is injective and, hence, that the cardinality of M is at least $2^{\aleph_0}$. To prove the claim, let C be the set of integers n such that $f_r(n) = f_s(n)$. We have just seen that C is nite; hence from part (b), $C \notin \mathcal{F}_a$ (because a non-trivial ultralter cannot contain a nite set) and $\mathcal{M} \models \neg \underline{C} a$. Also,

$$\mathcal{N} \models \forall v_0 (\underline{f_r} v_0 \simeq \underline{f_s} v_0 \Leftrightarrow \underline{C} v_0);$$

therefore

$$\mathcal{M} \models \forall v_0 (\underline{f_r} v_0 \simeq \underline{f_s} v_0 \Leftrightarrow \underline{C} v_0),$$

which shows that $\mathcal{M} \models \neg \underline{f_r} a \simeq \underline{f_s} a$.

So all denumerable models of T are isomorphic and T is $\aleph_0$-categorical.

(e) The model $\mathcal{N}$ has only one enrichment $\mathcal{N}'$ to an L'-structure that is a model of T'; it is the one in which the symbol X is interpreted by the whole set $\mathbb{N}$. Every model of T' is then isomorphic to $\mathcal{N}'$ (its L-reduct being isomorphic to $\mathcal{N}$).

However, T', which obviously has no nite models, is not complete; the formula $\forall v_0 X v_0$ is true in $\mathcal{N}'$, but we can nd a model of T' in which it is not true by choosing a proper elementary extension of $\mathcal{N}$ and interpreting X by the set $\mathbb{N}$.

This example illustrates that the hypothesis `κ is greater than or equal to card(L)' cannot be omitted from Vaught's theorem (Theorem 8.18).

Bibliography

First, we will suggest a list (no doubt very incomplete) of works concerning mathematical logic. These are either general treatises on logic or more specialized books related to various topics that we have presented. There is one exception: the book edited by J. Barwise, whose ambition was to produce, at the time it was published, a survey of the state of the art. Some of these titles are now out of print but we have included them nonetheless because they should be available in many university libraries.

J.P. Azra and **B. Jaulin**, *Recursivite*, Gauthiers-Villars, 1973.
J. Barwise (editor), *Handbook of mathematical logic*, North-Holland, 1977.
J.L. Bell and **A.B. Machover**, *A course in mathematical logic*, North-Holland, 1977.
J.L. Bell and **A.B. Slomson**, *Models and ultraproducts*, North-Holland, 1971.
E.W. Beth, *Formal methods*, D. Reidel, 1962.
C.C. Chang and **J.H. Keisler**, *Model theory*, North-Holland, 1973.
A. Church, *Introduction to mathematical logic*, Princeton University Press, 1996.
P. Cohen, *Set theory and the continuum hypothesis*, W.A. Benjamin, 1966.
H. Curry, *Foundation of mathematical logic*, McGraw-Hill, 1963.
D. van Dalen, *Logic and structures*, Springer-Verlag, 1983.
M. Davis, *Computability and unsolvability*, McGraw-Hill, 1958.
F. Drake, *Set theory*, North-Holland, 1979.
H.D. Ebbinghaus, **J. Flum** and **W. Thomas**, *Mathematical logic*, Springer-Verlag, 1984.
R. Frasse, *Cours de logique mathematique*, Gauthier-Villars, 1972.
J.Y. Girard, *Proof theory*, Bibliopolis, 1987.
P. Halmos, *Lectures on Boolean algebras*, D. Van Nostrand, 1963.
P. Halmos, *Naive set theory*, Springer-Verlag, 1987.
D. Hilbert and **W. Ackermann**, *Mathematical logic*, Chelsea, 1950.
K. Hrbacek and **T. Jech**, *Introduction to set theory*, Marcel Dekker, 1984.
T. Jech, *Set theory*, Academic Press, 1978.
S. Kleene, *Introduction to metamathematics*, Elsevier Science, 1971.
G. Kreisel and **J.L. Krivine**, *Elements de logique mathematique*, Dunod, 1966.
J.L. Krivine, *Theorie des ensembles*, Cassini, 1998.
K. Kunen, *Set theory*, North-Holland, 1985.
R. Lalement, *Logique, reduction, resolution*, Masson, 1990.
R.C. Lyndon, *Notes on logic*, D. Van Nostrand, 1966.
A.I. Mal'cev, *The metamathematics of algebraic systems*, North-Holland, 1971.
J. Malitz, *An introduction to mathematical logic*, Springer-Verlag, 1979.
Y. Manin, *A course in mathematical logic* (translated from Russian), Springer-Verlag, 1977.
M. Margenstern, *Langage Pascal et logique du premier ordre*, Masson, 1989 and 1990.
E. Mendelson, *Introduction to mathematical logic*, D. Van Nostrand, 1964.

P.S. Novikov, *Introduction a la logique mathematique* (translated from Russian), Dunod, 1964.
P. Odifreddi, *Classical recursion theory*, North-Holland, 1989.
J.F. Pabion, *Logique mathematique*, Hermann, 1976.
R. Peter, *Recursive functions*, Academic Press, 1967.
B. Poizat, *Cours de theorie des modeles*, Nur al-Mantiq wal-Ma'rifah (distributed by Oflib, Paris), 1985.
D. Ponasse, *Logique mathematique*, *OCDL*, 1967.
W. Quine, *Mathematical logic*, Harvard University Press, 1981.
W. Quine, *Methods of logic*, Harvard University Press, 1989.
H. Rasiowa and **R. Sikorski**, *The mathematics of metamathematics*, PWNñPolish Scientic Publishers, 1963.
A. Robinson, *Complete theories*, North-Holland, 1956.
A. Robinson, *Introduction to model theory and to the metamathematics of algebra*, North-Holland, 1974.
H. Rogers, *Theory of recursive functions and effective computability*, McGraw-Hill, 1967.
J.B. Rosser, *Logic for mathematicians*, McGraw-Hill, 1953.
J.R. Shoeneld, *Mathematical logic*, Addison-Wesley, 1967.
K. Shutte, *Proof theory*, Springer-Verlag, 1977.
W. Sierpinski, *Cardinal and ordinal numbers*, PWNñPolish Scientic Publishers, 1965.
R. Sikorski, *Boolean algebras*, Springer-Verlag, 1960.
R. Smullyan, *First order logic*, Springer-Verlag, 1968.
R.I. Soare, *Recursively enumerable sets and degrees*, Springer-Verlag, 1987.
J. Stern, *Fondements mathematiques de l'informatique*, McGraw-Hill, 1990.
P. Suppes, *Axiomatic set theory*, D. Van Nostrand, 1960.
P. Suppes, *Introduction to logic*, D. Van Nostrand, 1957.
A. Tarski, *Introduction to logic and to the methodology of deductive sciences*, Oxford University Press, 1965.
A. Tarski, **A. Mostowski** and **R. Robinson**, *Undecidable theories*, North-Holland, 1953.
R.L. Vaught, *Set theory*, Birkhauser, 1985.

Below, for the benet of the eclectic reader whose curiosity may have been piqued, we supplement this bibliography with references to books which, while related to our subject overall, are of historical or recreational interest.

L. Carroll, *The game of logic*, Macmillan, 1887.
M. Gardner, *Paradoxes to puzzle and delight*, W.H. Freeman, 1982.
K. Godel, *Collected works* (published under the supervision of S. Feferman), Oxford University Press, 1986.
J. van Heijenoort, *From Frege to Godel, a source book in mathematical logic (1879ñ1931)*, Harvard University Press, 1967.
A. Hodges, *Alan Turing: the enigma*, Simon & Schuster, 1983.
R. Smullyan, *What is the name of this book?*, Prentice Hall, 1978.
J. Venn, *Symbolic logic*, Chelsea, 1971 (rst edition: 1881).

Index